AF669837

Methods in Molecular Biology™

Series Editor
John M. Walker
School of Life Sciences
University of Hertfordshire
Hatfield, Hertfordshire, AL10 9AB, UK

For further volumes:
http://www.springer.com/series/7651

Functional Genomics

Methods and Protocols

Second Edition

Edited by

Michael Kaufmann and Claudia Klinger

Private Universität, Witten/Herdecke gGmbH, Witten, Germany

Editors
Michael Kaufmann, Ph.D.
Witten/Herdecke University
Faculty of Health
School of Medicine
Center for Biomedical Education and Research
Institute for Medical Biochemistry
The Protein Chemistry Group
58448 Witten
Stockumer Str. 10
Germany
mika@uni-wh.de

Claudia Klinger, Ph.D.
Witten/Herdecke University
Faculty of Health
School of Medicine
Center for Biomedical Education and Research
Institute for Medical Biochemistry
The Protein Chemistry Group
58448 Witten
Stockumer Str. 10
Germany
cklinger@uni-wh.de

ISSN 1064-3745 e-ISSN 1940-6029
ISBN 978-1-61779-423-0 e-ISBN 978-1-61779-424-7
DOI 10.1007/978-1-61779-424-7
Springer New York Dordrecht Heidelberg London

Library of Congress Control Number: 2011940130

Printed on acid-free paper

Humana Press is part of Springer Science+Business Media (www.springer.com)

Preface

The Life Sciences are undergoing more than ever an accelerating evolution currently culminating in the -omics era characterized by the development of a multitude of high-throughput methods that are now getting to be routinely applied in the modern biochemistry lab. While the basic principles of classic analytical methods, such as Northern or Western blot analysis, are still dominating, the individual methods have advanced and continuously morphed into sophisticated techniques, such as expression profiling of whole genomes via DNA microarrays or the use of delicate protein chips to specifically detect thousands of macromolecules simultaneously during one single experiment. Those innovative techniques are capable of delivering tremendous amounts of data accompanied by the need of only trace amounts of samples and at a minimum in both personnel and material costs. The progress in almost every aspect of computer hardware technology obeys Moore's law, i.e., computer performance still grows exponentially at doubling times in the range of months rather than years. In fact, these advances are an indispensable prerequisite to handle data sets typically obtained by today's procedures applied in the field of Functional Genomics.

Now, after almost a decade has passed by since the first edition of this book has been released, the pace in progress of biochemical and biotechnological high-throughput methodologies ultimately requires the release of an updated version. Compared to the first edition, the scope of this book has been extended considerably, now no longer just dealing with DNA microarrays as the pioneering technology that then initiated the establishment of the formerly new field of Functional Genomics. Instead, due to the methodological expansion of Functional Genomics, other high-throughput techniques, for instance those involved in analyzing proteins and metabolites, are also included.

Functional Genomics can be distinguished from Comparative Genomics by its focus on the dynamic aspects of the transcriptome, proteome, and metabolome, respectively. Nevertheless, it is noticeable that in the literature the two disciplines are frequently mentioned in the same breath which prompted us to open this volume with a chapter about Bioinformatics, although with a strong focus on computational tools suitable to make functional predictions. In contrast to most other publications in the field, the following paragraphs are structured with attention to the nature of the biochemical target molecules rather than the different laboratory methods under consideration, i.e., each chapter contains separate discussions about the analysis of DNA, RNA, proteins, and metabolites. Although we are aware that this strategy cannot completely exclude redundancies, we feel that they at least can be reduced to a minimum. Overall, each individual contribution is intended to be self-contained and largely independent from the other chapters of the book. Ideally, each chapter can be seen as a unit of its own which consequently reduces the importance of the order of chapters.

The book is useful for all scientists who plan to establish or extend one of the technologies described here in their own labs. Although short introductions of the basic principles of each procedure are not omitted, the focus of each chapter lies mainly on the practical aspects of each method enabling the reader to easily acquire all the equipment and materials needed and to successfully perform the experiments autonomously. As often as possible original lab protocols are included, making it easier to reproduce the respective procedures.

Finally, we would like to thank all the contributors for their time, patience, and endurance that undoubtedly was necessary to do such an excellent work. Regarding the reader, we hope that this book will satisfy its intention of being one of the pieces helping him to perform his experiments successfully, which, with respect to everyday lab experience, unfortunately is often an exception rather than the rule in a scientist's real life.

Witten, Germany

Michael Kaufmann
Claudia Klinger

Contents

Contributors

MARTIN F. ALMSTETTER • *Institute of Functional Genomics, University of Regensburg, Josef-Engert-Str. 9, Regensburg 93053, Germany*
RUISHENG AN • *Department of Entomology, The Ohio State University, 1680 Madison Ave, Wooster, OH 44691, USA*
PHILIPP BAASKE • *NanoTemper Technologies GmbH, Floessergasse 4, Münich 81369, Germany*
MIGUEL A. PEREZ-AMADOR • *Pérez-AmadorInstituto de Biología Molecular y Celular de Plantas, CSIC-Universidad Politécnica de Valencia, Ciudad Politécnica de la Innovación, Ingeniero Fausto Elio s/n, 46022, Valencia, Spain*
ARMIN BAIKER • *Bavarian Health and Food Safety Authority, Oberschleissheim, Germany*
BOGDAN BARZ • *Department of Physics and Astronomy, University of Missouri, Columbia, MO, USA*
DIETER BRAUN • *Ludwig-Maximilians-Universiät München, System Biophysics, Münich, Germany*
NICOLAS P. BURTON • *Inspiralis Ltd, Norwich Bioincubator, Norwich Research Park, Colney, Norwich NR4 7UH, UK*
IOURI CHEPELEV • *Laboratory of Molecular Immunology, National Heart, Lung and Blood Institute, National Institutes of Health, Bethesda, MD 20892, USA*
YOUNG HAE CHOI • *Pharmacognosy Department/Metabolomics, Gorlaeus Laboratories, Institute of Biology, Leiden University, P.O. Box 9502, RA Leiden 2300, The Netherlands*
VALLI DE RE • *Experimental and Clinical Pharmacology Unit, CRO Centro di Riferimento Oncologico, IRCCS National Cancer Institute, via F. Gallini 2, Aviano (PN) 33081, Italy*
STEFAAN DERVEAUX • *Center for Medical Genetics, Ghent University, De Pintelaan 185, Ghent 9000, Belgium*
CLAUS DESLER • *Center for Healthy Aging, University of Copenhagen, Blegdamsvej 3, Copenhagen 2200, Denmark*
KATJA DETTMER • *Institute of Functional Genomics, University of Regensburg, Josef-Engert-Str. 9, Regensburg 93053, Germany*
KIRSTEN DÖRRIES • *Institute of Pharmacy, Ernst-Moritz-Arndt-University Greifswald, Friedrich-Ludwig-Jahn-Str. 17, Greifswald 17487, Germany*
STEFAN DUHR • *NanoTemper Technologies GmbH, Münich, Germany*
JON AMBÆK DURHUUS • *Center for Healthy Aging, Faculty of Health Sciences, University of Copenhagen, Copenhagen N 2200, Denmark*
ISVETT JOSEFINA FLORES-SANCHEZ • *Institute of Biological Chemistry, Washington State University, Pullman, WA, USA*
SHAN GOH • *Department of Pathology and Infectious Diseases, Royal Veterinary College, University of London, London, UK*

LIAM GOOD • *Department of Pathology and Infectious Diseases, Royal Veterinary College, University of London, London, UK*
JOHN M. GREALLY • *Albert Einstein College of Medicine, Price 322, 1301 Morris Park Avenue, Bronx, NY 10461, USA*
PARWINDER S. GREWAL • *Department of Entomology, The Ohio State University, 1680 Madison Ave, Wooster, OH 44691, USA*
ROMAN HÄUSER • *Institute of Toxicology and Genetics, Karlsruhe Institute of Technology (KIT), Eggenstein-Leopoldshafen 76344, Germany*
BO HANSEN • *Santaris Pharma, Kogle Alle 6, Horsholm DK-2970, Denmark*
ARJANG HASSIBI • *Institute for Cellular and Molecular Biology, University of Texas, 1 University Station C8800, Austin, TX 78712-0323, USA*
BABAK HASSIBI • *Electrical Engineering Department, California Institute of Technology, Pasadena, CA 91125, USA*
ZHIQUAN HE • *Department of Computer Science, University of Missouri, Columbia, MO, USA*
ANJA HOEG • *Santaris Pharma, Kogle Alle 6, Horsholm DK-2970, Denmark*
FLORIAN HOLLFELDER • *Department of Biochemistry, University of Cambridge, 80 Tennis Court Road, Cambridge, CB2 1GA, UK*
MORAN JERABEK-WILLEMSEN • *NanoTemper Technologies GmbH, Münich, Germany*
YUN-JIN JIANG • *Institute of Molecular and Genomic Medicine, National Health Research Institutes, 35 Keyan Road, zhunan Town, Miaoli County 35053, Taiwan*
QIANG JING • *Departments of Genetics (Computational Genetics) and Center for Epigenomics, Albert Einstein College of Medicine, 1301 Morris Park Avenue, Bronx, NY, USA*
SHAM S. KAKAR • *Department of Physiology and Biophysics and James Graham Brown cancer Center, University of Louisville, Clinical and Translational Building, Room 322, Louisville, KY 40202, USA*
MIRIAM KALTENBACH • *Department of Biochemistry, University of Cambridge, 80 Tennis Court Road, Cambridge, CB2 1GA, UK*
TROELS KOCH • *Santaris Pharma, Kogle Alle 6, Horsholm DK-2970, Denmark*
IOAN KOSZTIN • *Department of Physics and Astronomy, University of Missouri, Columbia, MO, USA*
JOHNATHAN LAI • *Santaris Pharma, Kogle Alle 6, Horsholm DK-2970, Denmark*
GERNOT LÄNGST • *Universität Regensburg, Biochemistry III, Regensburg, Germany*
MICHAEL LALK • *Institute of Pharmacy, Interfaculty Institute for Genetics and Functional Genomics, University of Greifswald, F.-L.-Jahnstr. 15, Greifswald D-17487, Germany*
MANUEL LIEBEKE • *Biomolecular Medicine, Department of Surgery and Cancer, Faculty of Medicine, Imperial College London, London SW7 2AZ, UK*
AMELIE S. LOTZ-HAVLA • *Department of Molecular Pediatrics, Dr. von Hauner Children's Hospital, Ludwig-Maximilians-University, Münich 80337, Germany*
HIDEAKI MAEDA • *Measurement Solution Research Center, National Institute of Advanced Industrial Science and Technology, Tosu, Saga, Japan*
M. CARMEN MARQUES • *Instituto de Biología Molecular y Celular de Plantas (IBMCP), Universidad Politécnica de Valencia (UPV) and Consejo Superior de Investigaciones Científicas (CSIC), CPI 8E, Ingeniero Fausto Elio s/n, Valencia 46022, Spain*

ANTHONY MAXWELL • *Department of Biological Chemistry, John Innes Centre, Colney, Norwich NR4 7UH, UK*
ANDREW MCLELLAN • *Departments of Genetics (Computational Genetics) and Center for Epigenomics, Albert Einstein College of Medicine, 1301 Morris Park Avenue, Bronx, NY, USA*
PIETER MESTDAGH • *Center for Medical Genetics, Ghent University, De Pintelaan 185, Ghent 9000, Belgium*
HANNA MEYER • *Institute of Pharmacy, Interfaculty Institute for Genetics and Functional Genomics, University of Greifswald, F.-L.-Jahnstr. 15, Greifswald D-17487, Germany*
MASAYA MIYAZAKI • *Measurement Solution Research Center, National Institute of Advanced Industrial Science and Technology, 807-1 Shuku, Tosu, Saga 841-0052, Japan*
ANIA C. MUNTAU • *Department of Molecular Pediatrics, Dr. von Hauner Children's Hospital, Ludwig-Maximilians-University, Münich 80337, Germany*
NOBUTAKA NAKASHIMA • *Bioproduction Research Institute, National Institute of Advanced Industrial Science and Technology (AIST), 2-17-2-1 Tsukisamu-Higashi, Toyohira-ku, Sapporo 062-8517, Japan*
AGATA NOTARPIETRO • *Experimental and Clinical Pharmacology Unit, CRO Centro di Riferimento Oncologico, IRCCS National Cancer Institute, via F. Gallini 2, Aviano (PN) 33081, Italy*
PETER J. OEFNER • *Institute of Functional Genomics, University of Regensburg, Josef-Engert-Str. 9, Regensburg 93053, Germany*
HENRIK OERUM • *Santaris Pharma, Kogle Alle 6, Horsholm DK-2970, Denmark*
RON OPHIR • *Institute of Plant Sciences, Agricultural Research Organization, Volcani Research Center, Bet Dagan 50250, Israel*
SIVA KUMAR PANGULURI • *Department of Anatomical Sciences and Neurobiology, University of Louisville, 500 S Preston Street, HSC A-Tower, room 1001, Louisville, KY 40202, USA*
KRISZTIÁN PAPP • *Immunology Research Group, ELTE-MTA, Pazmany P.s. 1C, Budapest H-1117, Hungary*
JÓZSEF PRECHL • *Immunology Research Group, ELTE-MTA, Pazmany P.s. 1C, Budapest H-1117, Hungary*
LENE JUEL RASMUSSEN • *Center for Healthy Aging, Faculty of Health Sciences, University of Copenhagen, Copenhagen N 2200, Denmark*
JOSÉ LUIS RIECHMANN • *Division of Biology, California Institute of Technology, Pasadena, CA 91125, USA*
VIRAJ R. SANGHVI • *Department of Microbiology and Immunology, Institute for Molecular Medicine and Infectious Disease, Drexel University College of Medicine, 245 North 15th Street, Philadelphia, PA 19102, USA*
YI SHANG • *Department of Computer Science, University of Missouri, Columbia, MO, USA*
AMIR SHERMAN • *Institute of Plant Sciences, Agricultural Research Organization, Volcani Research Center, Bet Dagan 50250, Israel*
MARIA PAOLA SIMULA • *Experimental and Clinical Pharmacology Unit, CRO Centro di Riferimento Oncologico, IRCCS National Cancer Institute, via F. Gallini 2, Aviano (PN) 33081, Italy*

Roy D. Sleator • *Department of Biological Sciences, Cork Institute of Technology, Bishopstown, Cork, Ireland*
Harris S. Soifer • *Department of Oncology, Montefiore Medical Center, Albert Einstein College of Medicine, 111 East 210th Street, Hofheimer 1st Floor, Bronx, NY 10467, USA*
Laura F. Steel • *Department of Microbiology and Immunology, Institute for Molecular Medicine and Infectious Disease, Drexel University College of Medicine, 245 North 15th Street, Philadelphia, PA 19102, USA*
C.A. Stein • *Albert Einstein College of Medicine, Albert Einstein-Montefiore Cancer Center, Montefiore Medical Center, 111 E. 210 St. Bronx, NY 10467, USA*
Viktor Stein • *Department of Biochemistry, University of Cambridge, 80 Tennis Court Road, Cambridge, CB2 1GA, UK*
Thorsten Stellberger • *Max-von-Pettenkofer-Institut, Lehrstuhl Virologie, Ludwig-Maximilians-Universität (LMU), Münich, Germany*
Masako Suzuki • *Departments of Genetics (Computational Genetics) and Center for Epigenomics, Albert Einstein College of Medicine, 1301 Morris Park Avenue, Bronx, NY, USA*
Tomohiro Tamura • *Bioproduction Research Institute, National Institute of Advanced Industrial Science and Technology (AIST), Tsukuba, Ibaraki, Japan*
James A. Taylor • *Department of Biological Chemistry, John Innes Centre, Colney, Norwich NR4 7UH, UK*
Giuseppe Toffoli • *Experimental and Clinical Pharmacology Unit, CRO Centro di Riferimento Oncologico, IRCCS National Cancer Institute, via F. Gallini 2, Aviano (PN) 33081, Italy*
William Ka Fai Tse • *Craniofacial Developmental Biology Laboratory, Center for Regenerative Medicine, Massachusetts General Hospital, Harvard Medical School, 185 Cambridge Street, Boston, MA 02114, USA*
Peter Uetz • *Center for the Study of Biological Complexity, Virginia Commonwealth University, PO Box 842030, 1015 Floyd Ave. Richmond, VA 23284, USA*
Jo Vandesompele • *Center for Medical Genetics, Ghent University, De Pintelaan 185, Ghent 9000, Belgium*
Robert Verpoorte • *Pharmacognosy Department/Metabolomics, Gorlaeus Laboratories, Institute of Biology, Leiden University, P.O. Box 9502, RA Leiden 2300, The Netherlands*
Haris Vikalo • *Electrical and Computer Engineering Department, University of Texas, Austin, TX 78712, USA*
María G. Vizoso Pinto • *Department of Virology, Max von Pettenkofer-Institute, Pettenkoferstr. 9a, Münich 80336, Germany*
Albrecht von Brunn • *Max-von-Pettenkofer-Institut, Lehrstuhl Virologie, Ludwig-Maximilians-Universität (LMU), Pettenkoferstr. 9a, Münich 80336, Germany*
Mu Wang • *Department of Biochemistry and Molecular Biology, Indiana University School of Medicine, 635 Barnhill Drive, MS 4053, Indianapolis, IN 46202, USA*

QINGGUO WANG • *Department of Computer Science, University of Missouri, Columbia, MO, USA*
INA WEIDENFELD • *Molecular, Cellular, and Developmental Biology, University of Colorado at Boulder, Campus Box 347, Boulder, CO 80309, USA*
DONG XU • *Department of Computer Science, University of Missouri-Columbia, 201 Engineering Building West, Columbia, MO 65211, USA*
HIROSHI YAMAGUCHI • *Measurement Solution Research Center, National Institute of Advanced Industrial Science and Technology, Tosu, Saga, Japan*
JINSAM YOU • *Department of Biochemistry and Molecular Biology, Indiana University School of Medicine, 635 Barnhill Drive, MS 4053, Indianapolis, IN 46202, USA*
JINGFEN ZHANG • *Department of Computer Science, University of Missouri, Columbia, MO, USA*
KARINA ZILLNER • *Universität Regensburg, Biochemistry III, Regensburg, Germany*

Part I

Bioinformatics

Chapter 1

Prediction of Protein Tertiary Structures Using MUFOLD

Jingfen Zhang, Zhiquan He, Qingguo Wang, Bogdan Barz, Ioan Kosztin, Yi Shang, and Dong Xu

Abstract

There have been steady improvements in protein structure prediction during the past two decades. However, current methods are still far from consistently predicting structural models accurately with computing power accessible to common users. To address this challenge, we developed MUFOLD, a hybrid method of using whole and partial template information along with new computational techniques for protein tertiary structure prediction. MUFOLD covers both template-based and ab initio predictions using the same framework and aims to achieve high accuracy and fast computing. Two major novel contributions of MUFOLD are graph-based model generation and molecular dynamics ranking (MDR). By formulating a prediction as a graph realization problem, we apply an efficient optimization approach of Multidimensional Scaling (MDS) to speed up the prediction dramatically. In addition, under this framework, we enhance the predictions consistently by iteratively using the information from generated models. MDR, in contrast to widely used static scoring functions, exploits dynamics properties of structures to evaluate their qualities, which can often identify best structures from a pool more effectively.

Key words: Protein structure prediction, Multidimensional scaling, Molecular dynamics simulation

1. Introduction

Protein tertiary structure often provides a basis for understanding its function. Experimental approaches for protein structure determination, such as X-ray crystallography (1) and Nuclear Magnetic Resonance (NMR) techniques (2), are typically expensive and time-consuming. The increase of the structures in Protein Data Bank (PDB) (3) cannot keep up with the increase of proteins characterized in high-throughput genome sequencing (4). Compared to experimental approaches, computational methods, i.e., to predict the native structure of a protein from its amino acid sequence, are much cheaper and faster. As significant progress has been made over

Michael Kaufmann and Claudia Klinger (eds.), *Functional Genomics: Methods and Protocols*,
Methods in Molecular Biology, vol. 815, DOI 10.1007/978-1-61779-424-7_1, © Springer Science+Business Media, LLC 2012

the past two decades, computational methods are becoming more and more important for studying protein structures in recent years.

The foundation to predict the protein structure by computational methods relies on two sets of principles: the laws of physics and the theory of evolution. The protein folding theory based on the laws of physics states that at physiological conditions (temperature, ion concentration, etc.), a protein folds into its native structure with a unique, stable, and kinetically accessible minimum of free energy (5). The theory of evolution gives us the other guidance for structure prediction: (1) proteins with similar sequences usually have similar structures and (2) protein structures are more conserved than their sequences (6). When the structure of one protein in a family of proteins with similar sequences/structures has been determined by experiment, the other members of the family can be modeled based on their alignments to the known structure.

According to the above foundations, computational prediction methods can be classified into three categories: (1) ab initio prediction (7–10), (2) comparative modeling (CM) (11–13), and (3) threading (14–17). Ab initio methods assume that native structure corresponds to the global free energy minimum accessible during the lifespan of the protein and attempt to find this minimum by an exploration of many conceivable protein conformations. The prediction results are unreliable due to (1) the huge conformational search space and (2) the limitations of the currently used scoring functions. Both CM and threading are template-based methods. CM, using sequence comparison, is a successful category of prediction methods. With the increasing accumulation of experimentally determined protein structures and the advances in remote homology identification, CM has made continuing progress. However, when the sequence identity drops below 30%, the accuracy of CM sharply decreases because of substantial alignment errors. Threading is based on sequence-structure comparison and measures the fitness of the target sequence into templates. A special category of threading called mini-threading, obtains matches between a query sequence and short structure fragments in PDB to build local structures, which are then assembled into final models that require a significantly smaller computational search space than ab initio methods. Thus, minithreading has a better chance to achieve high prediction accuracy than CM in cases when no evolutionarily relationship is available between the target and template sequences.

Although significant progress has been made, existing computational methods are still far from consistently providing accurate structural models with reasonable computing time. Currently, the most popular optimization methods used in structure prediction such as genetic algorithms and Monte Carlo simulations are time-consuming so as to generate structural models often far from the global optimal solution of a scoring function. In addition, widely used scoring functions are generally not accurate enough to identify the best structure from the generated structure pool. Hence, although a number of prediction servers such as Modeller (12), HHpred (13), I-TASSA (18), and Rosetta (19) have been developed, protein structure prediction has not been widely applied in molecular biology studies other than homology modeling with structural templates of high-sequence identity, due to low prediction accuracies and long computing times.

To address the above issues, we proposed a hybrid method, MUFOLD (see Note), by using whole and partial template information to cover both template-based and ab initio predictions using the same framework. The framework generates structural models very fast so that it can assess and improve the model quality more directly than sequence alignment only. Two major novel contributions of MUFOLD are fast graph-based model generation and molecular dynamics ranking (MDR).

On the one hand, instead of using the Monte Carlo method to sample the conformation space, we have tried to find suitable templates and fragment structures in PDB to estimate the spatial constraints between residues in the target sequence, which decreases the search space. At the same time, we bypassed the energy functions and formulated the structure prediction problem as a graph realization problem. Then we applied an efficient optimization approach of MDS to speed up the prediction dramatically. In addition, under this graph-based framework, we can improve the distance constraints by iteratively using the information from the models and thus enhance the predictions consistently.

On the other hand, in contrast to widely used static scoring functions, we have proposed a ranking method, MDR, to exploit dynamics properties of structures to evaluate their qualities, which can often identify the best structures from a pool more effectively. This is a rare success in applications of molecular dynamics simulation for general protein structure predictions.

2. Materials

2.1. Alignment Tools

In MUFOLD we make use of sequence–profile alignment tools, e.g., PSI-BLAST (20), profile–profile alignment tool, e.g., HH-Search (21) and an in-house threading approach, PROSPECT (22), to search possible templates against PDB for target sequences.

2.2. Scoring Functions

Currently, structure quality assessment and model selection generally use the scoring functions in two categories (23): physics-based energy functions and knowledge-based statistical potentials. The knowledge-based statistical potentials are typically fast to calculate, easy to construct, and hence are most widely used in structure quality assessment. We investigated some state-of-the-art scoring functions, and finally chose OPUS (24), Model Evaluator (25), and Dfire energy (26) as the scoring functions to evaluate the quality of models.

2.3. Multidimensional Scaling

The multidimensional scaling (MDS) method is efficient for solving the graph realization problem. It starts with one or more distance matrices derived from points in a multidimensional space and finds a placement of the points in a low-dimensional space. In MUFOLD, we estimate the distances between Cα (or backbone) atoms for each pair of amino acids in the target sequence as distance matrix and then calculate the coordinates of the atom for each amino acid. We generate models using different techniques of MDS: classical metric MDS (CMDS) (27), weighted MDS (WMDS) (28), and split-and-combine MDS (SC-MDS) (29). In our study, we mainly use CMDS, which is the simplest MDS algorithm. CMDS minimizes the sum of least squared errors between the estimated distances and the actual distances in the output model for all pairs of points.

3. Methods

3.1. The Overview of MUFOLD

MUFOLD takes whole and partial template information for both template-based and ab initio predictions using the same framework toward achieving improved accuracies and fast computing in automated predictions. The overview of MUFOLD is presented in Fig. 1, which includes three main parts: (1) template selection and

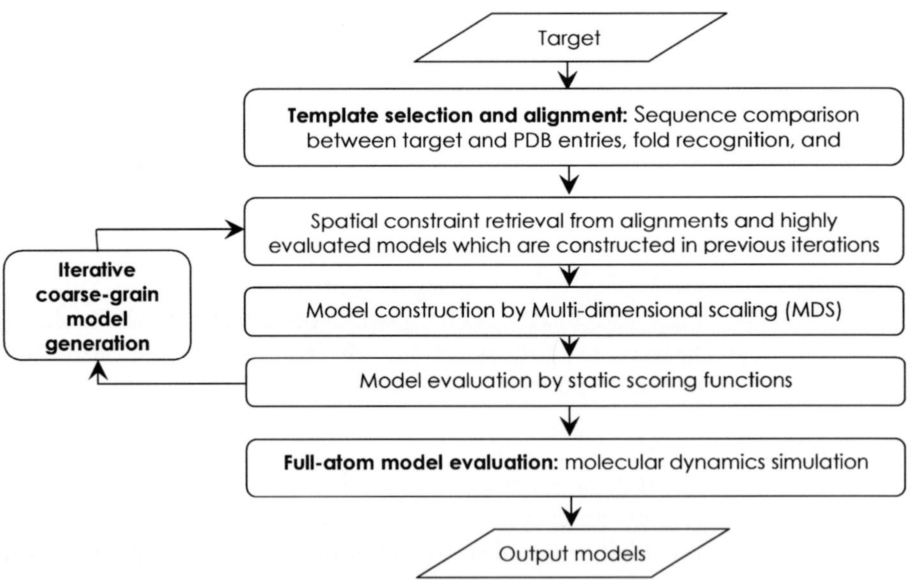

Fig. 1. Flowchart of the MUFOLD structure prediction method.

alignment, i.e., recognizing potentially useful templates/fragments in PDB for the target sequence and building alignments, (2) coarse-grain model generations and evaluations, including fast model generations using MDS techniques at the Cα or backbone level, evaluations of models through static scoring functions, and iterative improvement of selected models by integrating spatial constraints from sequence alignments and selected models, and (3) full-atom model evaluation through MD simulations.

3.2. Template Selection and Alignment

The first step of MUFOLD is to find suitable templates and alignments. Here, the "template" is a general concept including the global homologous templates, nonevolutionarily related (analogous) templates, and the locally compatible protein fragments from PDB. MUFOLD adaptively applies different strategies for various targets. In general, target sequences are classified into three categories: easy, medium, and hard.

"Easy" targets have significant hits by applying sequence–profile alignment in PSI-BLAST (20) against the PDB, i.e., there is at least one alignment hit that can cover more than 70% of the target sequence and with an *E*-value 10^{-3} or less. As homologous templates with high confidence alignments can be easily found for this case, it is intuitive that the sequence alignment can be used to obtain high-quality distance constraints directly. "Medium" targets have remote homologies obtained by using profile–profile alignment in HHSearch (21), i.e., there is at least one alignment hit with an *E*-value less than 10^{-2} (excluding "easy" targets defined above). These targets probably have the correctly identified fold information, but the alignments may be incorrect. Therefore, we try to obtain various alignments by applying different tools and parameters for the correct fold. Coupled with the optimization of MDS, we sample distance constraints and improve the constraints iteratively. "Hard" targets have analog structural templates in PDB that cannot be assigned even by profile–profile alignment. We use an in-house threading approach, PROSPECT (22), to search for possible templates. Although the top one hit may not represent the correct fold, the compatible protein fragments of top *n* (20–100) folds usually include the correct fold.

3.3. Coarse-Grain Model Generation

3.3.1. Graph-Based Model Generation Formulation

We formulate the structure prediction problem as a graph realization problem and then apply a MDS technique to solve it. The basic idea is to estimate the distances between Cα atoms for each pair of residues in the target sequence and then calculate the corresponding Cα coordinates by applying MDS. Assume there are *n* points (each representing the Cα atom of a residue) $X_k \in R^3, k = 1,\dots,n$ in a 3-D space. If we know the exact distances between some pairs of points, e.g., d_{ij} between residue *i* at X_i and residue *j* at X_j, then the graph realization problem is to determine the coordinates of the points from the partial distance constraints such that the distance between each pair of points matches the given distance constraint, $\|X_i - X_j\| = d_{ij}$ for all d_{ij}. If the distance constraints are inaccurate, usually there is no exact or unique solution to the over-determined system of equations. Instead, the problem is formulated as an optimization problem that minimizes the sum of squared errors as:

$$\min_{X_1 \ldots X_n \in R^3} \sum_{i,j=1,\ldots,n} \left(\|X_i - X_j\| - d_{ij}\right)^2 \quad (1)$$

The optimization problem of Eq. 1 is generally nonconvex with many local minima and MDS is very suitable for this optimization problem.

3.3.2. Spatial Distance Constraints

Since predicted distance constraints are often noisy, our strategy is to keep refining the initial models by sampling and improving the distance constraints (or contact maps) iteratively. The initial contact maps of a target protein are retrieved from alignments between the target sequence and various template proteins in PDB obtained in the above step of "template selection and alignment".

For a given alignment between the target and a template, we first estimate the pair-wise distance of the aligned residues in target by the distance of the corresponding residues in the template. Although we select multiple long templates and short fragments, there may still be residues in the target that are aligned to gaps or two residues that are not covered by any single hit simultaneously so that related pair-wise distances cannot be derived directly. For these missing distances, we estimate them by the shortest-path distance. We know that the adjacent Cα atom distance is about 3.8 Å, which means that any two Cα atoms can be connected at least through adjacent Cα atoms. There may be many different paths to connect two Cα atoms, we use the shortest path distance to estimate the unknown pair-wise distance. Although the shortest path often overestimates the distance, it provides an initial complete contact map for calculating a model by MDS.

It should be mentioned that MDS generates two mirror models for any given contact map. Technically, we superimpose the model configuration to the template, and calculate the reflection factor of the superimposition. If the reflection factor equals to 1, it indicates that the configuration is correct; otherwise, it is the incorrect mirror.

3.4. Coarse-Grain Model Evaluation

Coarse-grain model generation using MDS leads to a large number of candidate structures. In MUFOLD, we apply static scorings to evaluate and select better models for the next iteration of model improvement. Specifically, the method consists of filtering and representative finding. At first, we calculate the scoring functions, such as OPUS, Model Evaluator, and Dfire energy for each model. These scoring functions are normalized to *z*-score and summed to filter out those models with lower sum value. Next, the remaining structures are grouped into clusters based on pair-wise similarity measured by RMSD. For each cluster, we find a representative model whose average RMSD to all the other models in the cluster is minimal. These representative models can be reported as final models or as the input models for the next iteration.

3.5. Iteratively Improved Coarse-Grain Model Generation

Although using multiple templates and fragments can generate models that are closer to the native structure than any template alone, inconsistent constraints from different alignments and distances estimated by the shortest path method may compromise the quality of the models. Our strategy is to refine and improve the constraints iteratively by combining the original constraints derived from the alignments ($D_{alignment}$) and the measured distances from the generated models (D_{model}) as: $D_{refine} = \lambda \times D_{alignment} + (1 - \lambda) \times D_{model}$, $0 \leq \lambda \leq 1$. There are different ways to set the value of λ. For example, a simple way is to set $\lambda = 0.5$ if $D_{alignment}$ is available, otherwise $\lambda = 0$. Another way is to set λ according to the confidence level of $D_{alignment}$. By performing this iterative generation, the quality of models often gets better and better, while many deficiencies in the models are fixed over iterations.

Figure 2 shows an example of iteratively improved coarse-grain model generation, where we show the original and improved contact maps and the corresponding models in (a)–(c), (d)–(f), respectively. In the image of contact map, we use colors to illustrate the distances between pair-wise residues, where the lighter the color is, the larger the distance is. We can observe the color changes within the red rectangle regions in Fig. 2a, d, which means the modification of the distance constraints. From the data showed in Fig. 1.2b, e, we can see the significant improvement of the models, for example, all of the quality score such as RMSD, GDT_TS, GDT_HA (30), and TM score (31) of the model against the native have been improved.

3.5.1. Full-Atom Model Evaluation: Molecular Dynamics Ranking

The coarse-grain model generation described above provides various structures with significantly different conformations. How to identify the one with the smallest RMSD compared to the unknown native structure is a highly challenging problem. Existing methods generally use static scoring functions (measurements from static conformations) to rank models. However, the dynamics properties of a model may reveal its structural quality better than static information. Near native models are always more stable than poor-quality models during simulated heating, i.e., the latter unfold at lower temperatures than the former. Thus, the quantitative assessment of relative stabilities of structural models against gradual heating provides an alternative way of ranking the structures' quality. Here,

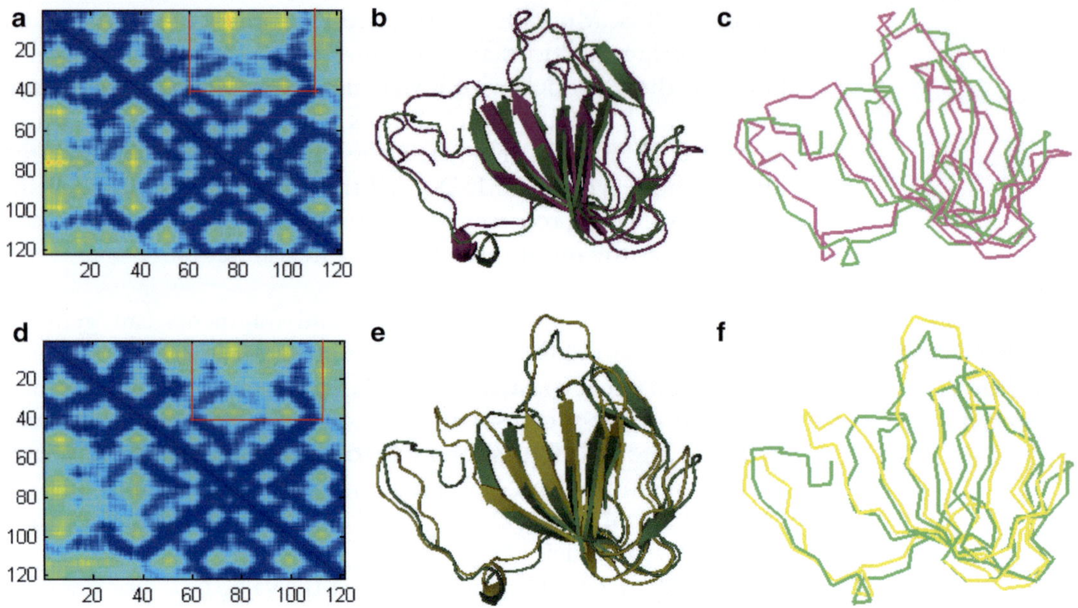

Fig. 2. An example of iterative coarse-grain model generation. (**a**) Original contact map, (**b**) model against the native (RMSD: 3.665 GDT_TS: 0.687 GDT_HA: 0.485 TM score: 0.737), (**c**) trace of Cα atoms against the native, (**d**) improved contact map, (**e**) new model against the native (RMSD: 3.211 GDT_TS: 0.784 GDT_HA: 0.595 TM score: 0.826), (**f**) trace of Cα atoms against the native for the new model.

we propose a novel MD-Ranking (MDR) method based on full-atom MD simulations (32) to evaluate and rank protein models according to their stabilities against external perturbations, e.g., change in temperature or externally applied forces. The basic idea is to build all-atom models from the coarse-grain models, optimize these models by energy minimization, gradually heat them through MD simulations, and then rank the models based on their structural changes during heating.

More specifically, first, an all-atom model is built for each of the top selected structures by the above coarse-grain model process. The coordinates of the missing backbone and side-chain heavy atoms are predicted by using the program Pulchra (33), and the hydrogen atoms are added by using psfgen, which is part of the VMD package (34). Next, the obtained structures are optimized by removing the bad contacts through energy minimization. Finally, the stability of a structure is tested by monitoring the change of its Cα RMSD (cRMSD) with respect to its initial structure during the MD simulation of a scheduled heating at a rate of 1 K/ps. The MD simulations are carried out in vacuum by coupling the system to a Langevin heat bath whose temperature can be varied (i.e., the dynamics of protein atoms is described by a Langevin equation). All energy minimizations and MD simulations

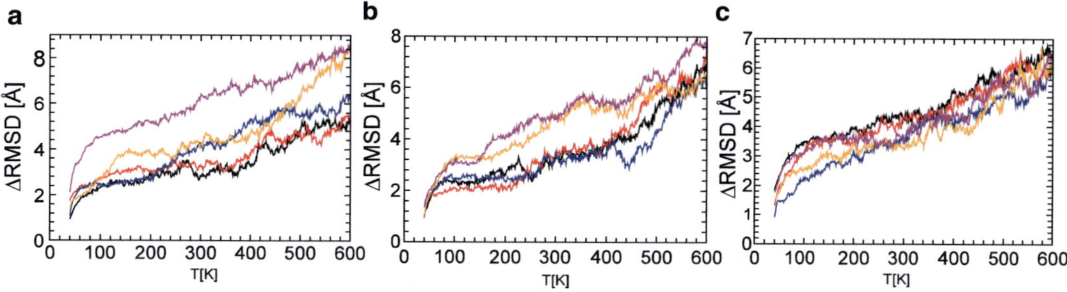

Fig. 3. Change in RMSD during the heating in MD simulations for three proteins. The *colored curves* correspond to the five different models with various RMSD values. The *bold faced models* have been top ranked by MDR. (**a**) *black* – **2.7 Å**, *red* – 3.9 Å, *blue* – 12.3 Å, *orange* – 12.6 Å, *magenta* – 12.9 Å; (**b**) *black* – 3.1 Å, *red* – 3.2 Å, *blue* – **3.3 Å**, *orange* – 9.9 Å, *magenta* – 9.9 Å; and (**c**) *black* – 3.3 Å, *red* – 4.5 Å, *blue* – **5.4 Å**, *orange* – 6.3 Å, *magenta* – 6.7 Å.

were performed by employing the CHARMM27 force field and the parallel NAMD2.6 MD simulation program (35).

Figure 3 shows three typical examples of MDR, i.e., plots of the changes in cRMSD during the heating MD simulations. For the first case (Fig. 3a), the data set contains a good model with RMSD < 3 Å to the native and many poor models. In this case, MDR can easily differentiate the best one from the others. When the best structure in the set has RMSD > 3 Å, the top ranked model of MDR is within 0.5 Å from the best one in most cases (Fig. 3b). In a few cases, however, when the quality of the models are similar and not good enough, the MDR method yielded only mediocre results, as shown in Fig. 3c, where the curves of different cRMSD changes mostly overlap with lack of discerning power. In summary, the performance of MDR varies for different cases while it is most efficient when the pool of models contains high-quality models (RMSD < 3 Å) besides poor ones.

4. Note

As a completely new framework for protein structure prediction, there are various limitations to address and new functionalities to implement for MUFOLD. MUFOLD currently can only handle protein monomers but not protein oligomers or complexes. We are improving the system in many aspects. For example, we are using multiple sequence alignment information to improve the distance constraints. Furthermore, the lack of solvent in the MD simulations may lead to errors of ranking, especially for structures that show comparable change in cRMSD during heating. Therefore, the MDR ranking method can be improved by considering a longer heating interval, using the GDT-TS or TM score instead of cRMSD, and including implicit solvent in the simulation, although adding implicit solvent may not be feasible in large-scale protein structure prediction due to a too long computational time. Like other tools, it is important to combine predicted structural models and wet-lab experiments to take advantage of the power of protein structure prediction.

Acknowledgments

This work has been supported by National Institutes of Health Grant R21/R33-GM078601. Major computing resource was provided by the University of Missouri Bioinformatics Consortium. We like to thank Jianlin Cheng, Yang Zhang, and Joel L. Sussman for helpful discussions.

References

1. Browne, W. J., North, A. C., Phillips, D. C., Brew, K., Vanaman, T. C., and Hill, R. L. 1969. A possible three-dimensional structure of bovine alpha-lactalbumin based on that of hen's egg-white lysozyme. J. Mol. Biol. 42:65–86.
2. K. Wuthrich, The way to NMR structures of proteins, *Nature Structural Biology* 2001; 8, 923–925.
3. Berman HM, Westbrook J, Feng Z, Gilliland G, Bhat TN, Weissig H, Shindyalov IN, Bourne PE. The protein data bank. Nucleic Acids Res. 2000; 28:235–242.
4. The UniProt Consortium. The Universal Protein Resource (UniProt). Nucleic Acids Res. 2008; 36:D190–D195.
5. Anfinsen, C.,"The formation and stabilization of protein structure". Biochem. *J.* 128 (4): 737–749.
6. Browne, W. J., North, A. C., Phillips, D. C., Brew, K., Vanaman, T. C., and Hill,R. L. 1969. A possible three-dimensional structure of bovine alpha-lactalbumin based on that of hen's egg-white lysozyme. *J. Mol. Biol.* 42:65–86.
7. HA. Monte carlo-minimization approach to the multiple-minima problem in protein folding, Proc. Natl. Acad. Sci. 1987; 84: 6611–6615.
8. Liwo A, Lee J, Ripoll DR, Pillardy J, Scheraga HA. Protein structure prediction by global optimization of a potential energy function. Proc. Natl.Acad. Sci. 1999; 96:5482–5485.
9. Simons KT, Strauss C, Baker D. Prospects for ab initio protein structural genomics. J. Mol. Biol. 2001; 306:1191–1199.
10. Zhang Y, Kolinski A, Skolnick J. TOUCHSTONE II: A New Approach to Ab Initio Protein Structure Prediction. Biophys. J. 2003; 85:1145–1164.
11. Bowie JU, Luthy R, Eisenberg D. A method to identify protein sequences that fold into a known three-dimensional structure. Science 1991; 253:164–170.
12. Sali A, Blundell TL. Comparative protein modelling by satisfaction of spatial constraints. J Mol Biol 1993; 234:779–815.
13. Soding J, Biegert A, Lupas A. The HHpred interactive server for protein homology detection and structure prediction. Nucleic Acids Research, 2005, 33:W244–W248.
14. Simons KT, Kooperberg C, Huang E, Baker D, Assembly of protein tertiary structures from fragments with similar local sequences using simulated annealing and Bayesian scoring functions, J. Mol. Biol. 1997; 268:209–225.
15. Xu Y, Xu D. Protein threading using PROSPECT: Design and evaluation. Proteins: Struct Funct Bioinformatics 2000; 40:343–354.
16. Inbar Y, Benyamini H, Nussinov R, Wolfson HJ. Protein structure prediction via combinatorial assembly of sub-structural units. Bioinformatics 2003; 19:158–168.
17. Skolnick J, Kihara D, Zhang Y. Development and large scale benchmark testing of the PROSPECTOR_3 threading algorithm. Proteins: Struct Funct Bioinformatics 2004; 56:502–518.
18. Zhang Y. I-TASSER server for protein 3D structure prediction. BMC Bioinformatics 2008; 9:40.

19. Kim DE, Chivian D, Baker D. Protein structure prediction and analysis using the Robetta server. Nucleic Acids Res. 2004; 32(2): 526–531.
20. Madden T, SchäfferA, Zhang J, Zhang Z, Miller W, Lipman D. Gapped BLAST and PSI-BLAST: a new generation of protein database search programs, Nucleic Acids Research 1997; 25(17): 3389–3402.
21. Soding J. Protein homology detection by HMM-HMM comparison. Bioinformatics 2005; 21:951–960.
22. Xu Y, Xu D. Protein threading using PROSPECT: Design and evaluation. Proteins: Struct Funct Bioinformatics 2000; 40:343–354.
23. Xu Y, Xu D, Liang J. Computational Methods for Protein Structure Prediction and Modeling, I, II, Springer-Verlag, 2006.
24. Wu Y, Lu M, Chen M, Li J, Ma J. OPUS-Ca: A knowledge-based potential function requiring only Ca positions. Protein Science 2007; 16:1449–1463.
25. Wang Z, Tegge A, Cheng J. Evaluating the absolute quality of a single protein model using structural features and support vector machines. Proteins: Struct Funct Bioinformatics 2009; 75:638–647.
26. Zhou H, Zhou Y. Distance-scaled, finite ideal-gas reference state improves structure-derived potentials of mean force for structure selection and stability prediction. Protein Science 2002; 11:2714–2726.
27. Borg I, Groenen P. Modern Multidimensional Scaling – theory and applications, Springer-Verlag, New York, 1997.
28. Torgerson WS, Multidimensional scaling of similarity, Psychometrika, 1965; 30: 379–393.
29. Tzeng J, Lu H, Li W. Multidimensional scaling for large genomic data sets. BMC Bioinformatics 2008; 9:179.
30. Zemla A. LGA: A method for finding 3D similarities in protein structures. Nucleic Acids Research 2003; 31:3370–374.
31. Zhang Y., Skolnick J., Scoring function for automated assessment of protein structure template quality. Proteins, 2004 57: 702–710.
32. Phillips J, Braun R, Wang W, Gumbart J, Tajkhorshid E, Villa E, Chipot C, Skeel R, Kale L, and Schulten K. Scalable molecular dynamics with NAMD. J Comput Chem 2005;26: 1781–1802.
33. Feig M, Rotkiewicz P, Kolinski A, Skolnick J, Brooks 3rd CL. Accurate reconstruction of all-atom protein representations from side-chain-based low-resolution models. Proteins: Struct Funct Bioinformatics 2000; 41(1):86–97.
34. Humphrey W, Dalke A, and Shulten K. VMD – Visual Molecular Dynamics. J. Molec. Graphics 1996; 14:33–38.
35. Phillips JC, Braun R, Wang W, Gumbart J, Tajkhorshid E, Villa E, Chipot C, Skeel RD, Kale L, and Schulten K. Scalable molecular dynamics with NAMD. J. Comput. Chem. 2005; 26(16):1781–1802.

Chapter 2

Prediction of Protein Functions

Roy D. Sleator

Abstract

The recent explosion in the number and diversity of novel proteins identified by the large-scale "omics" technologies poses new and important questions to the blossoming field of systems biology – What are all these proteins, how did they come about, and most importantly, what do they do?

From a comparatively small number of protein structural domains a staggering array of structural variants has evolved, which has in turn facilitated an expanse of functional derivatives. This review considers the primary mechanisms that have contributed to the vastness of our existing, and expanding, protein repertoires, while also outlining the protocols available for elucidating their true biological function. The various function prediction programs available, both sequence and structure based, are discussed and their associated strengths and weaknesses outlined.

Key words: Protein function, Homology-based transfer, Ontologies, Sequence and structure motifs, Evolution, Protein domains, Gene duplication, Divergence, Combination, Circular permutation

1. Introduction

While the famous quote from American architect, Louis Sullivan, that "form follows function" holds for man-made structures, in protein science the reverse is true – *function follows form.*

Data from the most recent large-scale sequencing projects has facilitated detailed descriptions of the constituent protein repertoires of more than 600 distinct organisms (1). Taking protein domains (clusters of 50–200 conserved residues) to represent units of evolution, as well as their more usual designation as structural/functional motifs, it is possible to accurately trace the evolutionary relationships of approximately half of these proteins (2).

Until recently, in the absence of any experimental evidence, homology-based transfer remained the gold standard for ascribing

Michael Kaufmann and Claudia Klinger (eds.), *Functional Genomics: Methods and Protocols*,
Methods in Molecular Biology, vol. 815, DOI 10.1007/978-1-61779-424-7_2, © Springer Science+Business Media, LLC 2012

a functional role to such newly identified proteins (3). Based on this approach, if a query protein shares significant sequence similarity (suggesting a common evolutionary origin) to a protein of known function, then the function of the latter may be transferred to the former (referred to as the query protein). However, as the databases continue to expand at an exponential rate, the utility of homology based prediction methods continues to contract, with fewer query proteins registering significant hits to known proteins. Herein, I review the current knowledge on protein evolution with a specific focus on how gene duplications, sequence divergence and domain combinations have shaped protein evolution. Furthermore, the most recent advances in the field of automated function prediction (AFP) are discussed, along with the future challenges and outstanding questions which still remain unanswered.

2. What Is Shaping Protein Structure?

2.1. Duplication

Of the animal genomes sequenced to date, the proportion of matched domains which are the result of duplications is estimated at between 93 and 97% (4). Indeed, the haemoglobins, which were the first homologous proteins to have their structure determined, are perhaps the best example of how duplication (and subsequent mutational events) has given rise to subtle structural and functional variations such as oxygen binding profiles (5). Furthermore, in addition to the generation of whole protein homologues, partial gene duplications resulting in domain duplication and elongation are also common features of protein evolution (6). In many cases, such enlargements have resulted from the addition of subdomains, variability in loop length, and/or changes to the structural core, such as beta-sheet extensions (7). Examples of such protein duplication events include cutinase and bovine bile-salt activated cholesterol esterase. While cutinase is the smallest enzyme of the α/β hydrolases, with five strands in the main beta-sheet (8), bovine bile-salt activated cholesterol esterase has 11 strands, and loop structures up to 79 residues in length (9).

2.2. Divergence

There are essentially two types of protein structural divergence: changes to the protein's surface or peripheral regions (e.g., surface loops, surfaces helices, and strands on the edges of ß-sheets) and the less common but far more detrimental modifications to the protein's interior or core (10). Indeed, it has been demonstrated that mutations in the protein surface are four times more biologically acceptable than those in the interior (1). In support of this is the observation that pairs of homologous proteins with identities of approximately 20% have been shown to exhibit up to 50% divergence in the peripheral regions alone (11).

In addition to subtle changes resulting from missense point mutations leading to single amino acid substitutions and the resulting gradual divergence in structure and function, more radical divergence of structure, mediated by domain shuffling (recombination or permutation) has also been reported (12). Circular permutations (CPs) in particular represent a specific form of recombination event that is characterized by the presence of the same protein subsequences in the same linear order but different positions of the N- and C-termini (13), in essence CP of a protein can be visualized as if its original termini were linked and new ones created elsewhere. First observed in plant lectins (14), a substantial number of natural examples of CP have been reported; indeed, some 120 protein clusters which appear to have segments of their sequences in different sequential order are reported in the Circular Permutation Database (15). In addition to natural evolutionary processes, artificial CPs have been engineered in an effort to study protein folding properties as well as the design of more efficient enzymes (16). A circularly permuted streptavidin, for example, has been designed to remove the flexible polypeptide loop that undergoes an open to closed conformational change when biotin is bound. The original termini have been joined by a tetrapeptide linker, and four loop residues have been removed, resulting in the creation of new N- and C-termini (16).

While domain shuffling may have dramatic effects on protein structure, protein homologues usually conserve their catalytic mechanisms, i.e., the relative positions of their functional active sites or catalytic residues may shift but they retain their functional activity. This usually occurs when divergence induces structural changes in the catalytic region, thus necessitating a reconfiguration of the position of the catalytic residues to maintain function (7). In several cases, while the functionally equivalent residues are located at non-homologous positions on the protein's 3D structure, the catalytic residues themselves are identical. An example of this is chloramphenicol acetyltransferase (PaXAT) and UDP-*N*-acetylglucosamine acyltransferase (LpxA); both of which contain an essential histidine residue thought to be involved in deprotonation of a hydroxyl group in their individual substrates. However, these residues are located at different points within the protein fold; in LpxA, the histidine is located in the core of the domain (17), whereas in PaXAT, it occurs in a loop extending from the solenoid structure.

Thus, two proteins may have quite divergent structures and/or sequences while retaining similar function; such proteins are said to be functional analogs. Such analogs may also arise as a result of convergent evolution; that is they do not diverge from a common ancestor but instead arise independently and converge on the same active configuration as a result of natural selection for a particular biochemical function. L-Aspartate aminotransferase and D-amino acid aminotransferase provide excellent examples of convergently evolved functional analogs. Despite having a strikingly similar arrangement of residues in their active sites, the two proteins have completely different architectures, differing in size, amino acid sequence, and the fold of the protein domains.

Conversely, certain proteins share significant sequence and/or structure similarity but differ in terms of substrate specificity or indeed catalytic function. An example of such structural analogs, which arise by means of divergent evolution from a single ancestor, include Human IL-10 (hIL-10), a cytokine that modulates diverse immune responses and the Epstein-Barr virus (EBV) IL-10 homolog (vIL-10). Although vIL-10 suppresses inflammatory responses like hIL-10, it cannot activate many other immune-stimulatory functions performed by the cellular cytokine (18).

2.3. Combination

While the evolutionary impact of duplication and divergence on protein sequence, structure and function is obvious, multidomain proteins are for the most part the result of gene combinations (19). Such combinations can give rise to domain recruitment and enlargement and can significantly affect both protein structure/stability and function. For example, in the case of domain recruitment the addition of an accessory domain may affect protein function by modulating substrate selectivity; achieved either by the addition of a binding site, or, by playing a purely structural role, shaping the existing active site to accommodate substrates of different shapes and/or sizes (7). For example, prokaryotic methionine aminopeptidase exists as a monomeric single-domain protein while creatinase, is a two-domain protein. The additional domain of the second subunit of creatinase caps the active site allowing the binding of the small molecule creatine (20).

3. What Is Protein Function?

Before commencing any discussion on protein function prediction we must first consider what is meant by "function". Biological function is highly contextual; different aspects of the function of a given protein may be viewed as occurring in different scales of space and time; from the almost instantaneous enzymatic reactions to the much slower overall biological process (21). Knowing which functional aspect is being investigated is thus extremely important and can only properly be achieved by the establishment of a standardized machine readable vocabulary.

Fortunately, significant progress has been made in the computer science arena in developing the theory and application of structured machine readable vocabularies, known as ontologies, which provide a formal explicit specification of a commonly used abstract model of the world (22). Ontologies not only allow formal definition of concepts but also enable the creation of software tools capable of reasoning about the properties and relationships of a domain. Formats such as the Resource Description Framework (RDF) and the Web Ontology Language (OWL) have been devised that allow ontological concepts to be persisted and communicated. RDF, for example, allows the creation of statements about a particular domain by the use of triples in the form of subject–predicate–object expressions. The subject and object represents a concept, whereas the predicate defines the relationship between them.

Detailed ontologies can be created by composing further defining concepts and relationships that model the domain of interest. Ontologies that define different aspects of proteins could be used to annotate biological data with functional facets and provide the basis of a framework for machine based reasoning.

The Gene Ontology (GO) (23) goes some way to achieving this goal, formulizing a definition of functional context and providing machine – legible functional annotation. GO has three "ontology trees" describing three aspects of gene product function: Molecular function, biological process and cellular location. By providing a standard vocabulary and defining relationships between terms, annotations can be computationally processed (24), thus providing a standard approach for programs to output their functional predictions.

Having defined biological "function" and the means of describing such functions we can now turn our attention to the various function prediction programs, and their associated strengths and weaknesses.

3.1. Protein Function Prediction Methods

Protein function prediction methods can be loosely divided into sequence and structure based approaches. Herein, we outline the current state of the art for sequence and structure based protein function prediction.

3.1.1. Sequence Based Approaches

Homology-Based Transfer

Homology-based transfer, using programs such as BLAST (25), is perhaps the most widely used form of computational function prediction method; assigning un-annotated proteins with the function of their annotated homologs. The rationale for this approach is based on the assumption that two sequences with a high degree of similarity most likely evolved from a common ancestor and thus must have similar functions.

While sequence similarity is undoubtedly correlated to functional similarity, exceptions have been observed on both ends of the similarity scale. Rost (26), for example, showed that even at high sequence similarity rates, enzymatic function may not necessarily be conserved, while Galperin et al. (27) observed that enzymes that are analogous on the basis of sequence dissimilarity are in fact homologous. While such errors are the exception rather than the rule, they may set the seed for further annotation errors; as more sequences enter the databases, more are annotated by homology-based transfer, thus helping to propagate and amplify the original single erroneous annotation (28, 29).

Furthermore, as the databases continue to expand, the utility of the homology-based transfer approach begins to break down. The recent explosion of large-scale metagenomic sequencing projects (30) has resulted in an unprecedented amount of novel sequences being deposited in the databases. As a direct consequence of this sequence expansion, the number of clustered similar proteins for which no single annotated reference sequence exists is expanding rapidly, eroding the foundations of the homology-based transfer approach. Indeed, it has been estimated that <35% of all proteins could be annotated automatically when accepting errors of ≤5%, while even allowing for error rates of >40% there is no annotation for >30% of all proteins (31).

Sequence Motifs

Typically of the 100–300 amino acids in a functional protein domain <10% constitute the protein's active sites (32). Therefore, homology-based transfer from a complete protein is often not necessary to predict a protein's function. All that is required is a sequence (or structure) based signature which is associated with a particular function. Such signatures may occur at a single position on the sequence or as a "fingerprint" composed of several such patters. A few databases are dedicated to motif searching; PROSITE (33), for example, is composed of manually selected biologically important motifs and has three types of signatures: patterns, rules, and profiles. Each signature represents a different automated method for searching motifs; while patterns and rules typically span only a few residues (e.g., A typical entry in PROSITE would be (ST)-x(2)-(DE), i.e., a Serine or Threonine, followed by any two residues, followed by Aspartate or Glutamate – the consensus sequence of a Casein kinase II phosphorylation site), profiles extend the similarity to the level of entire domains. Other well-known motif databases include BLOCKS (34) and PRINTS (35).

Genomic Context and Expression Based Prediction Methods

Genomic context based prediction, also referred to as phylogenomic profiling is a method for predicting protein function based on the observation that proteins with similar pedigrees (intergenomic profiles) are believed to have evolved in tandem and as such are likely to share a common function (36). Furthermore, in prokaryote genomes the loci of functionally related proteins tend to be colocated on the chromosome. Combining coevolution and colocation (chromosomal proximity) has given rise to a new generation of function-prediction algorithms such as Phydbac2 (37).

As an extension of colocation, genes involved in similar cellular functions also tend to be cotranscribed. Following this logic, unknown genes coexpressed with known genes may be functionally annotated by virtue of association. This "guilt by association" approach has given rise to an algorithm of the same name, developed by Walker et al. (38) for the analysis of gene expression arrays. Unlike the sequence motif based approach, which focuses on molecular function, annotation expression microarray based predictions are useful for annotation of the cellular aspect of protein function. Furthermore, given that most cellular processes are carried out by groups of physically interacting proteins, it is fair to assume that such interacting proteins have similar overall cellular functions. Thus, protein–protein interaction (PPI) data may also facilitate protein function annotation and several PPI databases are now available such as STRING – a database of known and predicted PPIs (39).

3.1.2. Structure Based Approaches

Given that protein structure is far more conserved than sequence, many proteins which exhibit little or no sequence similarities, due to evolutionary constraints still retain significant structure similarity (40). In this respect structure is a useful indicator of function; indeed most known protein folds are associated with a particular function or functional milieu (7). Programs that scan the Protein Data Bank (PDB) for structural similarity given a query sequence include, among others, FATCAT (41), PAST (42), and VAST (43). However, knowledge of 3D protein structure alone is not always sufficient to accurately infer function. Indeed, it is estimated that functional hypotheses can be made from 3D structures for only ~20–50% of hypothetical proteins (44, 45).

Rather than focusing on the protein as a whole, it is possible, and in some instances more desirable, to target 3D motifs associated with specific functions (e.g., binding sites or active sites). The rational for analysing structure motifs (or patterns) is analogous to that of sequence patterns – to identify unique signatures indicative of a particular function. Libraries of 3D motifs with known function have begun to evolve (46), one example of which is PROCAT (47), a database of 3D enzyme active sites that can be queried for specific functional signatures. In addition, hybrid motifs incorporating information from sequence and structure, as well as from the literature, have also been used to predict protein function (48).

4. Conclusions and Future Prospects

Herein, I have discussed how mechanisms such as gene duplication, sequence divergence and domain combinations (49) have shaped protein evolution and how the retention of sequence and/or structural domains has facilitated the tracking of this evolutionary process through the millennia. I have also introduced the far more complex issue of protein function elucidation wherein, in contrast to protein structure in which the data is either known or easily predicted, the multifaceted and ambiguous nature of biological function makes its elucidation a far more complex endeavor. The complexity of the problem is perhaps best illustrated by Jeffrey's (50) so called "moonlighting proteins" which perform several contextually different functions, ranging from the molecular to the cellular level. Thus, given the aggregate nature of protein function prediction, perhaps the best outcome will be achieved by adopting a multifaceted approach. For example, while biochemical function prediction is likely best served by focusing on sequence motifs, resolution of physiological function is better addressed at the genomic level, based for example on microarray expression data. Therefore, composite methods, employing a diversity of features to assess different functional aspects, are most likely to succeed. Examples of such aggregate functional prediction programs include InterPro, ProKnow and ProFunc, which utilize several data sources and/or algorithms to predict function.

However, despite the emergence of ever more sophisticated and versatile function prediction algorithms; the proper assessment of such programs still remains a significant limitation to the development of the field. Unlike assessment of protein structure, function prediction methods still lack a viable blind benchmark for which to assess program efficacy. This obstacle may eventually be overcome by emulating successful collaborative efforts of computational and experimental structural biologists in the form of CASP (Critical Assessment of Structure Prediction) for the benchmarking of protein structure.

References

1. Chothia, C., and Gough, J. (2009) Genomic and structural aspects of protein evolution, *Biochem J 419*, 15–28.
2. Sleator, R. D. (2010) An overview of the processes shaping protein evolution, *Science Progress 93*, 1–6.
3. Sleator, R. D., and Walsh, P. (2010) An overview of in silico protein function prediction, *Arch Microbiol 192*, 151–155.
4. Wilson, D., Pethica, R., Zhou, Y., Talbot, C., Vogel, C., Madera, M., Chothia, C., and Gough, J. (2009) SUPERFAMILY – sophisticated comparative genomics, data mining, visualization and phylogeny, *Nucleic Acids Res 37*, D380–386.
5. Blanchetot, A., Wilson, V., Wood, D., and Jeffreys, A. J. (1983) The seal myoglobin gene: an unusually long globin gene, *Nature 301*, 732–734.
6. Moore, A. D., Bjorklund, A. K., Ekman, D., Bornberg-Bauer, E., and Elofsson, A. (2008) Arrangements in the modular evolution of proteins, *Trends Biochem Sci 33*, 444–451.
7. Todd, A. E., Orengo, C. A., and Thornton, J. M. (2001) Evolution of function in protein superfamilies, from a structural perspective, *J Mol Biol 307*, 1113–1143.
8. Longhi, S., Czjzek, M., Lamzin, V., Nicolas, A., and Cambillau, C. (1997) Atomic resolution (1.0 A) crystal structure of Fusarium solani cutinase: stereochemical analysis, *J Mol Biol 268*, 779–799.
9. Chen, J. C., Miercke, L. J., Krucinski, J., Starr, J. R., Saenz, G., Wang, X., Spilburg, C. A., Lange, L. G., Ellsworth, J. L., and Stroud, R. M. (1998) Structure of bovine pancreatic cholesterol esterase at 1.6 A: novel structural features involved in lipase activation, *Biochemistry 37*, 5107–5117.
10. Gerstein, M., Sonnhammer, E. L., and Chothia, C. (1994) Volume changes in protein evolution, *J Mol Biol 236*, 1067–1078.
11. Chothia, C., and Lesk, A. M. (1986) The relation between the divergence of sequence and structure in proteins, *EMBO J 5*, 823–826.

12. Kawashima, T., Kawashima, S., Tanaka, C., Murai, M., Yoneda, M., Putnam, N. H., Rokhsar, D. S., Kanehisa, M., Satoh, N., and Wada, H. (2009) Domain shuffling and the evolution of vertebrates, *Genome Res* ***19***, 1393–1403.
13. Vogel, C., and Morea, V. (2006) Duplication, divergence and formation of novel protein topologies, *Bioessays* ***28***, 973–978.
14. Lindqvist, Y., and Schneider, G. (1997) Circular permutations of natural protein sequences: structural evidence, *Curr Opin Struct Biol* ***7***, 422–427.
15. Lo, W. C., Lee, C. C., Lee, C. Y., and Lyu, P. C. (2009) CPDB: a database of circular permutation in proteins, *Nucleic Acids Res* ***37***, D328–332.
16. Heinemann, U., Ay, J., Gaiser, O., Muller, J. J., and Ponnuswamy, M. N. (1996) Enzymology and folding of natural and engineered bacterial beta-glucanases studied by X-ray crystallography, *Biol Chem* ***377***, 447–454.
17. Wyckoff, T. J., and Raetz, C. R. (1999) The active site of Escherichia coli UDP-N-acetylglucosamine acyltransferase. Chemical modification and site-directed mutagenesis, *J Biol Chem* ***274***, 27047–27055.
18. Yoon, S. I., Jones, B. C., Logsdon, N. J., and Walter, M. R. (2005) Same structure, different function crystal structure of the Epstein-Barr virus IL-10 bound to the soluble IL-10R1 chain, *Structure* ***13***, 551–564.
19. Apic, G., Gough, J., and Teichmann, S. A. (2001) Domain combinations in archaeal, eubacterial and eukaryotic proteomes, *J Mol Biol* ***310***, 311–325.
20. Hoeffken, H. W., Knof, S. H., Bartlett, P. A., Huber, R., Moellering, H., and Schumacher, G. (1988) Crystal structure determination, refinement and molecular model of creatine amidinohydrolase from Pseudomonas putida, *J Mol Biol* ***204***, 417–433.
21. Godzik, A., Jambon, M., and Friedberg, I. (2007) Computational protein function prediction: are we making progress? *Cell Mol Life Sci* ***64***, 2505–2511.
22. Losko, S., and Heumann, K. (2009) Semantic data integration and knowledge management to represent biological network associations, *Methods Mol Biol* ***563***, 241–258.
23. Ashburner, M., and Lewis, S. (2002) On ontologies for biologists: the Gene Ontology – untangling the web, *Novartis Found Symp* ***247***, 66–80; discussion 80–63, 84–90, 244–252.
24. Smith, C. L., Goldsmith, C. A., and Eppig, J. T. (2005) The Mammalian Phenotype Ontology as a tool for annotating, analyzing and comparing phenotypic information, *Genome Biol* ***6***, R7.
25. Altschul, S. F., Madden, T. L., Schaffer, A. A., Zhang, J., Zhang, Z., Miller, W., and Lipman, D. J. (1997) Gapped BLAST and PSI-BLAST: a new generation of protein database search programs, *Nucleic Acids Res* ***25***, 3389–3402.
26. Rost, B. (2002) Enzyme function less conserved than anticipated, *J Mol Biol* ***318***, 595–608.
27. Galperin, M. Y., Walker, D. R., and Koonin, E. V. (1998) Analogous enzymes: independent inventions in enzyme evolution, *Genome Res* ***8***, 779–790.
28. Bork, P. (2000) Powers and pitfalls in sequence analysis: the 70% hurdle, *Genome Res* ***10***, 398–400.
29. Gilks, W. R., Audit, B., de Angelis, D., Tsoka, S., and Ouzounis, C. A. (2005) Percolation of annotation errors through hierarchically structured protein sequence databases, *Math Biosci* ***193***, 223–234.
30. Sleator, R. D., Shortall, C., and Hill, C. (2008) Metagenomics, *Lett Appl Microbiol* ***47***, 361–366.
31. Rost, B., Liu, J., Nair, R., Wrzeszczynski, K. O., and Ofran, Y. (2003) Automatic prediction of protein function, *Cell Mol Life Sci* ***60***, 2637–2650.
32. Friedberg, I. (2006) Automated protein function prediction – the genomic challenge, *Brief Bioinform* ***7***, 225–242.
33. Hulo, N., Bairoch, A., Bulliard, V., Cerutti, L., Cuche, B. A., de Castro, E., Lachaize, C., Langendijk-Genevaux, P. S., and Sigrist, C. J. (2008) The 20 years of PROSITE, *Nucleic Acids Res* ***36***, D245–249.
34. Henikoff, J. G., Greene, E. A., Pietrokovski, S., and Henikoff, S. (2000) Increased coverage of protein families with the blocks database servers, *Nucleic Acids Res* ***28***, 228–230.
35. Attwood, T. K., Bradley, P., Flower, D. R., Gaulton, A., Maudling, N., Mitchell, A. L., Moulton, G., Nordle, A., Paine, K., Taylor, P., Uddin, A., and Zygouri, C. (2003) PRINTS and its automatic supplement, prePRINTS, *Nucleic Acids Res* ***31***, 400–402.
36. Eisenberg, D., Marcotte, E. M., Xenarios, I., and Yeates, T. O. (2000) Protein function in the post-genomic era, *Nature* ***405***, 823–826.
37. Enault, F., Suhre, K., and Claverie, J. M. (2005) Phydbac "Gene Function Predictor": a gene annotation tool based on genomic context analysis, *BMC Bioinformatics* ***6***, 247.
38. Walker, M. G., Volkmuth, W., Sprinzak, E., Hodgson, D., and Klingler, T. (1999) Prediction of gene function by genome-scale expression analysis: prostate cancer-associated genes, *Genome Res* ***9***, 1198–1203.

39. Zhao, X. M., Chen, L., and Aihara, K. (2008) Protein function prediction with high-throughput data, *Amino Acids 35*, 517–530.

40. Watson, J. D., Laskowski, R. A., and Thornton, J. M. (2005) Predicting protein function from sequence and structural data, *Curr Opin Struct Biol 15*, 275–284.

41. Ye, Y., and Godzik, A. (2004) FATCAT: a web server for flexible structure comparison and structure similarity searching, *Nucleic Acids Res 32*, W582–585.

42. Taubig, H., Buchner, A., and Griebsch, J. (2006) PAST: fast structure-based searching in the PDB, *Nucleic Acids Res 34*, W20–23.

43. Gibrat, J. F., Madej, T., and Bryant, S. H. (1996) Surprising similarities in structure comparison, *Curr Opin Struct Biol 6*, 377–385.

44. Laskowski, R. A., Watson, J. D., and Thornton, J. M. (2003) From protein structure to biochemical function? *J Struct Funct Genomics 4*, 167–177.

45. Goldsmith-Fischman, S., and Honig, B. (2003) Structural genomics: computational methods for structure analysis, *Protein Sci 12*, 1813–1821.

46. Jones, S., and Thornton, J. M. (2004) Searching for functional sites in protein structures, *Curr Opin Chem Biol 8*, 3–7.

47. Wallace, A. C., Laskowski, R. A., and Thornton, J. M. (1996) Derivation of 3D coordinate templates for searching structural databases: application to Ser-His-Asp catalytic triads in the serine proteinases and lipases, *Protein Sci 5*, 1001–1013.

48. Di Gennaro, J. A., Siew, N., Hoffman, B. T., Zhang, L., Skolnick, J., Neilson, L. I., and Fetrow, J. S. (2001) Enhanced functional annotation of protein sequences via the use of structural descriptors, *J Struct Biol 134*, 232–245.

49. Chothia, C., Gough, J., Vogel, C., and Teichmann, S. A. (2003) Evolution of the protein repertoire, *Science 300*, 1701–1703.

50. Jeffery, C. J. (2003) Moonlighting proteins: old proteins learning new tricks, *Trends Genet 19*, 415–417.

Chapter 3

Genome-Wide Screens for Expressed Hypothetical Proteins

Claus Desler, Jon Ambæk Durhuus, and Lene Juel Rasmussen

Abstract

A hypothetical protein (HP) is defined as a protein that is predicted to be expressed from an open reading frame, but for which there is no experimental evidence of translation. HPs constitute a substantial fraction of proteomes of human as well as of other organisms. With the general belief that the majority of HPs are the product of pseudogenes, it is essential to have a tool with the ability of pinpointing the minority of HPs with a high probability of being expressed.

Key words: Hypothetical proteins, *In silico*, Pseudogenes

1. Introduction

The number of fully sequenced prokaryotic and eukaryotic organisms is ever increasing. From each organism, it is possible to identify open reading frames in the genome and thereby predict the total number of protein-encoding genes in the organism. For a greater proportion of these predicted proteins, translation of the protein has been verified and the function of the protein is likely to have been experimentally characterized. However, for the remaining group of genes predicted to encode proteins, translation has not been demonstrated and the proteins themselves have not been characterized. This group of proteins is accordingly defined as hypothetical.

Although many hypothetical proteins (HPs) most likely are predicted products of pseudogenes, there is a reasonable probability that a number of the HPs are truly novel and can perform uncharacterized biological functions. HPs can for that reason add knowledge to and/or constitute the key points missing for the understanding of biological pathways of an organism, specific biological mechanisms or pathologic conditions. Therefore, it makes good sense to mine

Michael Kaufmann and Claudia Klinger (eds.), *Functional Genomics: Methods and Protocols*,
Methods in Molecular Biology, vol. 815, DOI 10.1007/978-1-61779-424-7_3, © Springer Science+Business Media, LLC 2012

the HPs of an organism for translatable candidates. Screening HPs using in vitro and/or in vivo experiments can prove to be very laborious and an initial screening using *in silico* methods will be very helpful in finding the most probable candidates for subsequent in vitro and in vivo analyses.

The purpose of this chapter is to demonstrate how it is possible to screen HPs using a combination of *in silico* methods originally intended for the prediction of functions of proteins assumed to be expressed. As the field of bioinformatics is quickly progressing and different *in silico* models are often improved, replaced or abandoned, the focus of this chapter is on how to select the best models and how to combine different models to best screen HPs for translatable candidates. Examples are given with current models but the strategy will still be valid, even though models present at writing will become replaced or heavily modified.

1.1. Rationale of In Silico Selection Strategy

To this date, no single *in silico* model exist for the prediction of translatable candidates amongst HPs. Instead, it has been demonstrated that it is possible to device a selection strategy that can successfully sort HPs according to their probability of being translatable proteins (1). This strategy is based on *in silico* models normally used to make descriptive predictions of characterized proteins with unknown function. As the models are developed for use with characterized proteins, they are unable to differentiate between translatable proteins and the predicted product of a pseudogene. Therefore, if a HP is predicted by an *in silico* model to have a specific attribute, this does not necessarily increase the probability of this HP to be expressed. However, we have demonstrated that if more than one *in silico* model can predict independent attributes of a hypothetical protein, the probability of the protein to be translatable is increased (1) (see Fig. 1).

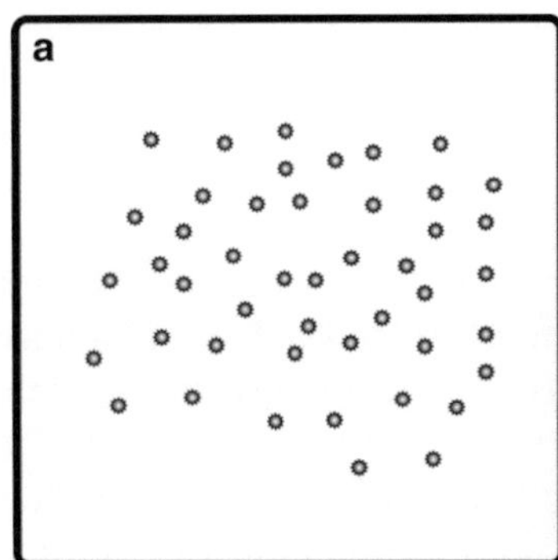

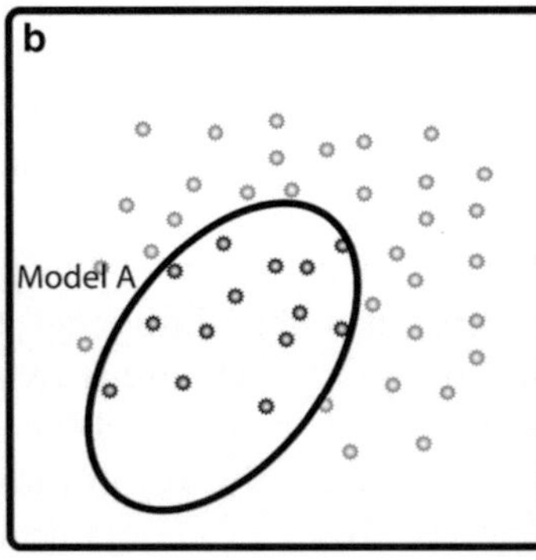

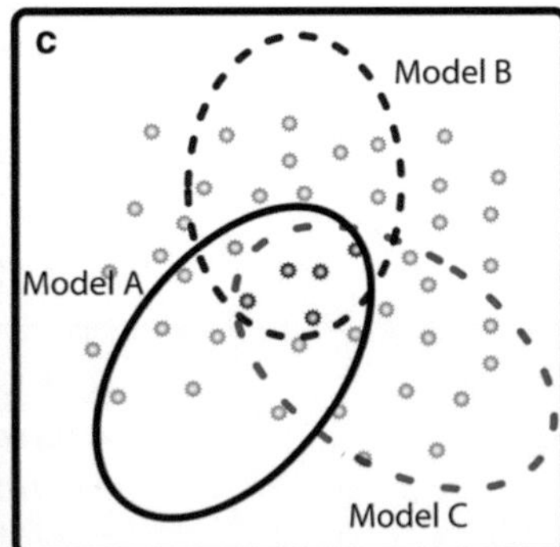

Fig. 1. (**a**) A collection of hypothetical proteins (HPs). (**b**) By using the in silico model A, it is possible to predict a specific biological attribute in a subset of the HPs. This does not necessarily mean that these HPs are translatable. The prediction model has been designed for analysis of proteins known to be translatable and cannot distinguish between translatable proteins and pseudogenes. (**c**) By analyzing the HPs with complementing in silico models B and C, a subset of HPs can be selected as having biological properties predicted by more than only one in silico model. This subset of proteins is more likely to contain translatable ones.

2. Selecting Components for an In Silico Selection Strategy

The focus of this chapter is on protein properties that can readily be predicted or identified by a good selection of contemporary and, most likely, future *in silico* models. In this chapter, four different protein properties are described. These include the following: existence of orthologs or paralogs, presence of protein domains, presence of subcellular targeting signals, and comparison of tertiary structure with known structures. For each property, a list of related *in silico* models is reviewed for usage, selected models are introduced for the reader, and the premises for the predictions made by the model are explained. To demonstrate the setup and usage of an *in silico* selection strategy, selected models are used to analyze a dataset of HPs. The dataset is an extract of a database of proteins extracted from GenBank in August 2006. At the time of extraction, all proteins were defined as hypothetical. We have selected a subset of this dataset and reinvestigated their present annotated status (2011). The entries of the 2006 dataset were divided into two groups according to their individual status in 2011: (A) characterized proteins and (B) proteins discovered to be pseudogenes and therefore removed by GenBank (see Table 1). By screening the dataset with the different *in silico* prediction models,

Table 1
Proteins defined as hypothetical in 2006 and their corresponding status in 2011

Proteins defined as hypothetical in 2006	Status of protein in 2011
NP_000009	Characterized protein
NP_000373	Characterized protein
NP_057164	Characterized protein
NP_060358	Characterized protein
NP_060616	Characterized protein
NP_001013750	Removed by GenBank
XP_496960	Removed by GenBank
NP_001001677	Removed by GenBank
NP_001004331	Removed by GenBank
XP_379036	Removed by GenBank
XP_939886	Removed by GenBank

Characterized proteins have been demonstrated to be expressed in vivo or in vitro. Removed proteins, are HPs that have been discovered to be the predicted product of a pseudogene we can use the results according to the selection strategy and predict which of the HPs are translatable candidates. By comparing our predicted result with the experimentally determined status of each of the proteins of the dataset in 2011, it is possible to comment on the fidelity of our selection strategy. The two proteins NP_057164 and NP_001004331 have been selected to serve as examples, and throughout the chapter, these proteins will be analyzed in greater detail than the remainder of the proteins listed in Table 1. NP_057164 is an example of a protein that was annotated as a HP in 2006, and in 2011 has been experimentally characterized. NP_001004331 was annotated as a HP in 2006, but have later been discovered to be a pseudogene. The proteins constituting the dataset used have been selected for pedagogic purposes, but they are all part of a much larger dataset previously published (http://www.biomedcentral.com/content/supplementary/1471-2105-10-289-S1.xls). The reader is encouraged to choose another dataset than the selected one and perform the screening to get a more independent view of the fidelity of the presented selection strategy.

2.1. Prediction Model: Existence of Orthologs or Paralogs

Alignment of a HP of interest can be used to yield information by comparison to existing proteins. Orthologs and paralogs or proteins with similar conserved domains could shed light on the function of the HP. Orthologous genes are similar genes in different species separated by specification, while paralogs are homologous sequences separated by gene duplication.

2.1.1. Models Available for Prediction

The basic local alignment search tool (blast) is a frequently used bioinformatical tool for comparing sequence similarity (2). Blast uses a heuristic algorithm that detects relationships among sequences sharing only isolated regions of similarity as opposed to global alignment, which is used to compare sequences similar in length by aligning every residue in the sequences compared. Because blast has been designed for speed, there is a risk of loss of sensitivity to low sequence similarity. A range of blast variants exist enabling more specific and/or sensitive analyses. The blast applications are listed on the blast homepage (http://www.ncbi.nlm.nih.gov/BLAST) where they are categorized as basic and specialized blasts (see Table 2). Basic blast tools useful for proteins include blastp, which is a protein–protein search, using a protein query against a protein database, and tblastn that searches a translated nucleotide database in all six reading frames using a protein query.

2.1.2. Advantages and Disadvantages

Blast is a fast and relatively easy procedure to find similar proteins based on comparison of amino acid sequences, emphasizing speed over sensitivity. However, there is a risk of false positives if accepting to low cutoff values.

Table 2
Selected basic local alignment search tools (blast) listed with specifications (http://blast.ncbi.nlm.nih.gov/Blast.cgi)

Tool	Comments
Protein blast	Search protein database using a protein query with the following algorithms: blastp, PSI-blast, or PHI-blast
Blastx	Search protein database using a translated nucleotide query
Tblastn	Search translated nucleotide database using a protein query
Tblastx	Search translated nucleotide database using a translated nucleotide query

2.1.3. How to Use: Blast

Retrieve FASTA file from protein of interest and insert into query sequence in the blast tool on NCBIs homepage. Choose blastp when searching proteins (http://blast.ncbi.nlm.nih.gov/). Consider which database and organism your query should be aligned with. If nothing is chosen, it will perform the search within nonredundant protein sequences across organisms by default. Choose the blastp (protein–protein) algorithm and press blast. A list of sequences with significant alignment (if any) will turn up showing query coverage, similarity and identified conserved domains.

2.1.4. What Constitutes a Positive Hit?

Two protein sequences can in general be regarded as having close homology if the percentage is above 30%, while proteins sharing 20–30% identity are less certain. The sequence length has to be taken into consideration as smaller peptides have a higher risk of alignment by chance. It should be noted that identity values only provide tentative guidance for possible homology.

2.2. Prediction Model: Protein Domains

Even though the different proteins of an organism can have very different and unique properties, the composition of each individual protein is not necessarily unique. Distinct protein parts have been demonstrated to be reoccurring throughout a large number of proteins. These parts are called protein domains and are defined as conserved parts of protein sequence and structure that are functionally independent of the protein it is occurring in.

The existence of protein domains represents a huge evolutionary advantage. Instead of having to invent every protein-encoding gene by means of random nucleotide substitutions, it is possible to evolve new genes by collecting different protein domains that to a large extend are able to stably fold independently when translated. Many of the protein domains have very distinct properties that can be of use in a variety of proteins. Such as zinc fingers that often mediate the binding of RNA or DNA, or ATP binding domains,

Table 3
A selection of programs capable to identify protein domains in a given protein

Model	URL
pFam	http://www.sanger.co.uk/Pfam/
Prosite	http://www.expasy.org/prosite/
SMART	http://www.smart.embl-heidelberg.de/
Superfamily	http://www.supfam.mrc-lmb.cam.ac.uk/SUPERFAMILY/hmm.html

which have the ability to bind and hydrolyze ATP. It is, therefore, clear that the combination of a very reduced set of protein domains can result in a wide variety of protein encoding genes with very diverse properties. To this day, more than 10,000 different protein domains have been identified.

2.2.1. Models Available for Prediction

Due to their conserved nature, protein domains can be easily identified. The range of available models identifies protein domains, by comparing different regions of the protein of interest with a database of already annotated protein domains (see Table 3).

2.2.2. Advantages and Disadvantages

The databases forming the basis of protein domain identification are most often of a reasonable quality. They have been compiled using multiple sequence alignments and hidden Markov models, and have often been refined by manual curation, therefore identification of protein domains is precise.

Not all protein domains are equally complex and a protein region for example can be identified as a transmembrane domain if the region spans between 25 and 35 amino acids and have satisfactory hydrophobic qualities. Similarly, coiled coil regions and signal peptides can be very simplistic. The less complex a protein domain is, the less the chance of the structure contributing to the properties of the protein. When identifying protein domains in a HP, it is therefore important that one considers the complexity of the identified protein domains. If a HP only has simple domains as transmembrane domains, coiled coil regions or signal peptides, or similar simple structures, it is recommended that the queried protein is regarded as not having any protein domains.

Advantages	Disadvantages
• Accuracy can be very high	• User have to evaluate the complexity of found protein domains

2.2.3. How to Use: SMART and Prosite

To demonstrate the identification of protein domains within a hypothetical protein, the SMART and Prosite programs (3–5) have been chosen as examples (http://smart.embl-heidelberg.de/). The SMART program is currently available in two versions: normal or genomic mode. In normal SMART, the database of protein domains is compiled from all available proteomes from Swiss-Prot, SP-TrEMBL, and Ensembl. In genomic SMART, the database of protein domains is compiled only from the proteomes of completely sequenced genomes. The database used in normal SMART is the most comprehensive, and will be used for our purposes.

Retrieve FASTA file from protein of interest and insert into query sequence on the SMART homepage (http://smart.embl-heidelberg.de/). Notice that the program has several options available. Make sure that you have checked the option "PFAM domains," as this allows for the consecutive use of the PFAM database together with the SMART database of protein domains. Press the "Sequence SMART" button and await the result of the program.

Identified protein domains within the queried protein will be graphically presented if found. One of the strengths of the SMART program is the user friendliness of the program. By clicking on the graphical representation of a protein domain it is possible to obtain detailed information of the protein domain, including, in which species the domain is found, literature related to the domain and structure of the domain. This information can be used to evaluate the complexity of the found protein domain.

The Prosite program is very similar to SMART in both use and output, but benefits from using a different database. The two programs can, therefore, supplement each other.

2.2.4. What Constitutes a Positive Hit?

Even though the presence of multiple complex protein domains exponentially increases the chance of the queried hypothetical protein to be a translatable protein, we have found that even the presence of a single protein domain increases the chance of the protein to be expressed. The user should, however, ignore proteins of low complexity, as these regions are not unique protein domains. When analyzing the HP NP_057164 with the SMART program, a glyoxalase Pfam domain is identified, while analysis of HP NP_001004331 only yields the identification of a coiled coil region. This corresponds very clearly with the current annotation of the proteins where NP_057164 has been experimentally characterized and NP_001004331 found to be a pseudogene.

2.3. Prediction Model: Subcellular Targeting Signals

With the exception of proteins encoded by the mitochondrial genome, eukaryotic proteins are translated in the cytosol from their corresponding mRNA. Many proteins are transported to specific parts of the cell where they function in context of the subcellular compartment. The sub-cellular localization of proteins can be facilitated by specific targeting peptides. There are two types of targeting peptides, the presequences and the internal targeting signals. Presequences are often localized at the N-terminal whereas internal targeting signals can be distributed throughout the whole protein sequence.

2.3.1. Models Available for Prediction

Prediction models for subcellular targeting signals primarily use two approaches: Either they evaluate the N-terminal region of the investigated protein for the presence of presequences or they search the entire protein for domains found in proteins known to localize to a specific cellular compartment, and are therefore believed to be internal targeting signals. For a eukaryotic cell, localization in up to 12 different compartments can be predicted.

2.3.2. Advantages and Disadvantages

Not all translatable proteins in a hypothetical population will have an equal probability of being identified using models predicting subcellular localization. Using this type of prediction models primarily applies for the search of translatable candidates that upon translation will localize to a cellular compartment. Therefore, translatable candidates not encoding targeting signals will most likely not be identified. Usage of this type of prediction models as a part of a selection strategy will consequently be most beneficial if the purpose of the screen is to find HPs that have a high probability of being translated AND localized to a cellular compartment of interest.

Not all targeting signals of the different cellular compartments are equally well characterized and understood. As an example, targeting for mitochondrial localization is fairly well characterized, while the targeting signals determining localization of proteins to the nuclear envelope have not to the same degree been elucidated. It is, therefore, important to understand the criteria on which a prediction model derives its predictions. As an example, most mitochondrial precursor proteins posses N-terminal presequences that generally have a length of 6–85 amino acid residues, enriched in Arg, Ser, and Ala, while negatively charged amino acids are rarely present. These N-terminal presequences are well characterized and are used by TargetP to predict mitochondrial localization with an accuracy of 90% (6) (see Table 4). By contrast, the prediction model pTarget screens for putative protein domains that have been related to a specific cellular localization but not necessarily for complete targeting signals. Even though the reported accuracy is between 68 and 87% (7) (see Table 4), this does not necessarily mean that the prediction model is well suited for screening of HPs. If a HP turns out to be a pseudogene, it will most likely consist of duplications from other proteins. Even though several protein domains related to specific cellular locations are found within a HP, the complete targeting signal is not guaranteed to be complete and functional.

Table 4
A selection of subcellular localization prediction programs for eukaryotic proteins reported to have a medium to high prediction accuracy

Model	Number of localization sites	Reported accuracy (%)
BaCelLo (http://gpcr.biocomp.unibo.it/bacello/)	4–5	67–76
MITOPRED (http://bioapps.rit.albany.edu/MITOPRED/)	1	85
MultiLoc2 (http://www-abi.informatik. uni-tuebingen.de/Services/MultiLoc2)	11	75
PA-SUB (http://www.cs.ualberta.ca/~bioinfo/PA/Sub/)	11	81–94
pTarget (http://bioapps.rit.albany.edu/pTARGET/)	9	68–87
TargetP (http://www.cbs.dtu.dk/services/TargetP/)	3	90
WoLF PSORT (http://wolfpsort.org/)	12	80

Listed are the numbers of compartments each program can predict as a target, and the reported accuracy of the prediction

Advantages	Disadvantages
• Accuracy can be very high • Several prediction models available	• Only identifies translatable proteins that will be localized to a cellular compartment • Some targeting signals are better characterized than others

2.3.3. How to Use: TargetP

To demonstrate the identification of protein domains within a HP, the TargetP program has been chosen as an example (http://www.cbs.dtu.dk/services/TargetP/). Retrieve FASTA file from protein of interest and insert into query sequence on the TargetP homepage. Make sure that you have selected the correct organism group (nonplant or plant). Choose "no cutoffs: winner-takes-all" which is also the default setting and finally, submit your HP and await the result of the program.

As the result of the programs analyses, the user will be met with a text output. Of the different panes of output the two most important are denominated "Loc" and "RC". Loc is short for predicted localization, ranging from "C" for chloroplast (if a plant protein is investigated), "M" for mitochondrion, "S" for secretory pathway and "–" and "*" for proteins that are not predicted to be targeted for any of the three cellular compartments. RC is short for reliability class. The reliability class is a number from 1 to 5 where 1 indicates the strongest prediction. TargetP also provides a prediction score, but this number is arbitrary, and it is much easier and less confusing to just use the reliability classes.

2.3.4. What Constitutes a Positive Hit?

Prediction models for subcellular targeting signals often provide some measure of reliability of their prediction. In the case of TargetP, this is the RC score, whereas other models have other measures of reliability. The prediction models can be regarded in a simple way, in which case prediction to any cellular compartment can be regarded as a positive hit when a sensible cutoff value is selected. For TargetP such a cutoff is RC class 1 and 2. Alternatively, the reliability scores can be useful for giving a more detailed prediction of whether a HP can be predicted to be expressed instead of just a "yes" or "no".

When querying the protein NP_057164 in TargetP, the protein is predicted to localize to the mitochondria. The RC value of this prediction is 1, indicating that the protein fulfills most or all of the criteria searched for by the prediction program. By contrast, the pseudogene NP_001004331 was neither predicted to localize to the mitochondria nor to be secreted. This corresponds to its now known annotation as a pseudogene. However, it could have been a translatable protein that was soluble or localized to another cellular compartment.

2.4. Prediction Model: Comparison of Tertiary Structure

By analyzing tertiary structures of proteins, new functional regions might be revealed. The RSCB protein data bank (http://www.pdb.org/pdb/home/home.do) currently has approximately 72,000 PDB structures. HPs are, however, not crystallized and no PDB structure, therefore, exists.

Predictions of tertiary structure can be made by bioinformatical tools as I-TASSER (8, 9) (http://zhanglab.ccmb.med.umich.edu/I-TASSER/). The iterative threading assembly refinement (I-TASSER) server is an online program for protein structure and function predictions made by Yang Zhang's lab. It has been ranked as the number one server for protein structure prediction in the last three Critical Assessment of Techniques for Protein Structure Predictions (CASP) (http://predictioncenter.org/). The main goal of the CASP experiments is to test the accuracy of structure prediction on protein structures that are soon to be crystallized.

Starting from an amino acid sequence, I-TASSER first generates 3D structure models from multiple threading alignments and iterative structural assembly simulations. Threading is a bioinformatical approach, used for fold recognition in proteins where the tertiary structure has not been experimentally characterized and therefore not found in the RSCB protein data bank. Possible functions of the protein are then derived by structurally matching the 3D structure model with the tertiary structure of other known proteins.

The output from a server run contains full-length secondary and tertiary structure predictions, functional annotations on ligand-binding sites, enzyme commission numbers and gene ontology terms. An estimate of accuracy of the predictions is also provided based on the confidence score of the modeling. The target sequence is first threaded through a PDB structure library to search for the possible folds by four methods, with different combinations of the hidden Markov models and PSI-blast profiles plus other alignment algorithms. I-TASSER assembles full-length models from identified fragments, while unaligned regions are built by *ab initio* (from scratch) modeling. The remaining structure is predicted utilizing several bioinformatical tools to remove steric clashes and to refine the generated 3D structure model further. A generated model of a HP can be used to visualize the tertiary structure of the HP investigated, but more important, to find similar proteins that have been experimentally characterized and thereby suggest analogous functional properties of the HP (8, 9).

2.4.1. How to Use: I-TASSER

Submit sequence to the I-TASSER server, fill in mandatory fields and press run I-TASSER. Consider to specify a template with or without a chosen alignment and excluding specific templates. When the results are finished, the submitted sequence and the predicted secondary structure, the solvent accessibly, the top models in 3D and the top ten models used by I-TASSER are shown.

2.4.2. Advantages and Disadvantages

The predictions made by I-TASSER are very accurate according to CASP and the server accepts sequences up to 1,500 residues, which is more than accepted by most other programs. However, the predictions are time-consuming and only one sequence at a time can be submitted. The structural information provided by the 3D structure model can provide new insights by elucidating similarity to known proteins with well characterized biological functions. Moreover, analogy that is undetected by sequence comparison can be recognized, as well as binding motifs and catalytic centers.

2.4.3. What Constitutes a Positive Hit

A selection of HPs was analyzed by I-TASSER and two different HPs were chosen as examples. For a HP which has been identified as a true protein we again chose NP_057164 a human glyoxalase domain-containing protein. Up to five predicted 3D structure models of protein queried are shown after analysis by I-TASSER. Each 3D structural model is given a C-score. The *C*-score is a confidence score based on the significance of threading template alignments and the convergence parameters of the structure assembly simulations. A high *C*-score signifies high confidence and the *C*-score is usually in the range from −5 to 2, from low to high confidence respectively. The highest scoring model of NP_057164 has a *C*-score of −0.39 and has the highest similarity with the crystal structure of *Bacillus cereus* metalloprotein from the glyoxalase family followed by other glyoxalases. It is also the *B. cereus* metalloprotein that has the highest rank among the templates used to build the 3D structure. I-TASSER also presents predicted binding sites and gene ontology terms.

The chosen HPs, which now have been removed from NCBI have in general a very poor *C*-score. NP_001004331 was chosen as an example of a HP, where there is no support for the transcript of the protein. This HP has a *C*-score of −3.36 indicating a low confidence. The predicted protein models seem unfolded and there is no consensus of the used templates and the structurally closest PDBs. The validation of HPs as translatable protein candidates can be deduced by a high *C*-score. However, the simplicity of structure has also to be taken into account. As the PDB library is expanding the power of such bioinformatical tools strengthen too.

3. Devising In Silico Selection Strategy from Selected Models

A HP may have a, yet uncharacterized, role in a biological context or simply be the predicted result of a pseudogene and have no biological relevance. The purpose of this chapter is to demonstrate how an *in silico* selection strategy for finding translatable candidates amongst HPs can be devised. We have argued that such a selection strategy can be based on *in silico* models, normally used to make descriptive predictions of characterized proteins with unknown function. Through the chapter we have reviewed four different protein attributes that can be reliantly predicted or identified and we have discussed existing models for doing so. We have utilized a database of HPs dating from 2006 and reviewed their annotated status in 2011. Accordingly, we can verify our selection strategy by reviewing the proteins that were hypothetical in 2006, but which have later been experimentally characterized. Unfortunately, it is not possible to demonstrate the predictive properties of finding orthologs or paralogs for HPs on the dataset of proteins annotated as hypothetical in 2006. The blast procedure will evaluate the queried protein as per its annotation in 2011 and not as per its status in 2006.

By using the reviewed programs, SMART, Prosite, TargetP and I-TASSER on the database of proteins annotated as hypothetical in 2006 (see Table 5), it is evident that the HPs which have been demonstrated to be translatable proteins, to a much higher degree than pseudogenes, have protein properties which have been possible to be identified with the used programs. This demonstrates the validity of the *in silico* selection strategy. The programs used for the selection strategy either identifiy or predict well-defined properties of the queried proteins and their output can be easily identifiable as either positive or negative hits. By using other *in silico* models with the same attributes, it is possible to replace or supplement the reviewed programs and still get a reliable prediction

Table 5
Proteins annotated as hypothetical in 2006 were analyzed using a multitude of prediction models

	Protein domain		Subcellular targeting signals	Tertiary structure	
Accession no.	**SMART**	**Prosite**	**TargetP**	**I-TASSER**	**Protein status 2011**
NP_000009	+	+	+		Characterized protein
NP_000373	+	+	+		Characterized protein
NP_057164	+	–	+	+	Characterized protein
NP_060358	+	+	+		Characterized protein
NP_060616	+	–	+		Characterized protein
NP_001013750	+	+	+		Removed by GenBank
XP_496960	+	–	+	–	Removed by GenBank
NP_001001677	–	–	–	–	Removed by GenBank
NP_001004331	–	–	–	–	Removed by GenBank
XP_379036	–	–	–	–	Removed by GenBank
XP_939886	–	–	–	–	Removed by GenBank

+ indicate that the used model was able to identify a specific property, while – indicate nothing was predicted. Owing to the workload of each submitted protein prediction by I-TASSER only a limited set of HPs were chosen for analysis

of which HPs are translatable candidates and which are most likely the predicted product of pseudogenes.

A simple *in silico* selection strategy is, therefore, an obvious first step when screening hypothetical proteins for translatable candidates, highlighting the group of proteins where further in vitro or in vivo characterization will be most productive.

Acknowledgments

This work was supported by a grant from the NORDEA foundation.

References

1. Desler, C., Suravajhala, P., Sanderhoff, M., Rasmussen, M., and Rasmussen, L.J. (2009). In Silico screening for functional candidates amongst hypothetical proteins. *BMC Bioinformatics* **10**, 289.
2. Altschul, S.F., Gish, W., Miller, W., Myers, E.W., and Lipman, D.J. (1990). Basic local alignment search tool. *J Mol Biol* **215**, 403–410.
3. Schultz, J., Milpetz, F., Bork, P., and Ponting, C.P. (1998). SMART, a simple modular architecture

research tool: identification of signaling domains. *PNAS USA* **95**, 5857–864.

4. Letunic, I., Copley, R.R., Pils, B., Pinkert, S., Schultz, J., and Bork, P. (2006). SMART 5: domains in the context of genomes and networks. *Nucleic acids research* **34**, D257–260.
5. de Castro, E., Sigrist, C.J.A., Gattiker, A., Bulliard, V., Langendijk-Genevaux, P.S., Gasteiger, E., Bairoch, A., and Hulo, N. (2006). ScanProsite: detection of PROSITE signature matches and ProRule-associated functional and structural residues in proteins. *Nucleic acids research* **34**, W362–365.
6. Emanuelsson, O., Brunak, S., von Heijne, G., and Nielsen, H. (2007). Locating proteins in the cell using TargetP, SignalP and related tools. *Nature protocols* **2**, 953–971.
7. Guda, C. (2006). pTARGET: a web server for predicting protein subcellular localization. *Nucleic acids research* **34**, W210–213.
8. Roy, A., Kucukural, A., and Zhang, Y. (2010). I-TASSER: a unified platform for automated protein structure and function prediction. *Nature protocols* **5**, 725–738.
9. Zhang, Y. (2008). I-TASSER server for protein 3D structure prediction. *BM Bioinformatics* **9**, 40.

Chapter 4

Self-Custom-Made SFP Arrays for Nonmodel Organisms

Ron Ophir and Amir Sherman

Abstract

Successful genetic mapping is dependent upon a high-density set of markers. Therefore, tools for high-throughput discovery of genetic variation are essential. The most abundant genetic marker is the single-nucleotide polymorphism (SNP). However, except for model organisms, genomic information is still limited. Although high-throughput genomic sequencing technologies are becoming relatively inexpensive, only low-throughput genetic markers are accessible (e.g., simple sequence repeats). The use of sequencing for the discovery and screening of high-density genetic variation in whole populations is still expensive. Alternatively, hybridization of genomic DNA (gDNA) on a reference (either genome or transcriptome) is an efficient approach for genetic screening without knowing the alleles in advance (Borevitz et al. Proc Natl Acad Sci USA 104:12057–12062). We describe a protocol for the design of probes for a high-throughput genetic-marker discovery microarray, termed single feature polymorphism (SFP) array. Starting with consensus cDNA sequences (UniGenes), we use OligoWiz to design T_m-optimized 50-bp long oligonucleotide probes (Ophir et al. BMC Genomics 11:269, 2010). This design is similar to expression arrays and we point out the differences.

Key words: Single-feature polymorphism, Microarray, Probe design, Crop

1. Introduction

1.1. Single-Feature Polymorphism

One of the technologies used for the identification of genetic variation is single-feature polymorphism (SFP). This approach is based on the concept that target DNA that perfectly matches its probes binds with greater affinity than comparative target DNA with a mismatch. Thus, natural imperfections can be detected as a difference in signal intensity in microarray hybridization using labeled genomic DNA (gDNA) (1, 2). In its most simple experimental implementation, one would hybridize two gDNAs derived from two individuals to a SFP array. Based on the assumption that there is no difference in copy number, a specific probe would reveal any

Michael Kaufmann and Claudia Klinger (eds.), *Functional Genomics: Methods and Protocols*,
Methods in Molecular Biology, vol. 815, DOI 10.1007/978-1-61779-424-7_4, © Springer Science+Business Media, LLC 2012

genetic variation, from a single to a few nucleotides polymorphism, as well as small insertions and deletions (indels; only a fraction of the existing variation). SFPs can be used, with no further information, as markers, or they can be sequenced to identify the genetic difference and translated into conventional markers (single-nucleotide polymorphism (SNP) and insertions/deletions (indels)). In most cases, it appears that the cause of the SFP is a SNP, which is the most abundant type of genetic variation (especially in protein-coding regions). SFP technology draws its strength from the fact that it can be implemented in a variety of genetic applications, such as marker discovery and fine-mapping of traits, as well as for genome-wide association studies (3–5). In particular, SFP discovery has been relatively successfully implemented in model organisms such as *Arabidopsis* (6) and *Drosophila* (7), and in other whole-genome-sequenced organisms, such as rice and soybean (8, 9). To date, all hybridizations have been performed on high-density short oligonucleotides (Affymetrix arrays). However, these types of arrays are only available for a few organisms, they are expensive and they are not flexible in their design. Therefore, the ability to implement this technology on any custom array (Agilent, NimbleGen, and others) has the potential to create a very useful tool in many breeding programs for agricultural crops.

1.2. Probe Design

There are various types of usage for microarrays, including arrays for expression detection (10, 11) and copy-number variation detection (12), SNP chips for genotyping (13), whole-genome array (tiling array) for transcript mapping, i.e., identification of genomic positions of exons (14), and regulatory-factor identification by hybridization of chromatin immunoprecipitation (ChIP) products on a tiling array (15). However, despite the high variety of microarray formats, most of the literature on probe design has focused on expression arrays (16–19). Strictly speaking, the methodology for designing probes for microarrays concerns factors – melting temperature (T_m), cross-hybridization, probe folding, and low complexity (18) – that affect target affinity, with T_m and cross-hybridization having the strongest impact (20). Specifically, for expression arrays it is very important to set the probe orientation, i.e., design all probes as sense or antisense, due to the fact that many labeling protocols target only one strand of the cDNA. Moreover, it is recommended that probes close to the 3′ end be favored because the RNA labeling is performed with T7 attached to poly-A (18). The length of the probes varies from 25 bp (25 mer) to 60 bp (60 mer) and this affects the average T_m more than the probes' composition does. In most cases, these two size extremes are the most common. A standard SFP experiment is carried out almost exclusively on Affymetrix expression arrays (25 mer probes). The reason for this is that in SFP experiments signal differences reflect the genetic variation of the target (gDNA). Short-length probes, as in Affymetrix arrays, are thought to be more sensitive at detecting genetic variation. The disadvantages of Affymetrix arrays are that they are not T_m-optimized, which is another important factor in probe sensitivity, and they are available for only a limited repertoire of organisms. Here, we describe a probe design for SFP based on 50-bp long T_m-optimized probes as implemented in a previous study (24). The T_m optimization is performed by varying the probe length between 45 and 55 bp. Because the target for SFP array is gDNA and the labeling protocol is for comparative genomics hybridization (CGH), which labels dsDNA, the orientation of the probe is irrelevant.

2. Software and Data

2.1. Input Data

1. *UniGenes*: A fasta file of UniGene sequences. De novo RNA transcript sequencing of transcriptomes is achieved by two approaches. The classical approach: preparation of cDNA libraries and 400-bp long sequence tags, on average, termed expression-sequence tags (ESTs). The novel approach: direct sequencing of fragmented cDNA generating millions of sequence-reads of 500 bp, on average. These reads, or tags, are assembled to reconstruct the mRNA by generating longer sequence fragments, which are called contigs or UniGenes (see Note 1).
2. *Genomic DNA*: A fasta file of gDNA, with an entry for each chromosome.

2.2. Software

1. OligoWiz 2.0: Program for microarray olinucleotide design. OligoWiz software includes three programs. There is no need for compilation, just a computer preinstalled with PERL (see Note 2). The three programs are as follows: (1) ow2.format.pl for background preparation, (2) oligowiz2.pl for probe parameter calculation, and (3) OligoWiz-2.1.3.jar for probe viewing and selection.
2. R statistical language: Freely available from http://cran.r-project.org/ for any platform. This step is optional.
3. Megablast: Program for sequence search. NCBI megablast can be downloaded from http://www.ncbi.nlm.nih.gov/blast/megablast.shtml. This step is for expression arrays.

3. Methods

Of all of the factors involved in probe design, cross-hybridization and T_m have the strongest impact on hybridization (20). For cross-hybridization calculations, OligoWiz prepares a database in blast format. This database is prepared from the target sequences that hybridized on the array, termed background. For example, if the target hybridized to the array is gDNA, then the database should be compiled, ideally, from the whole-genome sequence. If the target is mRNA, then the whole transcriptome is sufficient as background. The probes are a subset of the background and they are designed to enrich only specific regions of it. Ideally, for SFP arrays, the background should be gDNA, which is the target hybridized to the array. If the whole genome sequence is not available, as is the case for most nonmodel crops, the background may be the full set of protein-coding sequences, assuming that the coding and noncoding genomic loci are quite different from each other. The probes are designed from the loci under study. For trait mapping, the protein-coding genes' genomic loci are preferred. In the following protocol, we assume that the genomic sequence is available chromosome by chromosome, as the input data for SFP probe design, and the whole set of transcriptome, as the input data for expression probe design (see Note 3).

3.1. Preparing the Background

For the background, one would like to have a unique set of the genome in fasta format (see Note 4), with each chromosome being one entry. For an expression array background, one should have a unique set of mRNA transcripts. If the source of the mRNA is EST assembly (UniGenes) and not curated mRNA, it should be treated with caution. Such an assembly tends to create contigs per splice variant rather than per gene. Therefore, a few entries of the same gene may be present in the background. Two such entries in the background would be considered a duplicated locus, and therefore, the probes designed to this locus are suspected of being cross-hybridized. To avoid this possibility, one would like to run all of the mRNA sequences against each other to reduce unnecessary duplication in the background.

The fasta file of unique sequence entries is the input to the ow2.format.pl program (see Note 5).

3.2. Calculating Probe-Design Factors

The next step is to calculate the probe-design factors, these factors being cross-hybridization, T_m, folding, relative position, and low complexity. Cross-hybridization is calculated based on the background database prepared by the ow2.format.pl program. The definition of a two-locus cross-hybridization is a similarity higher than 75% or a perfect match of a 15-bp long consecutive sequence (21–23). The latter parameter is, of course, irrelevant for designing short 25-mer probes. Moreover, for such short probes, the similarity definition for cross-hybridization increases to 85% or higher. Based on the literature, these are the default parameters, although they can be modified when running the oligowiz2.pl program. The T_m is the most important parameter for signal success. It is roughly correlated with probe length (see Note 6). Thus by choosing the desired probe length, one is actually setting the average T_m. To ensure that all probes on the array will be T_m-unified, it is possible to give a range for probe size. Varying probe length enables flexibility in probe composition and therefore narrowing of the delta of probe T_m from the average probe T_m. We found that setting the probe length within a 10-bp range, e.g., 60 ± 5 bp, gives a very small delta for probe T_m. Thus, in this step, running the "oligowiz2.pl" program with the minimal parameter set would be:

oligowiz2.pl -in Melon.cds.fsa -species Melon -length 50 -lmax 55 -lmin 45 (see Note 7), where "Melon.cds.fsa" is the fasta file of the sequences that the probes are being designed from, "Melon" is the background database name, and the probe length is set to 50 mer with variable length allowed from 45 to 55 bp.

3.3. Selecting Probes

To select and review probes, it is recommended that initially, the probe parameter file be compressed (see Note 8). After loading the probe parameters into OligoWiz (see Note 9), the probes are placed on the contigs and exported. While placing the probes on the contigs, OligoWiz calculates normalized scores that range between 0 (bad) and 1 (good). These probe-design factor scores are integrated by weighting the summation to one score, termed total score. The weighting is user-defined. Our approach is to set a higher weight for array cross-hybridization, T_m, and low-complexity factors, although low complexity is somewhat redundant with T_m and cross-hybridization. The weighting of the probe position is array-dependent. For SFP, the array position score is insignificant since it is desirable to cover as many genomic regions as possible. By contrast, when designing probes for expression array, we would like to choose one or two probes that are close to the 3′ end of the mRNA. Therefore, the position score should receive a higher weight. Probe folding is the least important factor due to the fact that the probes are short and in most cases, they do get a high score for folding (see Fig. 1, in which more than 95% of the probes get a score of 0.9 or higher).

Steps for creating probes tab file:

1. Open OligoWiz2.0 (see Note 10).
2. Go to File → Open OWZ data file.
3. Select the score weights for cross. Hyb: 5, Delta T_m: 5, Folding: 1, Position: 0, and Low complexity: 2.
4. Go to menu Oligos → Place oligos.
5. For 50 mer ± 5 oligos, type 55 (50 + 5) in the "Min distance between oligo".
6. Check the "unlimit…" checkbox. Alternatively, for an expression array, type 1 in "Max number oligos/sequence" window.
7. Press "Apply to all" button.

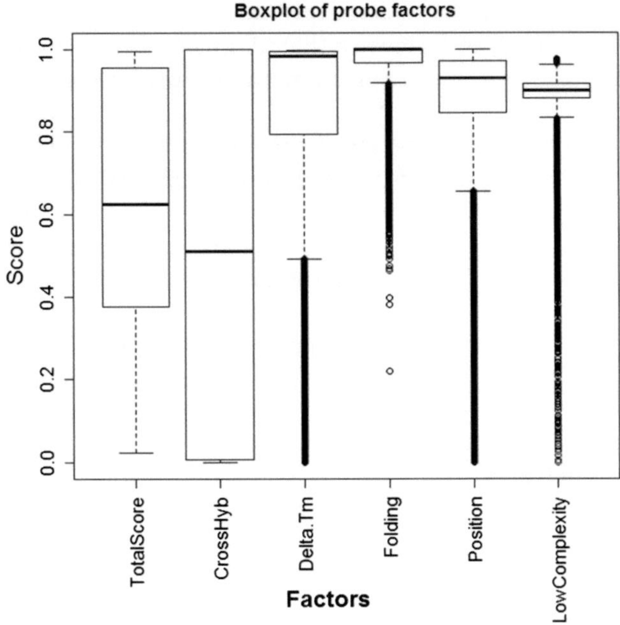

Fig. 1. Boxplot of probe-factor scores. A description of the distribution of each factor score is presented in a *boxplot*. The *thick line* in the box is the median, the upper limit of the box is the upper 25th percentile where the lower limit is the 75th percentile. The upper 5th and 95th percentiles are represented by the *upper* and *lower horizontal lines*, respectively. If the box limits and the upper and lower bounds are equally distant from the median, the distribution is symmetric (as is the case in normal distribution). Here, the total score is very close to being symmetric.

8. To regulate the number of probes to the array size, set score limits in the "minimum score allowed" window: 0.8 or higher is preferred (see Note 11).
9. Press "Export oligos" tab then press "OK".

3.4. Loading Probes to eArray

The last stage in creating a custom array is uploading or sending the probe sequences to a manufacturer. Although the outfile of the OligoWiz is tab-delimited, it is not standard due to the fact that the column headers are written in each row next to the values (see Note 11). The best way to load it into eArray (earray.chem.agilent.com) is to use the eArray minimal format. This format requires removing the OligoWiz output file hash (#) lines and then cutting out the two first columns. The fact that the probe-IDs (first column) are unique allows us to attach the rest of the probe factor information, usually assigned to other columns, later. Such actions can be done by any text editor or by the Linux command "grep '^#' -v Tilapia_1_oligos.tab | cut -f1,2 > To_earray.txt". Alternatively, it can be run by an R script (Fig. 2). In addition, this script draws a boxplot as in Fig. 1.

```
##############################################
# read oligowiz probe table assuming file name is
#  oligos.tab
##############################################
##############################################################
# parsing oligowiz table
##############################################################
Probes = scan(file="oligos.tab",what=character(0),blank.lines.skip=TRUE,comment="#",sep="\n")
Probes = lapply(Probes,strsplit,"\t")
Probes = lapply(Probes,sapply,function(x) gsub("^.*: ","",x))
#############################################################
# converting it to regular tab delimited table with header
# at the top
#############################################################
Table = sapply(Probes,function(x) x[,1])
Table = data.frame(t(Table),stringsAsFactors=FALSE)
colnames(Table) =
c("ProbeID","Sequence","Tm","Len","TotalScore","CrossHyb","Delta.Tm","Folding","Position","Low
Complexity")
#############################################################

score = c("TotalScore","CrossHyb","Delta.Tm","Folding","Position","LowComplexity")
num = c("Tm","Len","TotalScore","CrossHyb","Delta.Tm","Folding","Position","LowComplexity")
minimal = c("ProbeID","Sequence")
tmp = apply(Table[,num],2,as.numeric)
Table = data.frame(Table[,minimal],tmp)

###################################
# draw boxplot figure in jpeg format
###################################
jpeg(file="boxplot.jpg")
op=par(mar=c(8,4,3,2),font.main=2,cex.axis=1.2,cex.lab=1.5)
boxplot(Table[,score],ylab="Score",las=3)
mtext("Factors",side=1,line=-2,outer=TRUE,cex=1.5)
par(op)
dev.off()
#########################################################
# Wriet tab delimited file for eArray
#########################################################
write.table(Table[,minimal],file="probes.txt",sep="\t",row.names=FALSE)
```

Fig. 2. Converting and plotting probe factors. The R script reads an OligoWIZ output file, here named "oligos.tab". Next, the data are converted to a table which is plotted as a *boxplot* (see Fig. 1 for legend) and saved in a minimal two-column format. To save the whole table, modify, within the write.table() function, Table, [minimal] to Table. The figure is saved in jpeg picture format in a file called boxplot.jpg.

4. Notes

1. UniGenes can be downloaded from a database relevant to the species under study. In our case, the database is http://www.icugi.org/.
2. PERL is free with any Linux (free operating system) distribution. On Windows, it can be freely download from the "active state" site. However, we highly recommend running all programs on Linux. One can easily install Linux by downloading BioLinux.
3. The minimum data for the background, by definition, is that which the probes are designed from. If the whole genome

or whole transcriptome are not available, use the available whole-genomic information.

4. Fasta format is a text file in a simple format. Each entry has two lines: (1) the description, starting with ">" sign and (2) the sequence itself.
5. ow2.format.pl -genome melon_genome.fasta -dbname "Melon" -speciesname "C. melo". "Melon" is the name that will be specified in the program for probe-design factor calculation – oligowiz2.pl.
6. In most array platforms, the hybridization temperature is set to 10°C or more below the T_m.
7. The output of the oligowiz2.pl program is directed to the standard output (screen) and should be redirected to a file by the ">" sign as follows: "oligowiz2.pl -in Melon.cds.fsa -species Melon -length 50 -lmax 55 -lmin 45 > melon.owz".
8. Compress owz file (probe parameter file) as follows: "gzip -c melon.owz > melon.owz.gz". Compression saves memory and enables the loading of a larger number of probes.
9. To load probe parameters into the OligoWiz viewer, use command "java -jar OligoWiz-2.1.3.jar". It is possible to increase the program memory for loading big files by adding the -Xmx switch, as follows "java -Xmx2g -jar OligoWiz-2.1.3.jar". Here, the maximum memory size would be 2 GB.
10. When regulating the score limit, do not forget to check the "Replace existing oligos" radio button.
11. An example of an OligoWiz tab-delimited file: "lcl|amnone_c10005 No definition line found_288–343 ATGGACCCA-C TCATGGGTGGGGTGATCTCTGAAGACATGATTCAG T_m: 82.3, Len: 56, TotalScore: 0.638, Cross-hyb: 0.214, Delta T_m: 0.975, Folding: 0.769, Position: 0.864, and Low-complexity: 0.824".

References

1. Borevitz, J. O., Hazen, S. P., Michael, T. P., Morris, G. P., Baxter, I. R., Hu, T. T., Chen, H., Werner, J. D., Nordborg, M., Salt, D. E., Kay, S. A., Chory, J., Weigel, D., Jones, J. D., and Ecker, J. R. (2007) Genome-wide patterns of single-feature polymorphism in Arabidopsis thaliana, *Proc Natl Acad Sci USA 104*, 12057–12062.
2. Nordborg, M., Hu, T. T., Ishino, Y., Jhaveri, J., Toomajian, C., Zheng, H., Bakker, E., Calabrese, P., Gladstone, J., and Goyal, R. (2005) The pattern of polymorphism in Arabidopsis thaliana, *PLoS Biology 3*, e196.
3. Aranzana, M. J., Kim, S., Zhao, K., Bakker, E., Horton, M., Jakob, K., Lister, C., Molitor, J., Shindo, C., Tang, C., Toomajian, C., Traw, B., Zheng, H., Bergelson, J., Dean, C., Marjoram, P., and Nordborg, M. (2005) Genome-wide association mapping in Arabidopsis identifies previously known flowering time and pathogen resistance genes, *PLoS Genet 1*, e60.
4. Singer, T., Fan, Y., Chang, H. S., Zhu, T., Hazen, S. P., and Briggs, S. P. (2006) A high-resolution map of Arabidopsis recombinant inbred lines by whole-genome exon array hybridization, *PLoS Genet 2*, e144.

5. Zhu, T., and Salmeron, J. (2007) High-definition genome profiling for genetic marker discovery, *Trends Plant Sci 12*, 196–202.
6. Borevitz, J. O., Liang, D., Plouffe, D., Chang, H. S., Zhu, T., Weigel, D., Berry, C. C., Winzeler, E., and Chory, J. (2003) Large-scale identification of single-feature polymorphisms in complex genomes, *Genome Res 13*, 513–523.
7. Lai, C. Q., Leips, J., Zou, W., Roberts, J. F., Wollenberg, K. R., Parnell, L. D., Zeng, Z. B., Ordovas, J. M., and Mackay, T. F. (2007) Speed-mapping quantitative trait loci using microarrays, *Nat Methods 4*, 839–841.
8. Edwards, J. D., Janda, J., Sweeney, M. T., Gaikwad, A. B., Liu, B., Leung, H., and Galbraith, D. W. (2008) Development and evaluation of a high-throughput, low-cost genotyping platform based on oligonucleotide microarrays in rice, *Plant Methods 4*, 13.
9. Kaczorowski, K. A., Ki-Seung Kim, K. S., Diers, B. W., and Hudson, M. E. (2008) Microarray-Based Genetic Mapping Using Soybean Near-Isogenic Lines and Generation of SNP Markers in the Rag1 Aphid-Resistance Interval, *THE PLANT GENOME 1*, 89–98.
10. Cossins, A. R., and Crawford, D. L. (2005) Fish as models for environmental genomics, *Nat Rev Genet 6*, 324–333.
11. Sotiriou, C., and Piccart, M. J. (2007) Taking gene-expression profiling to the clinic: when will molecular signatures become relevant to patient care?, *Nat Rev Cancer 7*, 545–553.
12. Buckley, P. G., Mantripragada, K. K., Piotrowski, A., Diaz de St hl, T., and Dumanski, J. P. (2005) Copy-number polymorphisms: mining the tip of an iceberg, *Trends in Genetics 21*, 315–317.
13. Guo, Y., Tan, L. J., Lei, S. F., Yang, T. L., Chen, X. D., Zhang, F., Chen, Y., Pan, F., Yan, H., Liu, X., Tian, Q., Zhang, Z. X., Zhou, Q., Qiu, C., Dong, S. S., Xu, X. H., Guo, Y. F., Zhu, X. Z., Liu, S. L., Wang, X. L., Li, X., Luo, Y., Zhang, L. S., Li, M., Wang, J. T., Wen, T., Drees, B., Hamilton, J., Papasian, C. J., Recker, R. R., Song, X. P., Cheng, J., and Deng, H. W. (2010) Genome-wide association study identifies ALDH7A1 as a novel susceptibility gene for osteoporosis, *PLoS Genet 6*, e1000806.
14. Royce, T. E., Rozowsky, J. S., Bertone, P., Samanta, M., Stolc, V., Weissman, S., Snyder, M., and Gerstein, M. (2005) Issues in the analysis of oligonucleotide tiling microarrays for transcript mapping, *Trends Genet 21*, 466–475.
15. Wu, J., Smith, L. T., Plass, C., and Huang, T. H. (2006) ChIP-chip comes of age for genome-wide functional analysis, *Cancer Res 66*, 6899–6902.
16. Rouillard, J. M., Zuker, M., and Gulari, E. (2003) OligoArray 2.0: design of oligonucleotide probes for DNA microarrays using a thermodynamic approach, *Nucleic Acids Res 31*, 3057–3062.
17. Royce, T. E., Rozowsky, J. S., and Gerstein, M. B. (2007) Toward a universal microarray: prediction of gene expression through nearest-neighbor probe sequence identification, *Nucleic Acids Res 35*, e99.
18. Wernersson, R., Juncker, A. S., and Nielsen, H. B. (2007) Probe selection for DNA microarrays using OligoWiz, *Nat Protoc 2*, 2677–2691.
19. Wernersson, R., and Nielsen, H. B. (2005) OligoWiz 2.0 – integrating sequence feature annotation into the design of microarray probes, *Nucleic Acids Res 33*, W611–615.
20. Mueckstein, U., Leparc, G. G., Posckany, A., Hofacker, I., and Kreil, D. P. Hybridization thermodynamics of NimbleGen microarrays, *BMC Bioinformatics 11*, 35.
21. Kane, M. D., Jatkoe, T. A., Stumpf, C. R., Lu, J., Thomas, J. D., and Madore, S. J. (2000) Assessment of the sensitivity and specificity of oligonucleotide (50 mer) microarrays, *Nucleic Acids Res 28*, 4552–4557.
22. He, Z., Wu, L., Li, X., Fields, M. W., and Zhou, J. (2005) Empirical establishment of oligonucleotide probe design criteria, *Appl Environ Microbiol 71*, 3753–3760.
23. Hughes, T. R., Mao, M., Jones, A. R., Burchard, J., Marton, M. J., Shannon, K. W., Lefkowitz, S. M., Ziman, M., Schelter, J. M., Meyer, M. R., Kobayashi, S., Davis, C., Dai, H., He, Y. D., Stephaniants, S. B., Cavet, G., Walker, W. L., West, A., Coffey, E., Shoemaker, D. D., Stoughton, R., Blanchard, A. P., Friend, S. H., and Linsley, P. S. (2001) Expression profiling using microarrays fabricated by an ink-jet oligonucleotide synthesizer, *Nat Biotechnol 19*, 342–347.
24. Ophir, R., Eshed, R., Harel-Beja, R., Tzuri, G., Portnoy, V., Burger, Y., Uliel, S., Katzir, N., and Sherman, A. (2010) High-throughput marker discovery in melon using a self-designed oligo microarray., *BMC Genomics 11*, 269.

Part II

DNA Analysis

Chapter 5

Construction and Analysis of Full-Length and Normalized cDNA Libraries from Citrus

M. Carmen Marques and Miguel A. Perez-Amador

Abstract

We have developed an integrated method to generate a normalized cDNA collection enriched in full-length and rare transcripts from citrus, using different species and multiple tissues and developmental stages. Interpretation of ever-increasing raw sequence information generated by modern genome sequencing technologies faces multiple challenges, such as gene function analysis and genome annotation. In this regard, the availability of full-length cDNA clones facilitates functional analysis of the corresponding genes enabling manipulation of their expression and the generation of a variety of tagged versions of the native protein. The development of full-length cDNA sequences has the power to improve the quality of genome annotation, as well as provide tools for functional characterization of genes.

Key words: Library, cDNA, Citrus, Full-length, Normalized, SMART, DSN nuclease, Gateway technology

1. Introduction

Many methods for the construction of cDNA libraries have been developed in recent years. Conventional cDNA library construction approaches, however, suffer from several major shortcomings. First, the majority of cDNA clones are not full-length, mainly due to premature termination of reverse transcription or blunt-end polishing of cDNA ends prior to subcloning. A number of methods have been developed to overcome this problem and obtain cDNA library preparations enriched in full-length sequences, most of them based on the use of the mRNA cap structure (1–4). However, these methods require high quantities of starting material and complicated multistep manipulations of mRNA and cDNA intermediates, which often results in the degradation of mRNA and the isolation of short clones. The recently described SMART method (switching

Michael Kaufmann and Claudia Klinger (eds.), *Functional Genomics: Methods and Protocols*,
Methods in Molecular Biology, vol. 815, DOI 10.1007/978-1-61779-424-7_5, © Springer Science+Business Media, LLC 2012

mechanism at the 5′ end of the RNA transcript), exploits two intrinsic properties of Moloney murine leukemia virus (MMLV) reverse transcriptase, reverse transcription and template switching of blunt-ended cDNA copies, allowing an easy and efficient production of full-length clones (5). Second, the straightforward random sequencing of clones from standard cDNA libraries is inefficient for discovering rare transcripts, owing to the repeated occurrence of intermediately and highly abundant cDNAs, and a normalization process is often required. This process generally utilizes second-order reaction kinetics of re-association of denatured DNA, so that relative transcript concentration within the remaining single-stranded cDNA fraction is equalized to a considerable extent. A recently described method uses the properties of DSN nuclease to specifically cleave ds-DNA (in both DNA–DNA and DNA–RNA duplexes) allowing the separation of the normalized ss-fraction (6–8). Third, an adaptor-mediated cloning process is still a common approach for cDNA library construction, leading to undesirable ligation by-products and inserts of non-mRNA origin. Directional cloning using *Sfi*I endonuclease minimizes these problems, as it identifies variable target sequences and allows for designing adaptors with noncomplementary ends, thus avoiding their concatenation. As the *Sfi*I recognition sequence is very rare in eukaryotic genomes, the use of *Sfi*I also eliminates the need for methylation during cDNA synthesis (9). In the last place, gene discovery is facilitated by the ability to easily express proteins in both homologous and heterologous biological contexts and thus understanding gene function (10). This entails engineering of multiple expression constructs, which is time-consuming and laborious when using traditional ligase-mediated cloning methods. The recombinational cloning employed in the commercially termed Gateway technology (Invitrogen) exploits the accurate and site-specific recombination system utilized by bacteriophage lambda in order to shuttle sequences between plasmids bearing compatible recombination sites (11–13). This bypasses the need for traditional ligase-mediated cloning while maintaining orientation of the transferred DNA segment and yielding a high proportion of desired clones.

Herein, we describe how we took advantage of the SMART protocol, the DSN nuclease and the Gateway technology to maximize acquisition of full-length and rarely expressed cDNAs from citrus ready to use for functional analysis purposes (14).

2. Materials

2.1. Development of the Gateway-Based Cloning Vector

1. pENTR1A vector (Invitrogen).
2. Restriction enzymes: *Eco*RI, *Xho*I, *Sfi*I.

Table 1
Oligonucleotides used in this protocol

Name	Sequence	Step (section)
pENTR-SfiI-F	AATTCGGCCATTATGGCCTGCAGGATCC GGCCGCCTCGGCC	3.1.2
pENTR-SfiI-R	TCGAGGCCGAGGCGGCCGGATCCTGCA GGCCATAATGGCCG	3.1.2
SMART IV	AAGCAGTGGTATCAACGCAGAGT GGCCATTATGGCCGGG	3.3.1
CDSIII/3	ATTCTAGAGGCCGAGGCGGCC GACATG-d$(T)_{30}$NN	3.3.1
M1-5′	AAGCAGTGGTATCAACGCAGAGT	3.4.3
M1-3′	ATTCTAGAGGCCGAGGCGG	3.4.3
M2-5′	AAGCAGTGGTATCAACGCAG	3.4.4
M2-3′	ATTCTAGAGGCCGAGGCG	3.4.4
pENTR-F	GGCTTTAAAGGAACCAATTCAG	3.5.7
pENTR-R	GCAATGCTTTCTTATAATGCCAAC	3.5.7

*Sfi*I recognition sites (GGCCNNNNNGGCC) are underlined

3. Oligonucleotides: pENTR-SfiI-F, pENTR-SfiI-R, pENTR-F, and pENTR-R (Table 1).
4. JM110 *Escherichia coli* competent cells.
5. Shrimp Alkaline Phosphatase (SAP).
6. Qiaquick Gel Extraction kit (Qiagen).
7. TAE 1×.
8. Agarose.

2.2. Preparation of Poly(A+)-RNA

1. Oligotex mRNA kit (Qiagen). It includes the Oligotex resin, Binding Buffer (OBB), Washing Buffer (OW2), and Elution Buffer (OEB).
2. 3 M sodium acetate pH 5.2.
3. Ethanol 96%.
4. GlycoBlue.

2.3. Synthesis of Full-Length cDNAs for the Construction of a Full-Length Enriched Library

1. BD SMART PCR cDNA synthesis kit (BD Biosciences). This kit contains: 7 μl PowerScript Reverse Transcriptase, 200 μl 5× First-strand buffer, 100 μl 5′ PCR Primer IIA (12 μM), 70 μl dNTP mix, 200 μl DTT (20 mM), 5 μl Control Human Placental Total RNA (10 μg/μl), 1 ml deionized water. It also

includes an Advantage long distance PCR kit, containing: 30 μl 50× Advantage 2 Polymerase mix, 200 μl 10× Advantage 2 PCR buffer, 50 μl 50× dNTP mix, 30 μl Control DNA template, 30 μl Control Primer mix, and 2.5 ml PCR Grade water (see Note 1).

2. Oligonucleotides: SMART IV and CDSIII/3′ (Table 1).
3. Qiaquick PCR purification kit (Qiagen).
4. Qiaquick gel extraction kit (Qiagen).
5. T4 DNA ligase (2 U/μl).
6. *Sfi*I restriction enzyme.
7. Proteinase K (10 μg/μl).
8. One Shot MAX Efficiency DH5α-T1 Competent Cells.
9. Kanamycin, stock solution at 50 mg/μl in sterile water.

2.4. Normalization of cDNAs for the Construction of a Normalized Library

1. DSN nuclease (EVROGEN), including DSN enzyme (initially lyophilized and diluted after reception in 50 μl of DSN Storage Buffer to a final concentration of 1 U/μl); 100 μl of 10× Master Buffer; 500 μl of 2× DSN Stop Solution; 20 μl of DSN Control Template (100 ng/μl).
2. Hybridization Buffer: 200 mM HEPES pH 7.5 and 2 M NaCl.
3. Advantage 2 PCR kit (BD Biosciences).
4. Oligonucleotides: M1-5′, M1-3′, M2-5′, and M2-3′ (Table 1).

3. Methods

3.1. Development of the Gateway-Based Cloning Vector

1. Opening the vector.
   Digest 2 μg of purified pENTR1A plasmid (Invitrogen) with 10 U of *Eco*RI and *Xho*I in the appropriate buffer and a final volume of 50 μl by incubating at 37°C for 3 h, then add ten additional units of restriction enzymes and let at 37°C over night (Fig. 1). Dephosphorylate the vector by adding 20 U of SAP to the reaction and incubate at 37°C for 90 min. Then, incubate at 65°C for 45 min to quench the reaction. Run the resultant product in TAE 1× agarose electrophoresis. Purify the band corresponding to the vector with Qiaquick Gel Extraction Kit (see Note 2). Resuspend in water to a concentration of 10 ng/μl.
2. Introduction of the adapters for *Sfi*I recognition sites.
   Heat a 10-μM mix of the synthetic oligonucleotides pENTR-SfiI-F and pENTR-SfiI-R (Table 1) for 10 min at 70°C and let them anneal by slow cooling at room temperature. Digest the

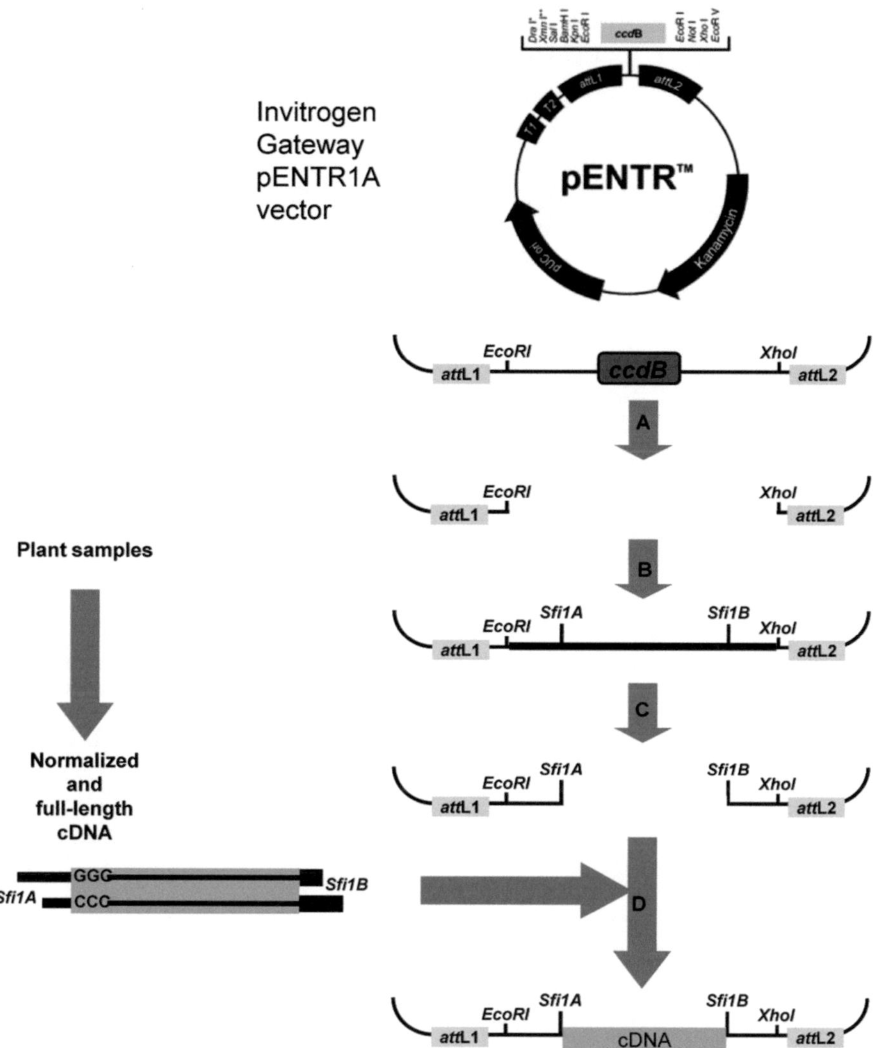

Fig. 1. Generation of the cloning vector pENTR-SfiI. The vector is a modification of the commercial pENTR1A Gateway vector (Invitrogen). Plasmid DNA is digested by *EcoR*I and *Xho*I and dephosphorylated with SAP (**a**). The SfiI adaptor (ds-DNA generated by the annealing of pENTR-SfiI-F and pENTR-SfiI-R oligonucleotides) is then ligated to the vector (**b**). Before ligation with the ds-cDNA, plasmid DNA is digested with *Sfi*I and dephosphorylated (**c**). Ligation with the normalized and full-length enriched cDNA generates the library (**d**).

adapters with 10 U of *EcoR*I and *Xho*I enzymes (see Note 3). Then ligate this double-stranded (ds) oligonucleotide to the opened pENTR1A (10–50 ng) with 10 U of T4 DNA ligase and incubate at 16°C over night. This plasmid constitutes the pENTR-SfiI vector (Fig. 1) (see Note 4).

3. Transformation of the JM110 *E. coli* competent cells. Transform home-made competent JM110 *E. coli* cells with 5 ng of pENTR-SfiI vector, and select transformants by plating onto

50 μg/μl LB-kanamycin plates. Check the construct by digestion/sequencing and make glycerol stocks (see Note 5).

4. Digestion and dephosphorylation of the pENTR1A-SfiI vector. Make a plasmid DNA prep using standard protocol. Add, in a 0.5-ml Eppendorf tube, 5 μg of the pENTR1A-SfiI vector (see step 3.1.3), 4 μl of buffer M 10×, and 2 μl (10 U/μl) of *Sfi*I restriction enzyme, in a final volume of 40 μl. Incubate immediately at 50°C for 4 h, add 2 μl of *Sfi*I and incubate at 50°C for additional 4 h or over night. Dephosphorylate the vector by adding 20 U of SAP to the reaction and incubate at 37°C for 90 min. Then, incubate at 65°C for 45 min to quench the reaction. Run the reaction in TAE 1× agarose electrophoresis, and purify the band corresponding to the vector with Qiaquick Gel Extraction Kit (see Note 2). Quantify and resuspend in water to a final working concentration of 2–5 ng/μl. Check the quality of the preparation by running a ligation of 5 ng of plasmid with T4 DNA ligase and transform (see Note 6).

3.2. Starting Material

1. Add RNase-free water to 200 μg of total RNA to a final volume of 250 μl. Poly(A)$^+$ RNA from different citrus tissues is purified using Oligotex mRNA Midi Kit (Qiagen) (see Note 7).
2. Add 250 μl of OBB and mix gently. Add 20 μl of Oligotex and mix gently. Denature this mix by heating at 70°C for 5 min (during this process, shake the tube every 2 min). Hybridize the samples at room temperature for 10 min.
3. Spin 2 min at maximum speed, add 400 μl of OW2 to the pellet, and resuspend with the pipette. Transfer the total volume to the provided column, spin 1 min at maximum speed, and remove the eluate. Wash again with 400 μl of OW2, and give an extra-spin for 1 min and transfer the column to a new tube.
4. Elute Poly(A)$^+$ RNA from the column by adding 75 μl of hot (70°C) OEB to the column, incubating it at 70°C for 2 min and spinning at maximum speed for 1 min. Transfer the eluate to a new tube, repeat the elution again, and pool both aliquots (150 μl final volume). Add 200 μl of RNase-free water to the elute to bring a final volume of 350 μl and quantify the whole volume in a spectrophotometer.
5. Precipitate Poly(A)$^+$ RNA by adding 35 μl of 3 M sodium acetate pH 5.2, 2 μl of glycoBlue and 900 μl EtOH 96%. Incubate at −80°C over night and recover the pellet by centrifuging for 15 min at maximum speed at 4°C. Wash the pellet with EtOH 70% and dry in SpeedVac. Finally, resuspend Poly(A)$^+$ RNA in RNase-free water to obtain a final concentration of 0.17 μg/μl.

3.3. Obtaining Full-Length ds-cDNA

1. First-strand cDNA synthesis, dC tailing, and template switching by reverse transcription (see Note 7b).

1. Combine the following reagents in a sterile 0.5 ml reaction tube: 3 μl Poly(A)$^+$ sample (0.17 ug/μl), 1 μl CDSIII/3′ oligonucleotide (10 μM), and 1 μl SMART IV oligonucleotide (10 μM) (Table 1).
2. Incubate the mix at 72°C for 2 min, cool the tube down on ice for 2 min, and add the following reagents to the reaction tube: 2 μl 5× first-strand buffer, 1 μl DTT (20 mM), 1 μl 50× dNTP (10 mM), and 1 μl PowerScript Reverse Transcriptase.
3. Incubate the tube at 42°C for 1 h to complete first-strand cDNA amplification (see Note 7c).

2. Second-strand synthesis by long-distance PCR (see Note 7d).
   1. Prepare a PCR mix containing the following components in the order shown: 80 μl deionized water, 10 μl 10× Advantage 2 PCR buffer, and 2 μl 50× dNTP Mix, 4 μl 5′ PCR primer IIA, and 2 μl 50× Advantage 2 polymerase mix. Finally, add 2 μl of the first-strand cDNA from the previous step to obtain a final reaction volume of 100 μl.
   2. Place the tube in the preheated (95°C) thermal cycler and commence thermal cycling using the following parameters: an initial preheating at 95°C for 1 min and additional 16 cycles of 5 s at 95°C, 5 s at 65°C, and 6 min at 68°C (see Note 8).

      Make three second-strand synthesis reactions for every full-length cDNA library you want to obtain. Therefore, a total of 300 μl of ds-cDNA is obtained (see Note 9).

3. ds-cDNA polishing. This step contains three procedures: (1) treatment with proteinase K to denature enzymes used in the previous steps, (2) amplification with T4 DNA polymerase to make ds-cDNA blunt-ended, (3) precipitation and concentration of ds-cDNA.
   1. Make 50 μl aliquots of the ds-cDNA obtained in the previous step in 0.5 ml Eppendorf tubes (six tubes). Add 4 μl of proteinase K (10 μg/μl) to each tube and incubate at 45°C for 1 h in order to eliminate the enzymes used in the previous steps that could interfere with the following reactions. Heat the tubes at 90°C for 10 min to inactivate the proteinase K. Then, chill the tubes in ice water for 2 min, add 3.5 μl (15 U) of T4 DNA polymerase and incubate at 16°C for 30 min. Afterward, heat the tubes at 72°C for 10 min to stop the reaction.
   2. Pool together the content of every two tubes and precipitate ds-cDNA by adding 55 μl ammonium acetate 4 M and 420 μl 95% ethanol to each tube. Mix thoroughly by inverting the tubes. Spin immediately at maximum speed

for 20 min at room temperature. Do not chill the tube before centrifuging as it could result in co-precipitation of impurities. Then, wash pellet with 80% ethanol and air dry to evaporate residual ethanol.

3. Collect the polished ds-cDNA contained in the three tubes by resuspension in water to obtain a single aliquot with a final volume of 66 μl. This cDNA is ready for digestion with *Sfi*I and ligation into the appropriate vector to produce the full-length enriched cDNA library.

3.4. Obtaining Normalized Full-Length ds-cDNA

1. Purification of ds-cDNA.
   Aliquot 100 μl of the ds-cDNA obtained in the step 3.3.2 (see Note 9) into two Eppendorf tubes, containing 50 μl each, and make two reactions of purification using the Qiaquick PCR Purification Kit (Qiagen), according to manufacturer's instructions. Elute cDNA from each column with 50 μl of 1 mM TE. Mix both elutes (100 μl) and concentrate in a SpeedVac to obtain a final concentration of 100 ng/μl approximately (see Note 9b).
2. Normalization step contains three procedures: (1) denaturing, (2) hybridization, and (3) degradation of re-natured ds-cDNA.
   1. Combine in a sterile tube 12 μl of ds-cDNA from the previous step and 4 μl of 4× Hybridization Buffer (see Note 10). Aliquot 4 μl of the reaction mixture into four 0.5 ml Eppendorf tubes.
   2. Denature ds-cDNA by incubating all the tubes at 98°C for 2 min and let rehybridize at 68°C for 5 h. Immediately, add 5 μl of hot Master Buffer 2× (68°C) and incubate the tubes at the same temperature for an additional 10 min (see Note 11).
   3. Dilute 1 μl of the DSN enzyme to 1/2 and 1/4 in DSN Storage Buffer. Add 1 μl (1 U) of DSN enzyme to the first tube, 1 μl of enzyme diluted to 1/2 (0.5 U) to the second, and 1 μl of the enzyme diluted 1/4 (0.25 U) to the third. Add 1 μl of Storage Buffer to the fourth tube to have a control of the normalization. Label each tube appropriately to avoid mistakes (see Note 12).
   4. Incubate at 68°C for 25 min. To quench the reaction add 10 μl of hot Stop Solution 2× to each tube and incubate at 68°C for additional 5 min. Finally, cool the tubes down on ice for several min and add 20 μl of water to each tube to obtain a final volume of 40 μl (see Note 13).
3. First round of amplification of the normalized cDNA and elucidation of the optimal number of cycles.
   1. Each DSN-treated cDNA from the previous step will be amplified separately.

Combine the following reagents in a tube to prepare a PCR Master Mix: 156 μl of water, 20 μl of 10× Advantage PCR buffer, 4 μl of 50× dNTP mix, 6 μl of primer M1-5′ (10 μM), 6 μl of primer M1-3′ (10 μM), and 4 μl of 50× Advantage Polymerase mix.

2. Aliquot 49 μl of this PCR master mix into four sterile 0.5 ml tubes and label them as in the previous step. Add 1 μl of the DSN-treated cDNA from the previous step to their corresponding reaction tube. Place the tubes in a preheated (95°C) thermal cycler and start thermal cycling using the following parameters: 7 s at 95°C, 10 s at 66°C, and 6 min at 72°C (see Note 14).
3. To establish the optimal number of cycles, transfer an aliquot (10 μl) of each PCR reaction to a clean tube after 7, 9, 11, 13 and 15 PCR cycles, obtaining a series of five tubes from every initial PCR reaction (20 tubes) (see Note 15).
4. Electrophorese 5 μl aliquots from each tube in a TAE 1.5× agarose gel to determine the efficiency of normalization (Fig. 2a) (see Note 16).

4. Second round of amplification of the normalized cDNA. Make a tenfold dilution of the reaction that best fit in the normalization parameters. If you plan to estimate the normalization efficiency you should also amplify control (non-normalized) cDNA in parallel. Prepare a PCR master mix by combining the following reagents in the order shown: 76 μl of sterile water, 10 μl of 10× Advantage PCR buffer, 2 μl of 50× dNTP mix, 4 μl of primer M2-5′ (10 μM), 4 μl of primer M2-3′ (10 μM), and 2 μl of 50× Advantage Polymerase Mix. Finally, add 2 μl of the tenfold dilution of the normalized cDNA (see Note 17).
5. cDNA polishing. Proceed as described in step 3.3.3.

3.5. Cloning cDNA into the Plasmid Vector

1. Digestion of ds-cDNA. Combine, in two independent reactions, the following reagents: 33 μl of ds-polished-cDNA (see Note 18), 4 μl of buffer M 10×, and 2 μl (10 U/μl) of *Sfi*I restriction enzyme. Incubate immediately at 50°C for 1 h, add 1 μl of *Sfi*I, and incubate at 50°C for additional 3 h.
2. Purification of the digested cDNA. Purify each digestion using the Qiaquick PCR purification kit (Qiagen), by eluting with 50 μl of Tris–HCl 1 mM. Combine both eluates and concentrate to a final volume of 30 μl.
3. Electrophoresis of the digested cDNA. Electrophorese the 30 μl of digested cDNA obtained in the previous step in TAE 1× agarose gel. Purify cDNA of the appropriate size by cutting the gel in blocks containing cDNAs longer than 1,000 bp and shorter than 5,000 bp (see Note 19).

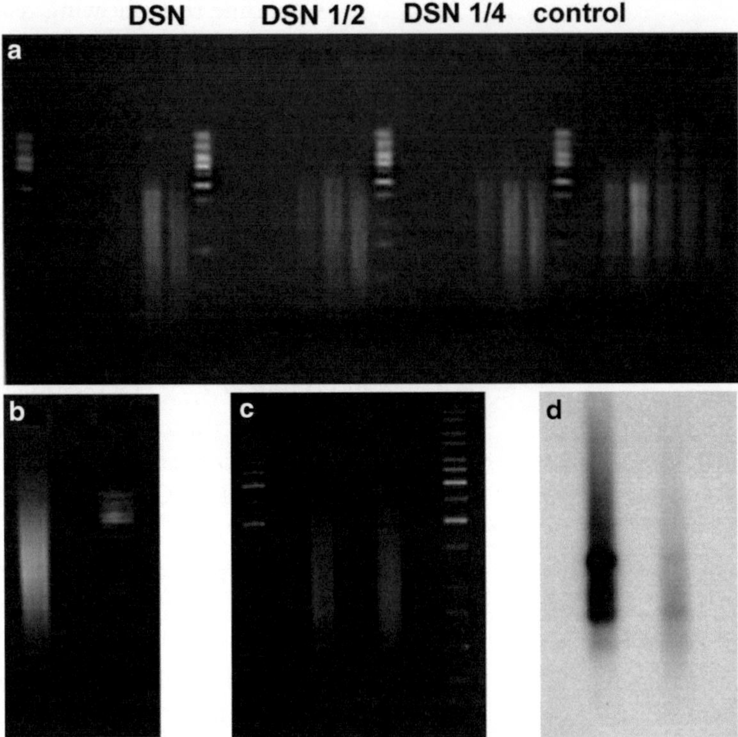

Fig. 2. Evaluation of the efficiency of normalization of cDNA libraries. (**a**) Gel electrophoretic analysis of 5 μl aliquots from the first amplification of the normalized cDNA taken at 7, 9, 11, 13, 15, and 17 PCR cycles (step 3.4.3). (**b**) Gel electrophoretic analysis of the normalized cDNA (as in step 3.5.3) utilized in the construction of the normalized cDNA library RVDevelopN (14). (**c**) Gel electrophoretic analysis of a non-normalized cDNA population (*left*) and a normalized cDNA population (*right*) from the same RNA sample. (**d**) Virtual northern of the cDNA smear blotted and hybridized with the highly abundant citrus clone *C32009H03*.

4. Purification of gel blocks. Purify cDNA from the gel blocks using the Qiaquick gel extraction kit (Qiagen) following manufacturer's instructions. Concentrate the eluted cDNA to a final volume of 20 μl (see Note 20).
5. Ligation into pENTR1A-SfiI vector. Combine, in a 0.5-ml Eppendorf tube, 5 ng of the digested pENTR1A-SfiI vector (see step 3.1.4), 6 μl deionized water, 2 μl cDNA from the previous step, 1 μl ligase buffer 10×, and 1 μl T4 DNA ligase. Incubate at 16°C over night. A control ligation without cDNA must be carried out in parallel (see Note 6).
6. Transformation of DH5α *E. coli* competent cells. We used One Shot MAXEfficiency DH5α-T1 competent cells (Invitrogen) and performed the transformation according to manufacturer's instructions, using 5 μl of the ligation. Plate the transformation onto two 50 μg/μl LB-kanamycin plates and grow at 37°C (see Note 21).

7. Check the quality of the library. Carry out colony PCR from 2 to 300 kanamycin-resistant colonies, using plasmid oligonucleotides pENTR-F and pENTR-R (Table 1). Check the size of the PCR fragments by electrophoresis (see Note 22).
8. Selection of recombinant clones. Select colonies, grow in LB-kanamycin media, and purify plasmid DNA using a 96-well plate format method (Eppendorf or Millipore). Sequence the corresponding cDNA inserts using plasmid oligonucleotide pENTR-F (Table 1) to generate an EST collection (see Note 23).

3.6. Virtual Northern

To get a better assessment of the normalization efficiency, carry out a virtual northern to estimate the relative concentration of a highly abundant clone in both the non-normalized and the normalized cDNA populations obtained from the second run of amplification (Fig. 2). Electrophorese, in a TAE 1.5× gel, equivalent quantities of cDNA corresponding to the non-normalized and normalized samples subjected to the second run of amplification. Transfer DNA to a nitrocellulose membrane, and run a standard Southern blot analysis. Obtain a probe of a highly abundant clone by carrying out a PCR of the corresponding cDNA (see Note 23).

4. Notes

1. This kit also provides seven CHROMA-SPIN-1000 columns and seven microfiltration columns (0.45 μm), but they are not used in this procedure. Please, note that the kit supplies oligonucleotides SMART IIA and 3′ SMART CDS primer II A, however, these oligonucleotides contain *Rsa*I cloning site. Since we are using the properties of *Sfi*I site for cloning purposes, we employ SMART IV and CDSIII/3′ oligonucleotides instead.
2. This process removes the ccdB gene which allows for negative selection of expression clones and lets two binding sites for *Sfi*I adapters.
3. The adapters are ready to be inserted in the opened pENTR1A vector. They provide two recognition sites (*SfiI*A and *SfiI*B, underlined in Table 1) which, once cut with *Sfi*I restriction enzyme, generate two nonsymmetrical ends ready for directional subcloning.
4. The developed pENTR1A-SfiI vector allows both, effective directional cloning by taking advantage of the nonsymmetrical cleavage of the *Sfi*I restriction enzyme and the ease of subcloning by the Gateway System.
5. Although the polylinker does not contain dam or dcm methylation-susceptible sequences, we observed that the digestion of

the vector with *Sfi*I was more efficient in plasmid obtained from JM110 *E. coli* cells.

6. It is very important to test the quality of the *Sfi*I-digested pENTR-SfiI vector prior to the ligation with the ds-cDNA, as well as every time you run a ligation with ds-cDNA. For a control ligation (without insert), mix 5 ng (1–2 μl) of *Sfi*I-digested vector in a 0.5-ml Eppendorf tube with 6 μl deionized water, 1 μl ligase buffer 10×, and 1 μl T4 DNA ligase. Incubate at 16°C over night. Transform One Shot MAXEfficiency DH5α-T1 competent cells using 5 μl of the ligation, plate onto two LB-kanamycin (50 μg/μl) plates, and grow at 37°C. Number of colonies is expected to be very low (less that 50 per plate). Larger numbers mean that probability to get nonrecombinant vectors during sequencing is high, which diminished the quality and efficiency of the library. A new vector preparation has to be obtained. According to our results only 1–5% of the colonies lacked an insert when sing cDNA as insert.
7. It is important to warm OEB at 70°C and Oligotex suspension at 37°C before starting the protocol.

7b. In this step the PowerScript Reverse Transcriptase (RT) provided in the BD SMART PCR Synthesis Kit synthesizes first-strand cDNA primed by CDSIII/3′, which contains a 30-mer oligo dT. This RT also promotes dC tailing (addition of three cytosines at the 3′ end of the cDNA when the first-strand reaches mRNA 5′ end). Furthermore, the addition of SMART IV oligonucleotide, which contains three guanines at its 3′ end, allows template switching needed for next steps.

7c. This first-strand cDNA (10 μl) can be stored at –20°C for up to 3 months.

7d. Double-stranded cDNA is generated with PCR catalyzed by a long-distance polymerase mixture which ensures processive second-strand synthesis and amplification while maintaining accurate size representation. This reaction uses the 5′ anchor primer, which is complementary to the *Sfi*A sequence and the CDS primer that contains the *Sfi*B sequence.

8. PCR parameters for long-distance PCR require short denaturing and annealing steps. Times are adjusted to a Perkin Elmer 9400 thermal cycler, if other apparatus are to be used parameters should be adjusted depending on the ramp rate needed to acquire the running temperature.
9. Three reactions are needed to prepare a full-length cDNA library. The normalization procedure requires ds-cDNA from this step as starting material. So, if you are planning to normalize the cDNAs two additional reactions should be performed, with one of them used in the normalization and the other kept intact in order to make future comparisons.

9b. Usually a final volume of 25 μl renders the appropriate concentration.

10. Heat Hybridization Buffer at 37°C for 10 min to dissolve any precipitate.

11. Note that during the process of normalization the temperature must be kept constant and tubes cannot be removed from the incubator for more than 30 s. Prepare and heat the dilutions and/or buffers to be used shortly before they are to be used.

12. For the degradation of the ds-fraction formed during re-association of cDNA using the DSN nuclease assay different enzyme concentrations in each tube, as the appropriate quantity cannot be known a priori.

13. This DSN-treated cDNA can be stored at −20°C for up to 2 weeks.

14. There is a well-known tendency of PCR to amplify shorter fragments more efficiently than longer ones. Thus, the cDNA sample should be somewhat biased toward longer cDNAs to obtain a natural length distribution upon cloning. This can be done by using a process of regulation of average length which combines the use of an enzyme mixture for long and accurate PCR, the design of primers with complementary sequences at their ends that tend to anneal to each other and compete with primer annealing, being this competition more pronounced in short molecules, and lowering the primer concentration to shift the equilibrium toward intramolecular annealing and therefore increase the suppression.

15. Choosing an optimal number of cycles ensures that the ds-cDNA will remain in the exponential phase of amplification. When the yield of PCR products stops increasing with every additional cycle, the reaction has reached its plateau. The optimal number of cycles for the experiment should be one or two cycles less than that needed to reach the plateau. It is better to use fewer cycles than too many.

16. The profile of an efficiently normalized and amplified cDNA is the one that (1) its overall signal intensity of the smear is similar to that shown for the control (not treated with DSN) but does not contain distinguishable bands, (2) the signal intensity of smear has reached its plateau, (3) the upper boundary of the cDNA smear do not exceed 4.5 kb. This amplified normalized cDNA can be stored at −20°C for up to a month (see Fig. 2).

17. To increase the cDNA concentration, perform three reactions of amplification for the normalized cDNA.

18. This cDNA proceeds from step 3.3.3 for full-length cDNA libraries or from step 3.4.4 for normalized libraries.

19. Elimination of fragments shorter than 1 kb allows enrichment of full-length clones and excludes those obtained in conventional libraries. On the other hand, we excluded fragments longer than 5,000 bp as full-length cDNAs do not seem to be larger than 4 kb.
20. The final concentration is approximately 10 ng/μl.
21. The library is completed. Approximately 40,000 kanamycin-resistant colonies (2,000 colonies per transformation, 2 transformations per ligation, 10 ligations per cDNA synthesis) can be obtained per assay.
22. Expected size of the PCR products (corresponding to the cloned cDNAs) range between 500 bp to 2 kb. Although cDNAs between 1 and 5 kb are purified from the agarose gel (see step 3.5.3), the average size of the cDNAs is smaller due to low cloning efficiency of large cDNAs.
23. Plasmid DNA preps, sequencing of ESTs, and Virtual northern are carried out by standard protocols.

Acknowledgments

The authors would like to thank to all participants in the Spanish Citrus Functional Genomic Project, specially to Drs. Javier Forment, Jose Gadea, and Vicente Conejero. This work was funded by grants from the Spanish Government GEN2001-4885-CO5-01 and GEN2001-4885-CO5-02.

References

1. Carninci P, Kvam C, Kitamura A, Ohsumi T, Okazaki Y, Itoh M, Kamiya M, Shibata K, Sasaki N, Izawa M, Muramatsu M, Hayashizaki Y, and Schneider C (1996) High-efficiency full-length cDNA cloning by biotinylated CAP trapper. *Genomics* **37**, 327–336.
2. Carninci P, Shibata Y, Hayatsu N, Sugahara Y, Shibata K, Itoh M, Konno H, Okazaki Y, Muramatsu M, and Hayashizaki Y (2000) Normalization and subtraction of CAP-trapper-selected cDNAs to prepare full-length cDNA libraries for rapid discovery of new genes. Genome Res. **10**, 1617–1630.
3. Suzuki Y, and Sugano S (2003) Construction of a full-length enriched and a 5'-end enriched cDNA library using the Oligo-capping method. *Methods Mol. Biol.* **221**, 73–91.
4. Clepet C, Le Clainche I, and Caboche M (2004) Improved full-length cDNA production based on RNA tagging by T4 DNA ligase. *Nucleic Acids Res.* **32,** e6.
5. Zhu YY, Machleder EM, Chenchik A, Li R, and Siebert PD (2001) Reverse transcriptase template switching: a SMART approach for full-length cDNA library construction. *Biotechniques* **30,** 892–897.
6. Zhulidov PA, Bogdanova EA, Shcheglov AS, Vagner LL, Khaspekov GL, Kozhemyako VB, Matz MV, Meleshkevitch E, Moroz LL, Lukyanov SA, and Shagin DA (2004) Simple cDNA normalization using Kamchatka crab duplex-specific nuclease. *Nucleic Acids Res.* **32,** e37.
7. Zhulidov PA, Bogdanova EA, Shcheglov AS, Shagina IA, Wagner LL, Khazpekov GL, Kozhemyako VV, Lukyanov SA, and Shagin DA (2005) A method for the preparation of normalized cDNA libraries enriched with full-length sequences. *Russian J. Bioorg. Chem.* **31,** 170–177.

8. Anisimova VE, Rebrikov DV, Zhulidov PA, Staroverov DB, Lukyanov SA, and Shcheglov AS (2006) Renaturation, activation, and practical use of recombinant duplex-specific nuclease from Kamchatka crab. *Biochem.-Moscow* **71**, 513–519.
9. Toru M, Matsui T, Heidaran MA, and Aaronson SA (1989) An efficient directional cloning system to construct cDNA libraries containing full-length inserts at high frequency. *Gene* **83**, 137–146.
10. Castelli V, Aury JM, Jaillon O, Wincker P, Clepet C, Menard M, Cruaud C, Quétier F, Scarpelli C, Schächter V, Temple G, Caboche M, Weissenbach J, and Salanoubat M (2004) Whole genome sequence comparisons and "Full-length" cDNA sequences: a combined approach to evaluate and improve Arabidopsis genome annotation. *Genome Res.* **14**, 406–413.
11. Earley KW, Haag JR, Pontes O, Opper K, Juehne T, Song KM, and Pikaard CS (2006) Gateway-compatible vectors for plant functional genomics and proteomics. *Plant J.* **45**, 616–629.
12. Karimi M, Inzé D, and Depicker A (2002) GATEWAY™ vectors for *Agrobacterium*-mediated plant transformation. *TRENDS Plant Sci.* 7, 193–195.
13. Hartley JL, Temple GF, and Brasch MA (2000) DNA cloning using *in vitro* site-specific recombination. *Genome Res.* **10**, 1788–1795.
14. Marques M C, Alonso-Cantabrana H, Forment J, Arribas R, Alamar S, Conejero V, and Perez-Amador MA (2009) A new set of ESTs and cDNA clones from full-length and normalized libraries for gene discovery and functional characterization in citrus. *BMC Genomics* **10**, 428.

Chapter 6

Assembling Linear DNA Templates for In Vitro Transcription and Translation

Viktor Stein, Miriam Kaltenbach, and Florian Hollfelder

Abstract

Cell-free expression systems provide straightforward access from genes to the corresponding proteins, involving fewer handling steps than in vivo procedures. A quick procedure to assemble a gene of interest into a linear DNA template together with 3′- and 5′-untranslated regions using a coupled uracil-excision–ligation strategy based on USER Enzyme and T4 DNA ligase. This methodology will be useful for repeated cycles of expression and in vitro selection, in which gene libraries are repeatedly assembled and their products and templates regenerated.

Key words: Linear DNA template assembly, USER friendly cloning, In vitro screening and selection, In vitro transcription/translation

1. Introduction

Combinations of cell-free protein expression systems with different screening and selection platforms have found widespread application in the fields of functional genomics and protein engineering for the exploration of sequence–structure–function relationships (1, 2). The ready commercial availability of different in vitro expression systems from various sources synergizes with the rapidly falling costs of DNA synthesis and the seemingly exponential mining of DNA sequences from natural repertoires. A key advantage of using cell-free systems is the rapid expression of protein product from a gene. In addition, working in vitro gives access to toxic gene products that cannot be synthesized in vivo and allows the use of linear, PCR-generated templates, which obviates the need to reclone genes into appropriate plasmids for every screening or selection cycle.

Michael Kaufmann and Claudia Klinger (eds.), *Functional Genomics: Methods and Protocols*,
Methods in Molecular Biology, vol. 815, DOI 10.1007/978-1-61779-424-7_6, © Springer Science+Business Media, LLC 2012

To function as a template for in vitro transcription/translation (IVTT), linear DNA templates require a promoter to drive transcription as well as 5′- and 3′-untranslated regions that carry essential regulatory motives for efficient translation (the exact nature of which depends on the particular IVTT system employed). Consequently, the accumulation of mutations in these regions is undesirable – e.g., in directed evolution these regions are typically not diversified during library creation. Therefore, they need to be attached to the diversified gene of interest (GOI) after mutagenesis or after every selection cycle.

Here, we present an efficient protocol that allows the assembly of a GOI with a tag and its flanking untranslated regions (UTRs) in approximately 90 min (Fig. 1) for subsequent in vitro expression. The assembly procedure obviates the need to reclone the GOI between successive selection cycles into circular plasmid backbones. Assembly is based on a coupled uracil excision–ligation

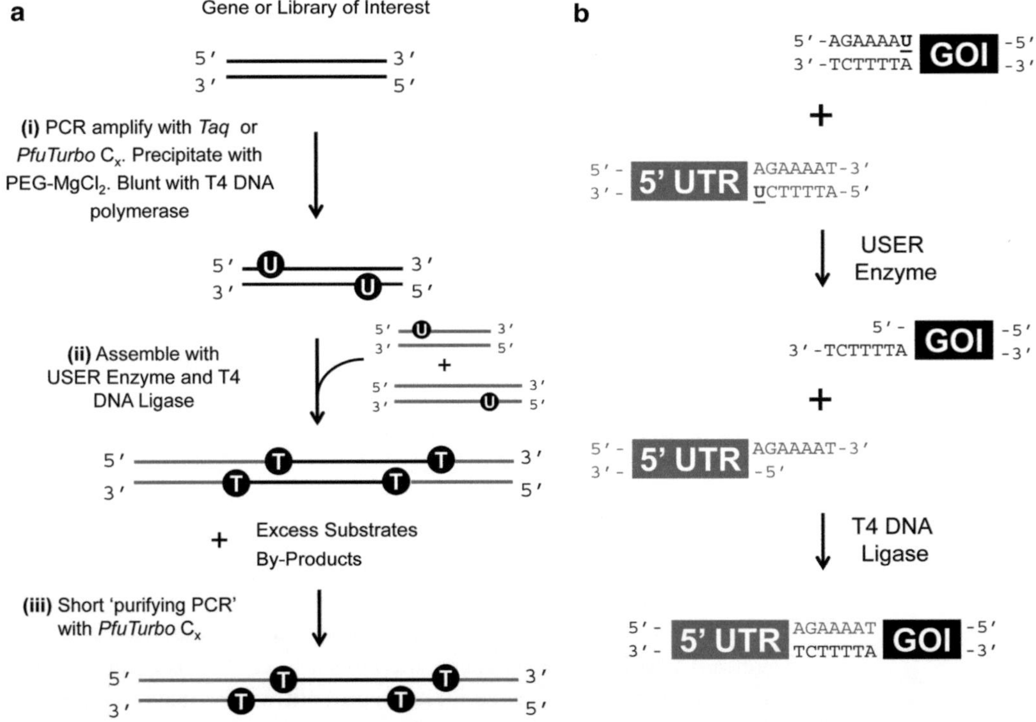

Fig. 1. (**a**) Assembly scheme. (i) GOI or a derivative library is amplified with primers that specifically incorporate uracil nucleotides close to both 5′-ends. (ii) Assembly of the GOI with its 5′- and 3′-untranslated regions including any constant protein-coding regions based on a coupled uracil excision–ligation strategy. (iii) Pure templates are obtained following a short "purifying PCR", which effectively removes excess substrates and partially assembled intermediates. (**b**) Mechanism of coupled uracil excision–ligation: first, USER enzyme catalyses the excision of uracil from DNA, thereby leaving a single base pair gap and a 3′-extension provided the 5′-portion can dissociate. Complementary overlapping 3′-extensions then direct the assembly of DNA fragments which are covalently sealed by T4 DNA ligase. Adapted from ref. (14).

strategy that is reminiscent of UDG (3–5) and USER enzyme cloning (6–8), but also includes T4 DNA ligase to generate DNA templates devoid of nicks. Briefly, USER (Uracil-Specific Excision Reagent) is a mixture of two enzymes: (1) uracil DNA glycosylase, which excises the uracil base from the DNA and (2) endonuclease VIII, which breaks the phosphodiester bond on both sides of the abasic site. In combination, USER generates a single nucleotide gap in the DNA. If the gap is positioned close enough to a nick or double stranded break, the intervening fragment will dissociate. The dissociation of this fragment in turn leaves a single-stranded overhang that is suitable for ligation with a complementary overhang from another DNA fragment. Templates that are functional for in vitro TS-TL are obtained following a short PCR run over ten amplification cycles and do not need to be purified further by agarose gel electrophoresis. The procedure is exemplified for a mutant of O^6-alkylguanine alkyltransferase (AGT) (9), which is assembled with an N-terminal peptide tag that can be biotinylated by biotin ligase (10) and its 5′- and 3′-UTRs that carry all the necessary sequences for IVTT including a T7 promoter, a ribosome-binding site, and a T7 terminator. We also highlight and discuss any adjustments to the protocol that need to be undertaken to process DNA fragments of different sizes and to optimize template yields.

2. Materials

2.1. Reagents

1. A plasmid DNA template that provides all regulatory components (e.g., promoters, 5′-UTRs, 3′-UTRs, tags, etc.) required to express proteins using an in vitro TS-TL system. We exemplify the procedure for regulatory elements derived from the pIVEX vector series (available from RiNA GmbH), which is suitable for the T7 RNA polymerase driven expression of proteins using *Escherichia coli*-based in vitro TS-TL.
2. The GOI or LOI that is to be expressed.
3. A thermostable DNA polymerase that is compatible with DNA templates that contain uracil residues such as Taq DNA polymerase (available with 10× NH_4 reaction buffer and 50 mM $MgCl_2$, Bioline) and PfuTurbo C_x Hotstart DNA polymerase (available with 10× PfuTurbo C_x reaction buffer, Agilent Technologies), see Notes 1 and 2.
4. 10 mM dNTPs (2.5 mM each).
5. Oligonucleotides as PCR primers to generate the 5′-UTR, GOI, and 3-′UTR resuspended in 10 mM Tris–HCl, pH 8.0 at a concentration of 100 μM (see Note 3). Except for the two

Table 1
Oligonucleotide sequences and loci of primer annealing

LMB-2-6 → Fw-AGT-U → Rec-pIV →

T7-Promoter 5'-UTR-Avi	AGT	3'-UTR T7-Terminator

← Rec-LMB ← Rv-AGT-U ← pIV-B1

LMB-2-6	5′-ATGTGCTGCAAGGCGATTAAGTTGGGTAACG-3′
Rec-LMB	5′-ATTTTCUGAGCCTCGAAGATGTC-3′
Fw-AGT-U	5′-AGAAAAUCGAATGGCACGAAGG-3′
Rv-AGT-U	5′-AACUCAGCTTCCTTTCGGGCTTTGTTAG-3′
Rec-pIV	5′-AGTUGGCTGCTGCCACCGCTGA-3′
pIV-B1	5′-GCGTTGATGCAATTTCTATGCGCACC-3′

Note: The highlighted U denotes the excision site of the USER enzyme mix.

outmost primers, oligonucleotides should contain suitable thymidine-to-uracil substitutions located up to 10 bp away from the 5′ end (see Subheading 3.1 and Table 1).

6. Spin column kit for purifying DNA from PCR reactions (e.g., QIAquick PCR purification kit, Qiagen) including elution buffer (10 mM Tris–HCl, pH 8.0).
7. DpnI restriction endonuclease and 10× NEBuffer 2 (NEB).
8. T4 DNA polymerase, 10× NEBuffer 2, 100× NEB BSA (NEB).
9. 30% (*w*/*v*) PEG-8000 (Sigma). To improve the longevity of the solution, we recommend passing it through a 0.25-μm sterilizing filter.
10. 30 mM $MgCl_2$. To ensure accurate concentrations, dilute from a commercially available stock of 1 M $MgCl_2$ (e.g., Sigma).
11. USER Enzyme (New England Biolabs).
12. T4 DNA ligase and 10× T4 DNA Ligase Buffer.

2.2. Equipment

1. PCR thermocycler.
2. Benchtop centrifuge.
3. Thermomixer.
4. Reagents and equipment for agarose gel electrophoresis.
5. Vortexer.
6. Spectrophotometer for determining DNA concentrations.

3. Methods

3.1. Primer Design

1. Identify possible splice sites as close as possible to the GOI (to avoid randomization outside the GOI during evolution cycles) to attach the desired promoter, 5′-UTR, protein based tags and 3′-UTR by uracil-excision–ligation mediated assembly. The sequence requirements for these splice sites are modest: only one adenine and one thymidine residue spaced apart by up to ten nucleotides in the 5′→3′ direction are necessary to ensure efficient dissociation of the excised regions. Splice sites are usually located in the 5′- and 3′-UTRs or in protein-based tags and linker regions that form part of the open reading frame (ORF), but not of the diversified region. To prevent nonspecific assembly reactions (e.g., by ligation of palindromic or mismatched single-stranded overhangs), great care should be taken in the design of splice sites where two DNA fragments are joined (see Note 4).
2. For a three-fragment assembly reaction, design three sets of primers to amplify the three DNA fragments that code for (a) the promoter, 5′-UTR and any N-terminal tags that form part of the open reading frame, (b) the gene or library of interest, and (c) the 3′-UTR and any C-terminal tags that form part of the open reading frame. The primer that covers any given splice site always features an adenine residue at its 5′-end and a uracil located up to 10 bp toward the 3′ end (Fig. 1 and Table 1). For guidelines on designing efficient PCR primers, see Note 5.

3.2. Preparation of Assembly Substrates

1. Amplify the 3′-UTR, the GOI and the 5′-UTR by PCR using PfuTurbo C_x. In the example given, we use primers LMB-2-6 and Rec-LMB for the 3′-UTR, Fw-AGT-U and Rv-AGT-U for the GOI and Rec-pIV and pIV-B1 for the 5′-UTR (Tables 1 and 2). For a reaction of 100 μl, prepare a PCR mix on ice as follows (see Note 6): 10 μl 10× PfuTurbo C_x reaction buffer (final concentration: 1×), 10 μl dNTP mix (final concentration: 250 nM each), 1 μl each of the forward and reverse primer (final concentration: 1 μM), 1 μl DNA template (e.g., 10 ng/μl plasmid DNA, Table 2), and 75 μl water. Finally, add 2 μl PfuTurbo C_x polymerase (2.5 U/μl stock) and place reaction tubes into a thermocycler. After an initial denaturation and heat activation step for 2 min at 95°C, cycle 30 times as follows: 30 s at 95°C, 1 min at the desired reannealing temperature (see Note 5), and 1 min/kbp at 72°C. Finish with a final extension step for 10 min at 72°C.
2. Purify the PCR products with the QIAquick PCR purification kit and elute with 50 μl elution buffer.

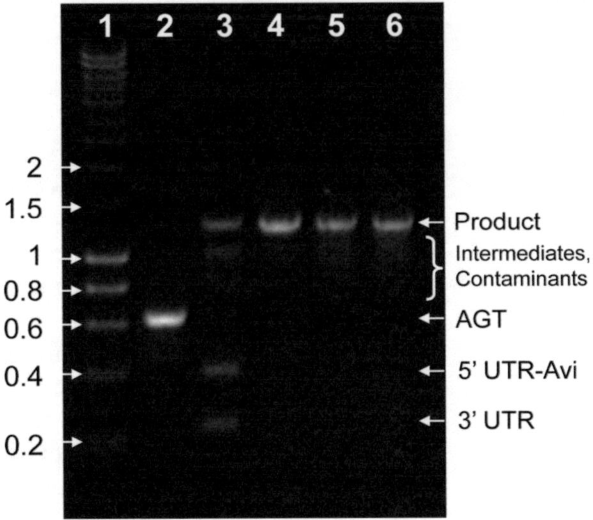

Fig. 2. Assembly of three DNA fragments by uracil excision–ligation. L1, hyperladder I in kilobase pairs; L2, substrate AGT amplified with PfuTurbo C_x DNA polymerase, precipitated with PEG–$MgCl_2$ and blunted with T4 DNA polymerase; L3, assembly of L2 with its UTRs + Avi-tag; L4, purifying PCR of L3 with 50 ng DNA per 100 μl PCR; L5, purifying PCR of L3 with 250 ng DNA per 100 μl PCR; L6, purifying PCR of L3 with 500 ng DNA per 100 μl PCR. Adapted from ref. (14).

Table 2
Summary of plasmid DNA templates used in the preparation of assembly substrates

Subheading (step)	DNA fragment	Size (bp)	Forward primer	Reverse primer	Reannealing temperature (°C)
3.2 (step 1)	AGT	638	Fw-AGT-U	Rv-AGT-U	62
3.2 (step 1)	T7-promoter 5′-UTR-Avi	410	LMB-2-6	Rec-LMB	62
3.2 (step 1)	3′-UTR T7-terminator	239	Rec-pIV	pIV-B1	62
3.3 (step 7)	Full-length DNA template for in vitro TS-TL	1,276	LMB-2-6	pIV-B1	68

3. To remove the plasmid DNA that served as a template in the PCR, incubate the purified DNA fragments (50 μl) with DpnI (1 μl or 20 U) in NEBuffer 4 for 1 h at 37°C (see Note 7).
4. Check the identity and purity of the PCR products by agarose gel electrophoresis. This quality control step is particularly critical for PCR protocols that have not been validated previously.

3.3. PEG–$MgCl_2$ Precipitation

1. To selectively remove any primer dimers that might still persist after purifying the PCR with the QIAquick kit, we generally purify all assembly substrates by precipitating them with 10% PEG-8000 and 10 mM $MgCl_2$ (see Note 8). To this end, combine equal amounts of DNA with 30% PEG-8000 and 30 mM $MgCl_2$ (e.g., 50 μl each). Slow and careful pipetting is required to dispense accurate amounts of the relatively viscous 30% PEG-8000 stock solutions. Mix thoroughly and vigorously by pipetting up and down until a homogeneous solution has been obtained. The concentration of the reagents in the final precipitation mix should be ~25–50 ng/μl DNA buffered in 3–5 mM Tris–HCl, pH 8.0 along with 10 mM $MgCl_2$ and 8–10% PEG-8000 depending on the size of the desired DNA fragment (see Note 9).
2. Centrifuge for 15 min at 16,000×*g* at room temperature. For large amounts of DNA, a pellet may be visible.
3. Carefully remove the supernatant with a pipette. Be careful not to disturb the DNA precipitate, which has accumulated on the wall of the tube that faces away from the centre of the centrifuge.
4. Resuspend in an appropriate volume of water or a mixture of 1:10 elution buffer/water. To ensure efficient recovery of the DNA, pipette the resuspension several times down the side of the tube over which the pellet has spread.

3.4. Blunting with T4 DNA Polymerase

1. Determine the DNA concentration using a spectrophotometer to assess the amount of T4 DNA polymerase that needs to be added in the blunting step (see Note 10).
2. For a 100 μl reaction, add the following substrates and enzymes to your DNA solution recovered in Subheading 3.3 step 5: 10 μl 10× NEBuffer 2 (final concentration: 1×), 4 μl dNTP mix (final concentration: 100 nM each), 1 μl 100× NEB BSA (final concentration: 1×) and fill up with water to approximately 98 μl. Precool the reaction mix at 12°C and then add 2 μl T4 DNA polymerase (3 U/μl stock) for 6 μg DNA. This is equivalent to 1 U/1 μg DNA (see Note 11).
3. Incubate for 15 min at 12°C. Do not exceed the reaction temperature, time and recommended amount of enzyme as this can lead to recessed 3′ ends as a result of the 3′→5′ exonuclease activity of T4 DNA polymerase.
4. Purify the DNA with the Qiaquick PCR purification kit and elute in a mixture of 1:10 elution buffer/water.

3.5. Assembly and Purification of DNA Fragments by Coupled Uracil Excision–Ligation

1. Determine the concentration of the DNA obtained in Subheading 3.4 step 4 using a spectrophotometer.
2. Calculate the molar amount of the DNA using the following formula:
   (a) Molar amount = mass (g)/(660 g/mol × template length in base pairs).

3. For a volume of 100 μl, set up the uracil-excision–ligation assembly reaction as follows: mix 3 pmol of the GOI with 4.5 pmol of the flanking 5′ UTR and 3′ UTR each (e.g., 15 μl from a 300 nM stock each). Then either measure or estimate the total amount of DNA that is present in the assembly reaction (see Note 12).
4. Add 10 μl 10× T4 DNA ligase buffer and fill up with a ddH_2O to a final volume of approximately 95 μl. To initiate the assembly reaction, add at least 1 μl USER enzyme per 1 μg DNA depending on the concentration of DNA (measured in Subheading 3.3 step 2).
5. Incubate at 37°C for 5 min.
6. Add 2 μl T4 DNA ligase (derived from a 400 U/μl stock) and incubate for a further 10 min at 37°C.
7. The assembly mixture can directly serve as template for the PCR described in the next step. If, however, a procedure is established for the first time or libraries are assembled, we recommend analyzing the assembly efficiency by agarose gel electrophoresis. In this case, it is necessary to purify the assembly reaction using the QIAquick kit before applying it onto the gel (see Note 13).
8. To "purify" the desired product, selectively amplify it from a mixture of the fully assembled template and partially assembled fragments by PCR. We use the two outmost primers LMB-2-6 and pIV-B1 in the example given. For a typical reaction, prepare a 100 μl PCR mix as follows: 10 μl PfuTurbo C_x reaction buffer (final concentration: 1×), 10 μl dNTP mix (final concentration: 250 nM), 1 μl each of the forward and reverse primer (final concentration: 1 μM final, Table 1), 1 μl DNA template (e.g., 1 μl assembly mix or 50 ng/μl spin column purified DNA, Table 2), and 75 μl water. Finally, add 2 μl PfuTurbo C_x polymerase (5 U/μl stock) and place reaction tubes into a thermocycler. After an initial denaturation and heat activation step for 2 min at 95°C, cycle ten cycles as follows: 30 s at 95°C, 1 min at the desired reannealing temperature (see Note 5), and 1 min/kbp at 72°C. Finish with a final extension step for 2 min at 72°C.
9. Purify the product DNA using the QIAquick kit and elute in a mixture of 1:10 elution buffer/water.
10. Analyze by agarose gel electrophoresis.
11. If necessary, precipitate the PCR product with PEG–$MgCl_2$ (see Subheading 3.3) to remove any primer dimers (see Note 14).

4. Notes

1. The majority of thermostable, proofreading DNA polymerases that are commercially available are derived from archaea and stall when they encounter uracil in the DNA template; they are thus not compatible with uracil-containing oligonucleotides. PfuTurbo C_x and PfuS7 are based on a mutant version V93Q where the uracil sensing function has been abolished (10, 11). We have established the procedure for PfuTurbo C_x, but PfuS7, a variant of Pfu DNA polymerase that has been engineered for greater processivity (11), should also be suitable for this method. Taq DNA polymerase, by contrast, is derived from a bacterial source and does not possess a uracil sensing function.
2. To prevent the accumulation of point mutations over the course of the assembly process, we highly recommend using a polymerase with a proofreading activity such as PfuTurbo C_x or PfuS7.
3. The composition of the PCR will depend on the type of the polymerase and any additional manufacturer's instructions and recommendations. In our experience, standard conditions work well and include 2 mM $MgCl_2$ and 1 μM oligonucleotides along with the recommended PCR buffer and a suitable chosen melting temperature, which will depend on the primer design. Depending on the nature of the DNA template, further optimization may have to be performed. However, no special precautions have to be taken per se.
4. Care should be taken in the design of splice sites to prevent separate sites from cross- or self-hybridizing; the latter occurs when the splice sites are palindromic leading to concatemerized DNA analogous to most restriction–digestion–ligation reactions. In our experience, all splice sites need to differ by at least 2 bp for a complementary overlap of 5–6 bp to prevent single-stranded extensions from cross- or self-hybridizing. Similarly, it must be ensured that the single-stranded extensions cannot fold onto themselves to prevent the formation of covalently closed loops, particularly if the extensions are longer than 6 bp. Calculating the base pair probabilities of the single-stranded extensions with a suitable folding program gives a good indication if secondary structures pose a problem, e.g., with the RNA-fold Web server (http://rna.tbi.univie.ac.at/), which conveniently displays base pair probabilities as a dot plot (12, 13).
5. When designing the PCR primers, the reannealing temperatures of the primers for amplification of the 3′-UTR, GOI, and 5′-UTR should match within one set (e.g., LMB-2-6 and Rec-LMB in case of the 3′-UTR). In addition, the annealing temperatures

of the two outmost primers for amplification of the full-length template (LMB-2-6 and pIV-B1) should match each other. Reannealing temperatures can be calculated using different online tools: e.g., the Oligo Analysis and Plotting Tool (http://www.operon.com) or the Sigma Genosys Oligocalculator (http://www.sigma-genosys.com/calc/DNACalc.asp). Note that different melting temperatures may be calculated depending whether an algorithm is based on a GC formula or thermodynamic parameters and to what extent the ion concentrations are taken into account.

6. The procedure is greatly facilitated by starting with a large amount of DNA, as each purification step necessarily reduces product yield. Starting amounts in the microgram range allow monitoring of each step by gel electrophoresis. However, once established, the protocol is also suitable for smaller amounts of DNA.

7. DpnI selectively digests methylated DNA of bacterial origin (unless it is derived from a DAM methylase negative strain such as JM110 or SCS110), but not synthetic DNA derived from PCR reactions. This is particularly important in the context of directed protein evolution to prevent repeated contamination with wild-type DNA.

8. Even though a DNA fragment may appear pure on an agarose gel, we generally find that for smaller DNA fragments primer dimers persist after spin column purification. The high *concentration* of these small fragments allows them to compete efficiently with larger assembly substrates, although their *mass* is small (explaining why they might not be visible on a gel). For this reason, we generally subject the DNA to a second purification step, which removes primer dimers: size-dependent PEG–$MgCl_2$ precipitation as described in Subheading 3.3. Should additional DNA fragments, e.g., larger unspecific PCR products still remain after the DNA precipitation, the assembly substrates may also be purified by agarose gel electrophoresis (e.g., using the Wizard SV Gel and PCR Clean-Up System). However, gel purification dramatically reduces yield. It may also interfere with downstream processes (e.g., inhibition of in vitro expression), which is why we recommend an additional ethanol precipitation of the DNA with sodium acetate should gel extraction be necessary. In our hands, however, spin column purification followed by PEG–$MgCl_2$ precipitation has proven an effective and sufficient means of purifying and preparing DNA fragments.

9. For optimal results, the concentration of PEG-8000 can be titrated by decreasing the concentration of PEG-8000 and selectively precipitating larger DNA fragments. As a rough guide,

we have successfully precipitated DNA fragments using different concentrations of PEG-8000 as follows: 300 bp and longer with 10% PEG, 415 bp and longer with 9.1% and 600 bp and longer with 8.3%.

10. Assembly by uracil-excision–ligation is highly dependent on defined 3′ extensions that mediate the assembly process. As the entire assembly reaction occurs in vitro, the method does not tolerate a significant fraction of single nucleotide gaps that can neither be ligated nor repaired by the endogenous repair machinery after transformation in *E. coli*. It is, therefore, critical that all assembly substrates have blunt ends. Blunting is absolutely essential for DNA fragments that have been prepared with Taq DNA polymerase. This is to remove the single 3′-adenine overhangs that are generated by its extendase activity. We also highly recommend to blunt all DNA fragments prepared with PfuTurbo C_x as in our experience the quality of blunt ends, and thus the efficiency of assembly, is variable (14).
11. Adjustments have to be undertaken where the DNA concentration varies. This may apply to the amount of enzymes that is added as well as the reaction volume. As a general guideline, we never exceed more than 2.5% glycerol in the final reaction mix; this is equivalent to no more than 5 μl of enzymes in a reaction volume of 100 μl assuming commercially available enzyme stocks are stored in 50% glycerol.
12. In our experience, it is critical for the efficiency of the assembly process to add at least 1 U USER enzyme per microgram DNA (as recommended by the manufacturer, NEB). It is, therefore, paramount to know the precise amount of DNA that is present in the assembly mix.
13. Applying the assembly reaction directly on an agarose gel generally results in smeary bands that do not correlate with the correct size of the DNA fragments. The reason for this is that either T4 DNA ligase or USER enzyme remain bound to DNA in the gel, which can lead to shifted bands.
14. Many in vitro DNA display systems require specific binding sites or chemical modifications to conjugate a protein to its coding DNA template: e.g., small molecules such as biotin, benzylguanine, or fluorouracil. These sites are generally introduced through specific primers modified at their 5′-end in the final "purifying" PCR step. An excess of primer dimers carrying these modifications can subsequently impair the conjugation reaction as they compete with the full-length DNA template for protein binding. In these cases, we recommend precipitating the fully assembled and amplified linear DNA templates with PEG-8000 and $MgCl_2$ to remove primer dimers.

References

1. Leemhuis, H., Stein, V., Griffiths, A.D. and Hollfelder, F. (2005) New genotype-phenotype linkages for directed evolution of functional proteins. *Curr Opin Struct Biol*, **15**, 472–478.
2. He, M. (2008) Cell-free protein synthesis: applications in proteomics and biotechnology. *New Biotechnol*, **25**, 126–132.
3. Rashtchian, A. (1995) Novel methods for cloning and engineering genes using the polymerase chain reaction. *Curr Opin Biotechnol*, **6**, 30–36.
4. Nisson, P.E., Rashtchian, A. and Watkins, P.C. (1991) Rapid and efficient cloning of Alu-PCR products using uracil DNA glycosylase. *PCR Methods Appl*, **1**, 120–123.
5. Smith, C., Day, P.J. and Walker, M.R. (1993) Generation of cohesive ends on PCR products by UDG-mediated excision of dU, and application for cloning into restriction digest-linearized vectors. *PCR Methods Appl*, **2**, 328–332.
6. Nour-Eldin, H.H., Hansen, B.G., Norholm, M.H., Jensen, J.K. and Halkier, B.A. (2006) Advancing uracil-excision based cloning towards an ideal technique for cloning PCR fragments. *Nucleic Acids Res*, **34**, e122.
7. Nour-Eldin, H.H., Geu-Flores, F. and Halkier, B.A. USER cloning and USER fusion: the ideal cloning techniques for small and big laboratories. *Methods Mol Biol*, **643**, 185–200.
8. Geu-Flores, F., Nour-Eldin, H.H., Nielsen, M.T. and Halkier, B.A. (2007) USER fusion: a rapid and efficient method for simultaneous fusion and cloning of multiple PCR products. *Nucleic Acids Res*, **35**, e55.
9. Gronemeyer, T., Chidley, C., Juillerat, A., Heinis, C. and Johnsson, K. (2006) Directed evolution of O6-alkylguanine-DNA alkyltransferase for applications in protein labeling. *Protein Eng Des Sel*, **19**, 309–316.
10. Connolly, B.A., Fogg, M.J., Shuttleworth, G. and Wilson, B.T. (2003) Uracil recognition by archaeal family B DNA polymerases. *Biochem Soc Trans*, **31**, 699–702.
11. Norholm, M.H. A mutant Pfu DNA polymerase designed for advanced uracil-excision DNA engineering. *BMC Biotechnol*, **10**, 21.
12. Hofacker, I.L. (2003) Vienna RNA secondary structure server. *Nucleic Acids Res*, **31**, 3429–3431.
13. SantaLucia, J., Jr. (1998) A unified view of polymer, dumbbell, and oligonucleotide DNA nearest-neighbor thermodynamics. *Proc Natl Acad Sci USA*, **95**, 1460–1465.
14. Stein, V. and Hollfelder, F. (2009) An efficient method to assemble linear DNA templates for in vitro screening and selection systems. *Nucleic Acids Res*, **37**, e122.

Chapter 7

Automated Computational Analysis of Genome-Wide DNA Methylation Profiling Data from HELP-Tagging Assays

Qiang Jing, Andrew McLellan, John M. Greally, and Masako Suzuki

Abstract

A novel DNA methylation assay, HELP-tagging, has been recently described to use massively parallel sequencing technology for genome-wide methylation profiling. Massively parallel sequencing-based assays such as this produce substantial amounts of data, which complicate analysis and necessitate the use of significant computational resources. To simplify the processing and analysis of HELP-tagging data, a bioinformatic analytical pipeline was developed. Quality checks are performed on the data at various stages, as they are processed by the pipeline to ensure the accuracy of the results. A quantitative methylation score is provided for each locus, along with a confidence score based on the amount of information available for determining the quantification. HELP-tagging analysis results are supplied in standard file formats (BED and WIG) that can be readily examined on the UCSC genome browser.

Key words: DNA methylation, Computational analysis, Bioinformatics, Pipeline

1. Introduction

Epigenetic regulation has been recognized to be essential for normal cellular function and development (1–3). Recently, many researchers have reported associations between epigenomic dysregulation and disease (4), with particular interest in its involvement in tumorigenesis (5–7). Cytosine methylation is one of the best-studied epigenomic modifications. It is well known that promoter hypermethylation is related to gene silencing (1, 8, 9), and recent studies show that increased DNA methylation of the gene body is associated with its transcription (10–12). Such findings highlight the importance of performing genome-wide DNA methylation assays to understand fully the role of DNA methylation in both normal and abnormal cellular states. The need to perform

Michael Kaufmann and Claudia Klinger (eds.), *Functional Genomics: Methods and Protocols*,
Methods in Molecular Biology, vol. 815, DOI 10.1007/978-1-61779-424-7_7, © Springer Science+Business Media, LLC 2012

rapid and accurate genome-wide DNA methylation analysis has driven the development of de novo analytical assays coupled with the algorithm and software development needed for performing data analysis. Recently, we have developed such an assay, which we call HELP-tagging (13).

The HELP-tagging method is based on parallel methylation-sensitive (HpaII) and methylation-insensitive (MspI) isoschizomer digestion of genomic DNA at CCGG restriction sites. After attaching a specially designed adaptor to the 5′ end of each cut restriction site, a 27-bp flanking genomic fragment is liberated by digestion with a type III restriction enzyme (EcoP15I), an adapter added to the other end of the molecule followed by library preparation and sequencing. Sequence tags are filtered for quality and then mapped back to the reference genome. DNA methylation is quantified by comparing the relative count of fragments generated by each of the two isoschizomers at every digested locus. The whole analysis is automated using a bespoke analytical pipeline, which utilizes our local high-performance computing facilities working as a module within our Wiki-based Automated Sequence Processor (WASP) system (http://wasp.einstein.yu.edu/). In this chapter, we explain the analysis pipeline in more detail.

2. Materials

2.1. DNA Extraction

1. Extraction buffer: 100 mM Tris–HCl, pH 8.0, 0.1 M EDTA, pH 8.0, 0.5% SDS, 20 μg/mL RNaseA.
2. Sodium Chloride–Sodium Citrate Buffer (SSC) buffer: prepare 20× stock with 3 M of Sodium Chloride and 300 mM Sodium Citrate, pH 7.0.

2.2. HELP Tagging

1. All restriction enzymes and modifying enzymes are available from New England Biolabs.
2. MEGAshortscript kit (Amibion).
3. Agencourt AMPure XP (AGENCOURT).

3. Methods

3.1. DNA Extraction

1. Approximately 5×10^6 cells are suspended in 10 ml of Extraction buffer and incubate for 1 h at 37°C.
2. Add 50 μl of proteinase K (20 mg/ml), mix gently and incubate in a 50°C water bath overnight.
3. Treat DNA three times with saturated phenol, then twice with chloroform, and then pour DNA into dialysis tube.

4. Dialyze the tube for 16 h at 4°C against three changes of 0.2× SSC buffer.
5. Concentrate DNA by coating the dialysis bags with polyethylene glycol (molecular weight 20,000).
6. The purity and final concentration of the purified DNA is checked by spectrometry.

3.2. HELP-Tagging Library Preparation

1. Five micrograms of genomic DNA are digested with HpaII and MspI in separate 200 μl reactions and treated by phenol/chloroform followed by ethanol precipitation.
2. The digested genomic DNA is ligated to a 5′ adapter that contains an EcoP15I recognition site and T7 promoter sequence.
3. The ligated products are digested with EcoP15I.
4. The EcoP15I digested fragments are end-repaired, and tailed with a single dA at the 3′ end.
5. After the dA tailing reaction a 3′ adapter containing the Illumina sequencing-primer binding site is ligated to the dA-tailed fragments using the Quick Ligation Kit.
6. After ligation, products are in vitro-transcribed, then the resulting mRNA reverse transcribed. The first strand cDNA produced is used as a template for PCR using the following conditions: 96°C for 2 min, then 18 cycles of 96°C for 15 s, 60°C for 15 s, and 72°C for 15 s followed by 5 min at 72°C for the final extension.

3.3. Illumina Sequencing

Illumina sequencing is performed on an Illumina GAIIx/HiSeq 2000 sequencer following the manufacturer's instructions. For this assay, single-end, 36–50 bp sequencing is required.

3.4. Input Data for HELP-Tagging Analysis

The Illumina Genome Analyzer uses proprietary software (CASAVA) (14) to perform image analysis and base calling. The output of the CASAVA pipeline required for HELP-tagging analysis is simply the raw sequence reads in Illumina's QSEQ format. At this stage, we are not interested in generating alignments because of the presence of ~8 bp of 5′ adapter sequence at the 3′ end of each read (13).

3.5. Library Quality Assessment and Prealignment Tag Processing

Within Illumina QSEQ files, each noncallable (unknown) base is represented as a period. These are converted to "N" characters when generating FASTA or FASTQ files. For all sequences, statistics are generated to determine the proportion and position of unknown bases to provide both an assessment of sequencing success and individual cycle efficacy, respectively. Library quality is assessed by determining the proportion and position of the 3′ adapter sequence within the reads. A high percentage of reads containing adapter sequence starting at around position 27 is indicative

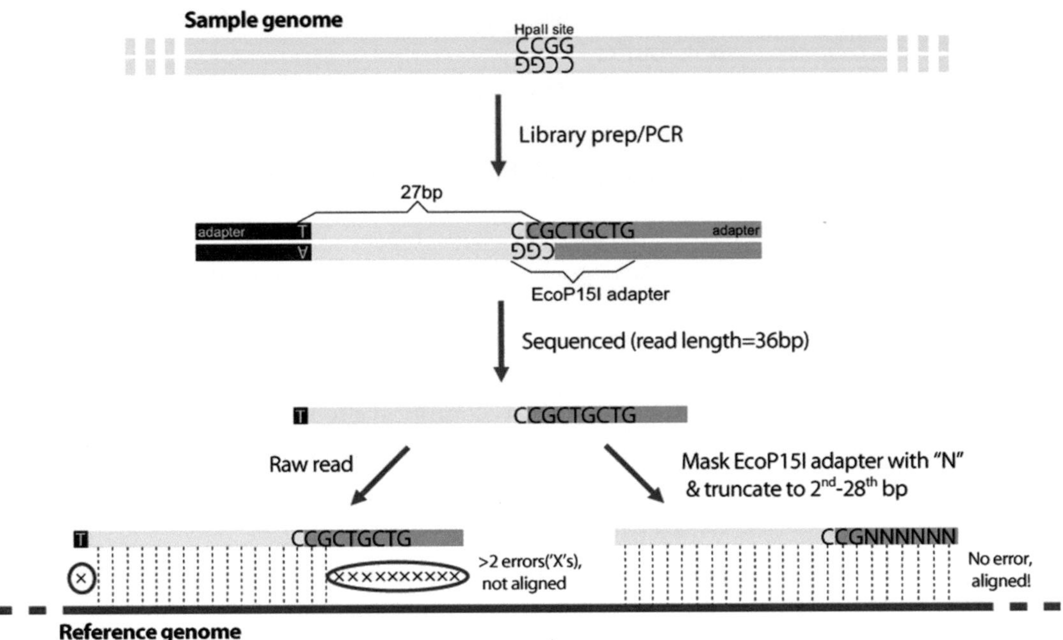

Fig. 1. Trimming and masking adaptor sequence from sequence tags. After sequencing, tags do not align back to the reference genome because of the presence of library adaptor sequence. However, after masking the adaptor sequence with "N"s and truncating to use bases 2–28 inclusively, it is possible to align the sequence tags to the reference genome.

of a high-quality library. Reads with excessive tracts of unknown bases and those failing to contain 5′ adapter sequence near the 3′ end are discarded (see Note 1).

Within remaining reads, the first base is removed, as it originates from 3′ adapter ligation, and the sequence originating from the 5′ adapter is masked by replacing the bases with "N"s so as to maintain a uniform tag size (Fig. 1). Illumina's proprietary alignment algorithm, ELAND, treats unknown ("N") bases as wild cards and therefore does not penalize them during alignment to the reference genome (14, 15). Some aligners, e.g., Bowtie (16), do not require tag length to be uniform and so the 5′ adapter sequence can be trimmed from the 3′ end of the sequences if using such an aligner.

3.6. Alignment

The trimmed and masked sequence tags are mapped back to a reference genome using the ELAND standalone program from CASAVA (see Note 2). During the alignment process, a maximum of two mismatches are allowed and indels are ignored. Statistics generated from the alignment included the number of sequence tags that are rejected due to there being too many matches to the reference genome (greater than a default value of 10 by ELAND algorithm), or for which there was no match at all (see Note 3).

3.7. HpaII Site Tag Counting and Angle Calculation

Counts of sequence tags aligned adjacent to every annotated HpaII/MspI site across the genome are assessed. Tags that aligned to more that one locus are given a proportional count relative to the number of mapped loci, e.g., a tag mapping to two loci is counted as 0.5 and one mapping to ten loci counted as 0.1. Cumulative counts at each annotated HpaII site are then normalized to represent a fraction of the total number of sequence tags aligned to annotated HpaII sites genome-wide.

In addition to assessing counts of sequence tags piled up adjacent to annotated HpaII/MspI sites, it is possible to observe sequence tags piling up at other genomic loci. This typically occurs where there is a SNP located within the CCGG site. Thus, all nonannotated HpaII/MspI loci, adjacent to which sequence tags are found to align, are compared to the Single Nucleotide Polymorphism Database (dbSNP) to discover putative polymorphic HpaII/MspI sites. The HELP-tagging analysis pipeline returns this information along with a comprehensive methylation analysis.

In order to quantify the level of methylation at each HpaII/MspI site, the normalized accumulative proportional (NAP) count for the HpaII digested sample can be compared to the NAP

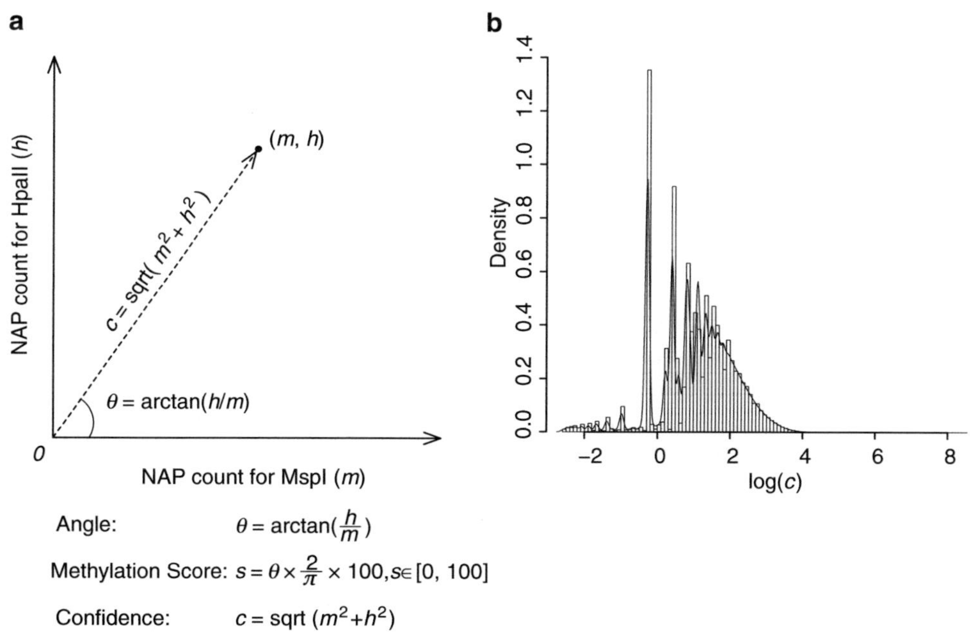

Angle: $\theta = \arctan(\frac{h}{m})$

Methylation Score: $s = \theta \times \frac{2}{\pi} \times 100, s \in [0, 100]$

Confidence: $c = \mathrm{sqrt}\ (m^2 + h^2)$

Fig. 2. Quantifying methylation at a CCGG locus and calculating a confidence score. (**a**) The NAP counts for HpaII- and MspI-digested samples represent the respective *x* and *y* coordinates for a vector representing a CCGG locus. The methylation score for this site is defined as the direction (*angle*) of the vector and this angle is linearly scaled in the range 0–100. Confidence in the methylation score is determined from the magnitude of the vector. (**b**) The probability density plot of the logarithm of vector magnitude shows an approximate normal distribution. Axis *x* is the logarithm of vector magnitude *c*, axis *y* is the density plot of log(*c*).

count for the MspI digested sample (Fig. 2). If the site is hypermethylated, the HpaII NAP count of this site should be less than the MspI NAP count since HpaII restriction enzyme is unable to cut DNA when the internal cytosine is methylated, whereas MspI enzyme is able to cut regardless of the cytosine methylation state (13). More specifically, the more that the HpaII NAP count approaches the MspI NAP count, the less methylated the CCGG site is. To quantify the methylation level of each CCGG locus, the NAP counts for HpaII and MspI digested samples can be represented as Cartesian coordinates for vectors projected in two-dimensional space where the *Y*-axis represents the HpaII NAP count and the *X*-axis represents the MspI NAP count. Represented as a vector, the direction (angle relative to the origin) corresponds to a quantification of hypomethylation and the magnitude represents a measure of the tag counts i.e. information content. Thus, the magnitude of each vector provides a level of confidence in the quantification (the greater the magnitude, the more tag counts contributed to the quantification and therefore the more confident we are in the result). The direction of each vector is calculated as the arc tangent of the ratio of the HpaII NAP count and the MspI NAP count. The final step in data processing is a linear scaling of these angular values to a range from 0 to 100, representing the percentage of unmethylation at that locus. These data are stored in a WIG format file for easy viewing of the data as a histogram within the UCSC genome browser. As fully methylated CCGG sites have an unmethylated value of 0, which would not appear on a genome browser view, the site is marked with a short tag underneath the *X*-axis (negative value) when the track is displayed within the genome browser.

An algorithm was developed to provide a level of confidence in methylation quantification based on vector magnitude (see explanation above), and it was decided to classify confidence as high, medium, or low based on the distribution of the logarithm of magnitudes (Fig. 3). A typical density plot of the logarithm of the vector magnitudes for data derived from an arbitrary HpaII digested sample showed that the data approximate a normal distribution (17). The three categories of confidence level are defined by using the mean value and standard deviation of this distribution (see Note 4). If the magnitude of a vector representing a particular CCGG site falls in the range of the mean value plus/minus the standard deviation, it would be categorized as medium confidence; if the magnitude is below or above this range, it would be categorized as being of low or high confidence respectively.

Fig. 3. (continued) *Rectangular boxes* represent internal process steps and *triangles* represent conditional evaluations. The *rounded boxes* represent alternate process steps producing useful statistics and additional results.

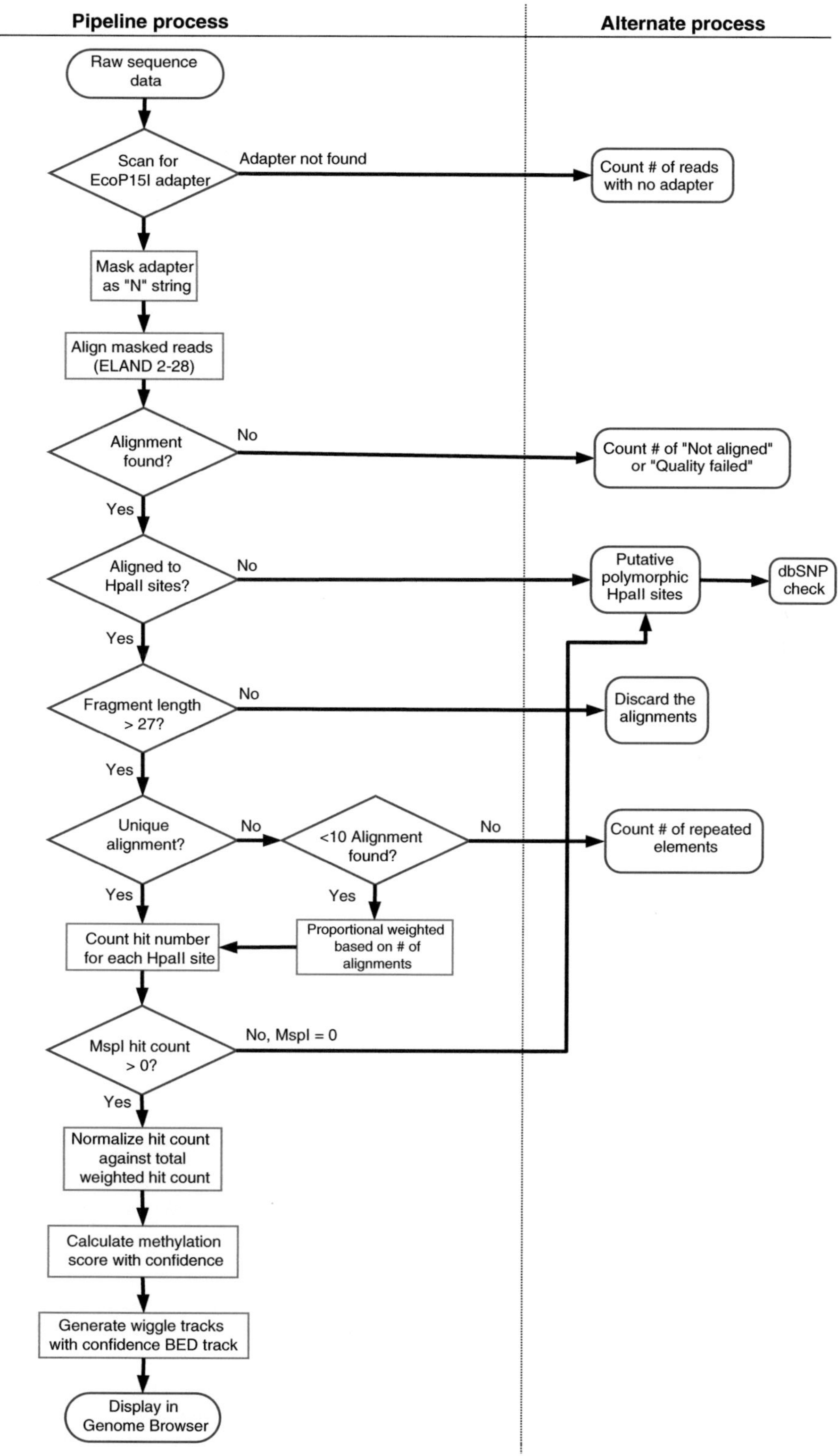

Fig. 3. Workflow for the HELP-tagging analytical pipeline. This flowchart shows the sequence of data processing performed by the HELP-tagging analysis pipeline. Steps in the *oval boxes* represent the input and output for the pipeline.

4. Notes

1. After applying the filtering by the occurrence for 5′ adaptor sequence, the number of remaining sequence reads should be at least twice the number of HpaII restriction sites genome-wide, to provide enough coverage for later comparison studies.
2. The ELAND alignment algorithm used in the analytical pipeline could be substituted by any other popular alignment algorithm such as Bowtie, BWA, or SOAP, without affecting most of other processing steps of the pipeline; however, the data format of the input/output files for the substitution algorithm may need to be modified.
3. The criterion of discarding repeated elements is more than ten alignments found which could be made more or less stringent. However, the methylation analysis result should not change too much as repetitive elements only make marginal contribution because of the use of proportional counting.
4. The log-normal distribution of confidence is approximate so the categorization of different confidence levels is also approximate. The purpose of providing these data is to provide the investigator with some reference for better interpretation of the methylation results at specific loci of interest.

Acknowledgments

We wish to thank Shahina Maqbool, Raul Olea, and Gael Westby of Einstein's Epigenomics Shared Facility for their contributions, and Einstein's Center for Epigenomics.

References

1. Bird, A. P., and Wolffe, A. P. (1999) Methylation-induced repression--belts, braces, and chromatin, *Cell 99*, 451–454.
2. Li, E., Bestor, T. H., and Jaenisch, R. (1992) Targeted mutation of the DNA methyltransferase gene results in embryonic lethality, *Cell* 69, 915–926.
3. Okano, M., Bell, D. W., Haber, D. A., and Li, E. (1999) DNA methyltransferases Dnmt3a and Dnmt3b are essential for de novo methylation and mammalian development, *Cell* 99, 247–257.
4. Costello, J. F., Fruhwald, M. C., Smiraglia, D. J., Rush, L. J., Robertson, G. P., Gao, X., Wright, F. A., Feramisco, J. D., Peltomaki, P., Lang, J. C., Schuller, D. E., Yu, L., Bloomfield, C. D., Caligiuri, M. A., Yates, A., Nishikawa, R., Su Huang, H., Petrelli, N. J., Zhang, X., O'Dorisio, M. S., Held, W. A., Cavenee, W. K., and Plass, C. (2000) Aberrant CpG-island methylation has non-random and tumour-type-specific patterns, *Nat Genet* 24, 132–138.
5. Greger, V., Passarge, E., Hopping, W., Messmer, E., and Horsthemke, B. (1989) Epigenetic changes may contribute to the formation and spontaneous regression of retinoblastoma, *Hum Genet* 83, 155–158.

6. Dobrovic, A., and Simpfendorfer, D. (1997) Methylation of the BRCA1 gene in sporadic breast cancer, *Cancer Res* 57, 3347–3350.
7. Gebhard, C., Schwarzfischer, L., Pham, T. H., Schilling, E., Klug, M., Andreesen, R., and Rehli, M. (2006) Genome-wide profiling of CpG methylation identifies novel targets of aberrant hypermethylation in myeloid leukemia, *Cancer Res* 66, 6118–6128.
8. Akiyama, Y., Watkins, N., Suzuki, H., Jair, K. W., van Engeland, M., Esteller, M., Sakai, H., Ren, C. Y., Yuasa, Y., Herman, J. G., and Baylin, S. B. (2003) GATA-4 and GATA-5 transcription factor genes and potential downstream antitumor target genes are epigenetically silenced in colorectal and gastric cancer, *Mol Cell Biol* 23, 8429–8439.
9. Suzuki, M., Sato, S., Arai, Y., Shinohara, T., Tanaka, S., Greally, J. M., Hattori, N., and Shiota, K. (2007) A new class of tissue-specifically methylated regions involving entire CpG islands in the mouse, *Genes Cells* 12, 1305–1314.
10. Suzuki, M. M., and Bird, A. (2008) DNA methylation landscapes: provocative insights from epigenomics, *Nat Rev Genet* 9, 465–476.
11. Backdahl, L., Herberth, M., Wilson, G., Tate, P., Campos, L. S., Cortese, R., Eckhardt, F., and Beck, S. (2009) Gene body methylation of the dimethylarginine dimethylamino-hydrolase 2 (Ddah2) gene is an epigenetic biomarker for neural stem cell differentiation, *Epigenetics* 4, 248–254.
12. Ball, M. P., Li, J. B., Gao, Y., Lee, J. H., LeProust, E. M., Park, I. H., Xie, B., Daley, G. Q., and Church, G. M. (2009) Targeted and genome-scale strategies reveal gene-body methylation signatures in human cells, *Nat Biotechnol* 27, 361–368.
13. Suzuki, M., Jing, Q., Lia, D., Pascual, M., McLellan, A., and Greally, J. M. (2010) Optimized design and data analysis of tag-based cytosine methylation assays, *Genome Biol* 11, R36.
14. Illumina. (2010) CASAVA Software Version 1.7 User Guide, Illumina Inc.
15. Cox, A. J. (unpublished) ELAND: Efficient Local Alignment of Nucleotide Data.
16. Langmead, B., Trapnell, C., Pop, M., and Salzberg, S. L. (2009) Ultrafast and memory-efficient alignment of short DNA sequences to the human genome, *Genome Biol* 10, R25.
17. Aitchison, J., and Brown, J. A. C. (1957) *The lognormal distribution, with special reference to its uses in economics*, University Press, Cambridge.

Part III

RNA Analysis

Chapter 8

Detection of RNA Editing Events in Human Cells Using High-Throughput Sequencing

Iouri Chepelev

Abstract

RNA editing can lead to amino acid substitutions in protein sequences, alternative pre-mRNA splicing, and changes in gene expression levels. The exact in vivo modes of interaction of the RNA editing enzymes with their targets are not well understood. Alterations in RNA editing have been linked to various human disorders and the improved understanding of the editing mechanism and specificity can explain the phenotypes that result from misregulation of RNA editing. Unbiased high-throughput methods of detection of RNA editing events genome-wide in human cells are necessary for the task of deciphering the RNA editing regulatory code. With the rapidly falling cost of genome resequencing, the future method of choice for the detection of RNA editing events will be whole-genome gDNA and cDNA sequencing. We describe a detailed procedure for the computational identification of RNA editing targets using the data from the deep sequencing of DNA and RNA from the peripheral blood mononuclear cells of a human individual with severe hemophilia A who is resistant to HIV infection. Interestingly, we find that mRNAs of the cyclin-dependent kinase CDK13 and the DNA repair enzyme NEIL1 undergo extensive A→I RNA editing that leads to amino acid substitutions in protein sequences.

Key words: RNA editing, Single nucleotide variants, High-throughput sequencing, Bioinformatics, Human immunodeficiency virus infection

1. Introduction

RNA editing is the posttranscriptional alteration of RNA sequences through the insertion, deletion or modification of nucleotides, excluding changes due to processes such as RNA splicing and polyadenylation (1). Such alterations in RNA sequences can bring about amino acid substitutions in protein sequences, alternative pre-mRNA splicing, and changes in gene expression levels (2). In higher eukaryotes, the most prevalent type of RNA editing is mediated by adenosine deaminase acting on RNA (ADAR) enzymes that convert adenosines to inosines (A→I editing) in double-stranded

Michael Kaufmann and Claudia Klinger (eds.), *Functional Genomics: Methods and Protocols*,
Methods in Molecular Biology, vol. 815, DOI 10.1007/978-1-61779-424-7_8, © Springer Science+Business Media, LLC 2012

RNA substrates (3). The three major types of A→I editing targets are protein-coding pre-mRNAs, repetitive elements such as Alu repeats located in exons or introns, and microRNA precursors (3). The exact in vivo modes of interaction of the RNA editing enzymes with their targets are unknown, but the base-paired RNA structures are believed to guide the enzymes to edit a single nucleotide with high specificity and efficiency (2). Alterations in RNA editing have been linked to various human disorders and the improved understanding of the editing mechanism and specificity can explain the phenotypic features that result from misregulation of RNA editing (4). Unbiased high-throughput methods of detection of RNA editing events genome-wide in normal and abnormal human cells are necessary for the task of deciphering the RNA editing regulatory code. Recent rapid developments in massively parallel DNA sequencing technologies (5, 6) have allowed the identification of RNA editing targets in human cells at 36,000 genomic loci by a targeted sequencing (7). With the rapidly falling cost of genome resequencing (8), the future method of choice for the detection of RNA editing events will be whole-genome genomic DNA (gDNA) and complementary DNA (cDNA) sequencing. Herein, we describe a detailed procedure for the computational identification of RNA editing targets using a dataset of raw sequence reads from the deep sequencing of DNA and RNA from one human individual.

2. Materials

2.1. Deep Sequencing Dataset

We obtained cDNA and gDNA raw sequencing data from the study published by Cirulli et al. (9). Let us briefly describe the samples from this work. The DNA and RNA were extracted from peripheral blood mononuclear cells (PBMCs) from an individual with severe hemophilia A, who is resistant to HIV infection. The DNA was prepared for sequencing according to Illumina's gDNA sample prep kit protocol. The total RNA was prepared according to the Illumina RNA-Seq protocol that involved globin reduction and polyA enrichment. For alternative RNA preparation procedures, see Note 1.

The paired-end reads for the gDNA and cDNA libraries are each around 75 bp long. There are 1,450 and 280 million reads in gDNA and cDNA libraries, respectively.

The raw sequencing data is in standard Sanger FASTQ format. A FASTQ file uses four lines per nucleotide sequence. An example entry for a single sequence in a FASTQ file is shown below:

```
@SRR037167.9742210
CCCGACGTTACATCATCTGCCCCGTTGTATGCAACA
+SRR037167.9742210
6-66+06(63+&0(&666+66(-6&+1(03&(.)&)
```

For each entry in a FASTQ file, the first line begins with a "@" character and is followed by a sequence identifier and an optional description. The second line is the raw sequence. The third line begins with a "+" character and is optionally followed by a description. The fourth line encodes Phred quality scores for the sequence from the second line, and contains the same number of symbols as letters in the sequence. Phred quality score Q of a base call is defined as $Q=-10\log_{10}p$, where p is the probability that the base call is incorrect. The Phred quality score Q encoded by an ASCII of the character x is given by $Q=\mathrm{ord}(x)-33$, where $\mathrm{ord}(x)$ is the decimal integer ASCII value corresponding to x. For example, the symbol "&" corresponds to the Phred score $Q=\mathrm{ord}(\&)-33=38-33=5$.

2.2. Computational Hardware

The handling and processing of large datasets generated from deep sequencing experiments is most convenient on Linux and Unix-based computers. The storage of the raw sequence dataset described in Subheading 2.1 alone requires more than 300 GB of disk space. There are several additional large processed files such as alignment files that need to be stored. In principle, the Unix piping utilities can be used to deal with compressed datasets to save some disk space at the expense of CPU time. Nevertheless, we recommend a computer with at least 2 TB of free disk space to comfortably work with the large datasets. It is also desirable that the computer has multiple processors/cores and at least 20 GB of RAM so that several computationally intensive jobs can be run concurrently.

2.3. Computational Software

Bowtie is an ultrafast, memory-efficient alignment program for aligning short DNA sequence reads to large genomes (10). For the human genome, Bowtie aligns more than 15 million 75 bp reads per CPU hour. The precompiled executable of Bowtie and prebuilt index of human hg18 genome can be downloaded from http://bowtie-bio.sourceforge.net.

SAMtools is a library and software package for parsing and manipulating alignments in the SAM/BAM formats (11). The source code for SAMtools is available from http://samtools.sourceforge.net/. SAMtools needs to be compiled using GNU C compiler (http://gcc.gnu.org/).

Picard comprises Java-based command-line utilities that manipulate SAM files. It is available from http://picard.sourceforge.net/. We use Picard to remove PCR amplification bias in alignment data.

3. Methods

The RNA editing analysis is conceptually simple as illustrated in Fig. 1. The cDNA and gDNA sequencing reads are aligned to a reference genome. The aligned reads are then passed into various filters. The aligned and filtered cDNA and gDNA reads are then fed into the

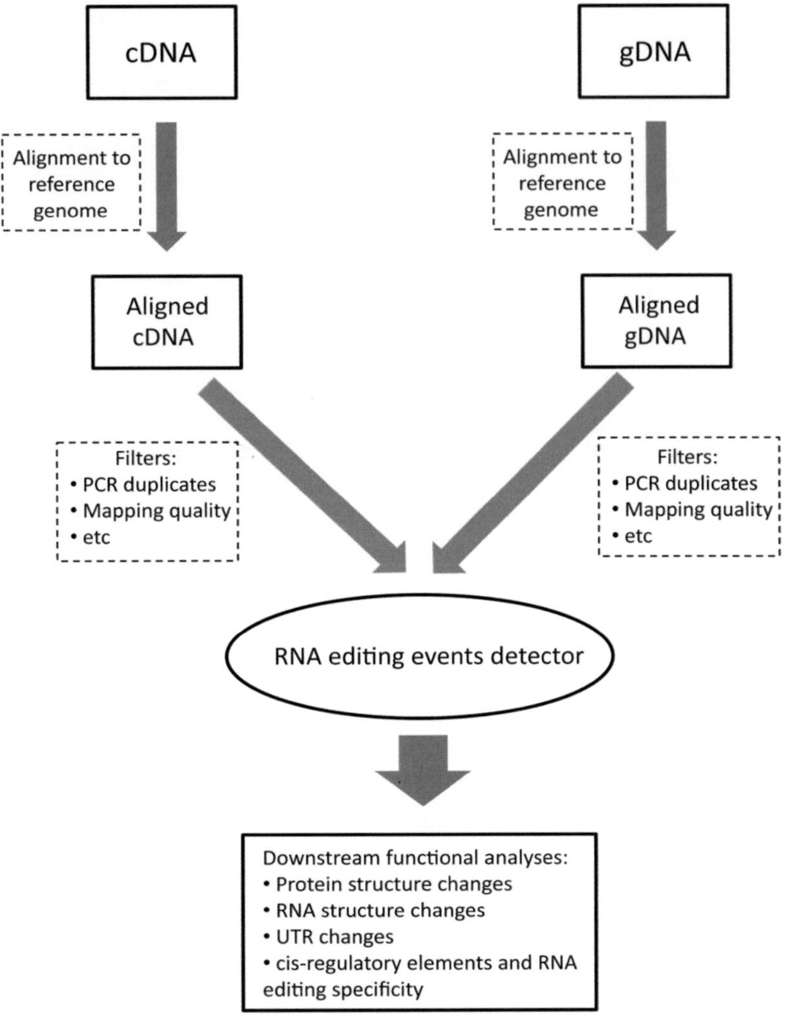

Fig. 1. Steps involved in detecting RNA editing events using cDNA and gDNA sequencing reads.

RNA editing events detector which outputs a list of candidate editing sites in the genome. The putative editing sites can then be analyzed for their functional consequences such as protein structure/function changes due to nonsynonymous base substitutions, changes in RNA structure and UTR. The identification of RNA editing sites genome-wide and in multiple cell types facilitates the decoding of the RNA editing regulatory code that involves a complex interaction of *cis*-regulatory sequences, base-paired RNA structures, and trans-acting elements.

3.1. Sequence Alignment

The Sequence Alignment/Map (SAM) format is a common alignment format that supports all sequence types (11). It is designed to scale to alignment sets of 10^{11} or more base pairs, which is typical for the deep sequencing of one human individual. Starting from the FASTQ file of raw reads data, we can align reads to hg18 reference human genome using Bowtie as follows:

```
bowtie hg18 -S -p 4 -n 2 -l 30 –e 70 –k 1 --best cDNA.fastq cDNA.sam
```

In the above command, the "-S" option sets the output alignment format to SAM. The "-p 4" allows the speed-up of the alignment process by using four processors/cores in parallel. The specification "-k 1 --best" guarantees that the best *valid* alignment per read will be reported. The *validity* of an alignment is specified as "-n 2 -l 30 –e 70". The latter specification means that (a) Alignments may have no more than two mismatches ("-n" option) in the first 30 bases ("-l" option) on the high-quality end of the read and (b) the sum of Phred quality values at all mismatched positions may not exceed 70 ("-e" option). See Note 2 for an alignment method more appropriate for cDNA sequences.

3.2. Postprocessing Alignments

3.2.1. Compressing Alignment Files and Indexing

The Binary Alignment/Map (BAM) format is the binary representation of SAM and keeps the same information as SAM (11). The BAM alignment file for the gDNA data from (9) requires around 90 GB disk space, which is equivalent to a compression rate of approximately 1.0 byte per input base. The command to generate BAM file from gDNA SAM file is:

```
samtools view -bS -o gDNA.bam gDNA.sam
```

The BAM alignment file should be sorted by coordinate for an efficient data processing and to avoid loading extra alignments into memory. A BAM file can be sorted as follows:

```
samtools sort gDNA.bam gDNA_sorted
```

The position sorted BAM file "gDNA_sorted.bam" can now be indexed to achieve fast random retrieval of alignments overlapping a specific genomic region as follows:

```
samtools index gDNA_sorted.bam
```

Let us also index the FASTA file "hg18.fa" of the reference human genome sequence using the command:

```
samtools faidx hg18.fa
```

We can now view alignments at any genomic location using a text-based viewer using "tview" option as follows:

```
samtools tview gDNA_sorted.bam hg18.fa
```

3.2.2. Duplicate Reads Filtering

In order to remove possible PCR amplification artifacts, it is reasonable to retain only one or a few reads that align to the same genomic position (12). We used Picard's "MarkDuplicates" function to properly remove duplicate reads. If multiple reads align to the same genomic location, Picard retains a single read with the best sequence quality. The following command removes duplicate reads from the sorted BAM file "cDNA_sorted.bam" and returns the sorted BAM file "cDNA_sorted_rmdup_picard.bam".

```
java -Xmx4g -jar MarkDuplicates.jar INPUT=cDNA_sorted.bam \
OUTPUT=cDNA_sorted_rmdup_picard.bam REMOVE_DUPLICATES=true
METRICS_FILE=MF.txt AS=true
```

In the above command, the –Xmx4g option sets the maximum java heap size to 4 GB.

3.2.3. Pileup Format and Variant Calling

The Pileup format describes the base-pair information at each chromosomal position. This format facilitates SNP/indel calling. An example pileup format file is shown below:

```
chr1  2012  T  6   ,C,...        CBBAAC
chr1  2586  T  8   CcGGCCC.      #BA##A;?
chr1  8745  A  5   c,...         ##AB?
chr1  8754  T  7   ,,...,c       ;9BA;##
chr1  8769  T  10  ,,...,,,.c    ;3AAA#/:B#
chr1  8772  t  11  ,,...,,,.,c   8>BB?#1;B#@
chr1  8773  c  12  ,,...,,,.,t,  A?BB9;=8@#6#
```

Here, each line consists of a chromosome, a 1-based coordinate, a reference base, the number of reads covering the site, the read bases, and the base qualities. At the read base column, a dot stands for a match to the reference base on the forward strand, a comma for a match on the reverse strand, "ACGTN" for a mismatch on the forward strand and "acgtn" for a mismatch on the reverse strand. The first line in the above example shows that (a) there are six sequence reads that cover position chr1:2,012 in the genome, (b) the reference nucleotide at this position is T, (c) five reads have matching nucleotides at this position with two of these reads aligning to the reverse strand of the genome and three to the forward strand, and (d) the remaining read aligns to the forward strand with the mismatch nucleotide C at this position. More details on the pileup format can be found in SAMtools manual. Given a FASTA file "hg18.fa" of human genome sequence and sorted BAM file "gDNA.bam" of aligned gDNA sequences, a huge pileup file "gDNA_pileup.txt" for all genomic locations covered by at least one sequencing read is generated as follows:

```
samtools pileup -f hg18.fa gDNA.bam > gDNA_pileup.txt
```

If option **-c** is applied to SAMtools, the IUPAC consensus base, Phred-scaled consensus quality, SNP quality, and root mean square mapping quality of the reads covering the site will be inserted between the "reference base" and the "number of reads covering the site" columns as in the following example:

```
chr1  95543  t  K  36  36  60  14  ..GG,.G...G.,.  B@@BB@BBBABCAC
chr1  98160  a  C  38  39  60  5   .cccc           #>;;#
chr1  98173  t  C  50  51  60  9   ccccCccac       B?@4/B9#.
```

3.2.4. Efficient Data Processing with UNIX Pipes

Whenever possible, generation of huge files should be avoided. UNIX pipes should be used for streaming the output of one program into the input of another program so that disk space usage is minimized. For example, after consensus base calling, we want to filter out sites with very high read coverage because such sites are error prone. We can use SAMtools to set maximum read depth using the "–D" option. We then apply a filter to keep only those sites that have mapping quality equal or greater than 20. SAMtools works well with UNIX pipes. Let "cDNA.bam" and "gDNA.bam" be duplicate-reads filtered sorted BAM files for cDNA and gDNA alignments, respectively. The variant calling and the two filters can be combined using pipes as follows:

```
samtools pileup -f hg18.fa –c gDNA.bam | samtools.pl varFilter -D100 | awk "$6>=20">gDNA_variants.txt
```

Similarly, instead of generating a full pileup file from cDNA alignment data, we can use pipes to keep only those sites that have read coverage of at least five and have at least one read with nucleotide mismatch as follows:

```
samtools pileup -f hg18.fa cDNA.bam | perl variant_site.pl > cDNA_pileup.txt
```

Here the script "variant_site.pl" is given by the following simple Perl code:

```
while(<>){
  @row = split /\t/;
  if ($row[4] =~ /[ACGTacgt]/){
    if ($row[3] >= 5){
      print $_;
    }
  }
}
```

3.3. Probabilistic Framework for Detecting RNA Editing Sites

3.3.1. Theory

At a given single-nucleotide position in the diploid human genome let the genotype be X_1/X_2. The genotype can be heterozygous ($X_1 \neq X_2$) or homozygous ($X_1 = X_2$). We look for the evidence of RNA editing only at homozygous sites in gDNA because such sites constitute an overwhelming majority of sites in the genome and because it is somewhat complicated to do RNA editing analysis of the heterozygous sites. So, let the genotype at the homozygous locus x be X_{gDNA}/X_{gDNA}. Let the nucleotide at the position x in the reference hg18 human genome be X_{hg18}. There are two possibilities: $X_{gDNA} = X_{hg18}$ or $X_{gDNA} \neq X_{hg18}$ .We only consider homozygous loci in gDNA where $X_{gDNA} = X_{hg18}$, since such loci constitute the overwhelming majority of homozygous loci in the human genome. See Note 3 for the $X_{gDNA} \neq X_{hg18}$ case.

Let the homozygous site x with the genotype X_{gDNA}/X_{gDNA} be located in a genomic region that is transcribed. In the absence of RNA editing the cDNA will have nucleotide $X_{cDNA}=X_{gDNA}$ at position x. In general, there will be two species of cDNA: a fraction f of cDNA will be unedited and have $X_{cDNA}=X_{gDNA}$ and a fraction $1-f$ of cDNA will be edited and have $X_{cDNA}\neq X_{gDNA}$ at position x. For the sake of simplicity, we only consider the most prevalent type of RNA editing: A→I editing. Inosine is interpreted as guanosine by the translational machinery, and therefore, A→I editing is functionally equivalent to an A→G conversion.

If the sequencing error rate were identically zero, the likelihood of observing $n(\mathrm{A})$ of A nucleotides and $n(\mathrm{G})$ of G nucleotides at the position x (the conditional probability of observed data given the unedited fraction f of RNA species), would be given by the binomial probability $P(\mathrm{D}\mid f)=f^{n(\mathrm{A})}(1-f)^{n(\mathrm{G})}$. The maximum likelihood estimate (MLE) of f is given by $f_{\mathrm{ML}}=n(\mathrm{A})/[n(\mathrm{A})+n(\mathrm{G})]$.

In reality the sequencing error rate is nonzero and the probability of base error, Phred probability, needs to be taken into consideration. Let D be the observed sequence data, which is generated by sampling RNA species and noisy sequencing measurements. If the maximum likelihood of data assuming nonzero fraction of edited RNA species, $\max_f P(\mathrm{D}\mid f)$, is much greater than the likelihood of the data assuming no RNA editing, $P(\mathrm{D}\mid f=1)$, we have a strong evidence for an RNA editing event (7).

If the base error probabilities are small, $P(\mathrm{D}\mid f\neq 1)$ can still be approximated by the binomial distribution mentioned above. Otherwise, one can proceed as follows. As a prerequisite, the reader is referred to Li et al. (13) for an introduction to a probabilistic theory of base error rates and variant calling. In the absence of any sequencing errors, there is still a variability in the number of observed A's and G's due to the sampling noise. Let us denote by R the unobserved "sequencing-error-free" data. $P(\mathrm{D}\mid f)$ can then be expanded as follows: $P(\mathrm{D}\mid f)=\Sigma_{\mathrm{R}} P(\mathrm{D}\mid \mathrm{R})\, P(\mathrm{R}\mid f)$. The conditional probability $P(\mathrm{D}\mid \mathrm{R})$ can be computed using Phred base error probabilities as in ref. 13. The conditional probability $P(\mathrm{R}\mid f)$ describes the sampling noise and is given by $f^{n(\mathrm{A})}(1-f)^{n(\mathrm{G})}$, where $n(\mathrm{A})$ and $n(\mathrm{G})$ are numbers of A's and G's in the data R.

The probability $P(\mathrm{D}\mid f=1)$ can be computed using Phred base error probabilities as follows. Let S_a and S_g be two sets of cDNA reads that contain reads with called bases "A" and "G" at a homozygous A/A genomic locus x, respectively. Since it is assumed that there is no RNA editing at the locus x, the base call "G" should be treated as a base calling error. If we denote by p the base error probabilities, we have $P(\mathrm{D}\mid f=1)=(\prod_{m\in Sg} p_m)(\prod_{k\in Sa}(1-p_k))$. The base error probability is related to Phred base quality score as $Q=-10\log_{10} p$.

3.3.2. Implementation

We now have almost everything at hand for detecting RNA editing sites. The pileup file "cDNA_pileup.txt" from Subheading 3.2.4 contains around 5.8 million sites that are covered by at least five cDNA reads and at least one read has single-nucleotide mismatch with hg18 reference human genome. As explained in Subheading 3.3.1 we restrict our analysis to homozygous gDNA loci that match the hg18 genome. In Subheading 3.2.4 we obtained "gDNA_variants.txt" file that contains homozygous and heterozygous sites in gDNA that have mismatches with the hg18 genome. We thus removed all these gDNA variant sites from the file "cDNA_pileup.txt". This procedure filtered out around 38,000 sites from the latter file. From the resulting list, we also removed sites that have gDNA reads coverage <10 because such sites can represent false negatives in gDNA variant discovery. The coverage of gDNA reads at a list of genomic sites can be computed using the "pileup -l" option in SAMtools. We then retrieved genomic coordinates of hg18 exons from Ensembl database (www.ensembl.org) and retained only the putative RNA editing sites in exons. The resulting filtered "cDNA_pileup_filtered.txt" file contains all necessary information for the calculation of likelihood ratios using the theory from Subheading 3.3.1.

Note that ASCII integer values of characters can be computed using the "ord" function in Perl. One additional thing to remember for an efficient computation of probabilities is that products of base error probabilities correspond to the sum of Phred quality values.

We set the cutoff for the log likelihood ratio to be 4: $\log_{10}[\max_f P(\mathrm{D} \mid f)/P(\mathrm{D} \mid f=1)] \geq 4$, and identified 7,955 A→G editing sites in human exons.

3.4. Functional Analysis of RNA Editing Sites

We obtained genomic coordinates of 5′ and 3′ UTR regions from the Ensembl database. 413 A→G editing sites are located in 5′UTR regions whereas 1,813 are in 3′UTR regions. We recommend BEDTools (14), a suite of utilities to work with BED format files, to identify editing sites that overlap genomic features such as UTR regions.

For the purposes of identifying nonsynonymous sites, miscellaneous information about hg18 transcripts such as transcription start and end sites, coding start and end sites, genomic strand, exon start and end sites, and exon frames was retrieved from the UCSC genome Table Browser (http://genome.ucsc.edu). We identified 1,860 nonsynonymous A→G editing sites. See Note 4 for a functional test of these sites.

To further narrow down the list of putative nonsynonymous editing sites to a very high-confidence list, we selected sites that fulfill the following criteria: (a) log likelihood ratio≥20, (b) the number of reads with "G" at the putative editing site is at least 10 i.e. $n(\mathrm{G}) \geq 10$, and (c) the editing level, defined as $100\ n(\mathrm{G})/[n(\mathrm{A})+n(\mathrm{G})]$, is at least 20%. There are 161 editing sites that fulfill these criteria. Interestingly, A→G editing at chr6: 32,822,103 in the HLA-DQA2 gene, which is known to interact with a number of HIV proteins (15), results in the amino acid substitution Q→R. A closer inspection reveals, however, that the putative RNA editing site in HLA-DQA2 is likely a false positive. The sequence around the putative RNA editing site in the HLA-DQA2 gene is identical to the sequence surrounding an A/G single nucleotide polymorphism at chr6:32,718,473 in the HLA-DQA1 gene.

The results of analysis of high-throughput data should always be carefully checked. Only after various confounding factors are excluded as possible explanation of the results, we can assume biological validity of conclusions. In the context of identification of RNA editing sites in mature microRNAs, it was noted that RNA sequences obtained from deep sequencing experiments could be inadvertently mapped to incorrect locations (16). Such cross-mapping of sequencing reads can lead to overrepresented mismatches at specific locations between the genome sequence and the RNA sequence, giving the appearance of RNA editing. The putative editing sites located in genes belonging to multigene families are more likely to be false positives. In order to reduce the number of cross-mapping events, the "uniqueness" of mapped reads can be controlled using the "-m" option in the Bowtie alignment program.

Interestingly, we found that 72% of CDK13 mRNAs undergo A→I editing at chr7:39,957,073 which results in the Q103R amino acid substitution in the protein product. CDK13, cyclin-dependent kinase 13, is known to interact with HIV-1 transactivator Tat protein and regulate viral mRNA splicing (17). CDK13 is also a known target of RNA editing in the brain (18). Intriguingly, we observed that 84% of mRNAs of NEIL1, an enzyme involved in base-excision repair of oxidative DNA damage (19), undergo A→I editing at chr15:73,433,139, which results in the K242R amino-acid substitution in the protein product. A very recent study (20) showed that the edited and the genome-encoded forms of NEIL1 have very distinct enzymatic properties, thus demonstrating the functional importance of RNA editing of NEIL1.

4. Notes

1. We used cDNA sequencing data from the polyA-enriched RNA sample. Since noncoding RNAs such as microRNAs and intronic RNAs are removed by the polyA enrichment procedure, no information about RNA editing of these RNA classes can be obtained. Specialized protocols should be used for the isolation and sequencing of specific classes of RNA as was, for example, done for the microRNAs in the study of Morin et al. (21).

2. For the sake of simplicity, we used Bowtie for aligning cDNA data and restricted our analysis to RNA editing sites in exons. Splice sites and surrounding regions were thus excluded from the analysis. To analyze RNA editing in the vicinity of splice sites, an algorithm tailored for aligning RNA sequencing data, such as TopHat (22), should be used instead.
3. As discussed in Subheading 3.3.1, for the sake of simplicity, we restricted RNA editing analysis to those homozygous sites in the sample genome that match the reference hg18 human genome. The variant homozygous sites can be analyzed as follows. We first extract the locations of homozygous variants from the file "gDNA_variants.txt" where homozygous sites correspond to lines with IUPAC symbols "ACGT" in the fourth column. Let us name the resulting space-separated two-column file as "pos.txt". We then pileup at these locations as follows:
4. Nonsynonymous base substitutions that result due to RNA editing may be computationally tested for functional importance using tools such as PolyPhen (23).

```
samtools pileup -f hg18.fa -l pos.txt cDNA.bam | perl variant_site.pl > cDNA_pileup.txt
```

Acknowledgments

I am grateful to Liz Cirulli and David Goldstein for providing the raw sequence data from their study (9). This work was supported by the Division of Intramural Research Program of the NIH, National Heart, Lung, and Blood Institute.

References

1. Gott, J.M. and Emeson, R.B. (2000) Functions and mechanisms of RNA editing. *Annu Rev Genet.* **34**:499–531.
2. Farajollahi, S. and Maas, S. (2010) Molecular diversity through RNA editing: a balancing act. *Trends Genet.* **26**(5):221–30.
3. Nishikura, K. (2010) Functions and Regulation of RNA Editing by ADAR Deaminases. *Annu Rev Biochem.* **79**, 321–349.
4. Maas, S., Kawahara, Y., Tamburro, K.M., Nishikura, K. (2006) A-to-I RNA editing and human disease. *RNA Biol.* **3**(1):1–9.
5. Metzker, M.L. (2010) Sequencing technologies – the next generation. *Nat Rev Genet.* **11**(1):31–46.
6. Wang, Z., Gerstein, M. and Snyder, M. (2009) RNA-Seq: a revolutionary tool for transcriptomics. *Nature Rev. Genet.* **10**, 57–63.
7. Li J.B., Levanon, E.Y., Yoon, J.K., Aach, J., Xie, B., Leproust, E., et al. (2009) Genome-wide identification of human RNA editing sites by parallel DNA capturing and sequencing. *Science.* **324**(5931):1210–3.
8. The 1000 Genomes Project Consortium. (2010) A map of human genome variation from population-scale sequencing. *Nature.* **467** (7319):1061–73.
9. Cirulli, E.T., Singh, A., Shianna, K.V., Ge, D., Smith, J.P., Maia, J.M., et al. (2010) Screening the human exome: a comparison of whole genome and whole transcriptome sequencing. *Genome Biol.* **11**(5):R57.
10. Langmead, B., Trapnell, C., Pop, M. and Salzberg, S.L. (2009) Ultrafast and memory-efficient alignment of short DNA sequences to

the human genome. *Genome Biol* 10:R25 (Software available at http://bowtie-bio.sourceforge.net).

11. Li H., Handsaker B., Wysoker A., Fennell T., Ruan J., Homer N., Marth G., Abecasis G., Durbin R. and 1000 Genome Project Data Processing Subgroup (2009) The Sequence alignment/map (SAM) format and SAMtools. *Bioinformatics*, **25**, 2078–9 (Software available at http://samtools.sourceforge.net/).
12. Chepelev, I., Wei, G., Tang, Q. and Zhao K. (2009) Detection of single nucleotide variations in expressed exons of the human genome using RNA-Seq. *Nucleic Acids Res.* **37**(16):e106.
13. Li, H., Ruan, J. and Durbin, R. (2008) Mapping short DNA sequencing reads and calling variants using mapping quality scores. *Genome Res.* **18**(11):1851–8.
14. Quinlan, A.R. and Hall, I.M. (2010) BEDTools: a flexible suite of utilities for comparing genomic features. *Bioinformatics.* **26**(6):841–2.
15. Fu, W., Sanders-Beer, B.E., Katz, K.S., Maglott, D.R., Pruitt, K.D. and Ptak, R.G. (2009) Human immunodeficiency virus type 1, human protein interaction database at NCBI. *Nucleic Acids Research.* **37**(Database issue):D417–22.
16. de Hoon, M.J., Taft, R.J., Hashimoto, T., Kanamori-Katayama, M., Kawaji, H., Kawano, M., et al. (2010) Cross-mapping and the identification of editing sites in mature microRNAs in high-throughput sequencing libraries. *Genome Res.* **20**(2):257–64.
17. Berro, R., Pedati, C., Kehn-Hall, K., Wu, W., Klase, Z., Even, Y., et al. CDK13, a new potential human immunodeficiency virus type 1 inhibitory factor regulating viral mRNA splicing. *J Virol.* **82**(14):7155–66.
18. Kiran, A. and Baranov, P.V. (2010) DARNED: a DAtabase of RNa EDiting in humans. *Bioinformatics.* **26**(14):1772–6.
19. David, S.S., O'Shea, V.L. and Kundu, S. (2007) Base-excision repair of oxidative DNA damage. *Nature.* **447**(7147):941–50.
20. Yeo, J., Goodman, R.A., Schirle, N.T., David, S.S. and Beal, P.A. (2010) RNA editing changes the lesion specificity for the DNA repair enzyme NEIL1. *Proc Natl Acad Sci USA.* **107**(48):20715–9.
21. Morin, R.D., O'Connor, M.D., Griffith, M., Kuchenbauer, F., Delaney, A., Prabhu, A.L., et al. (2008) Application of massively parallel sequencing to microRNA profiling and discovery in human embryonic stem cells. *Genome Res.* **18**(4):610–21.
22. Trapnell, C., Pachter, L. and Salzberg, S.L. (2009) TopHat: discovering splice junctions with RNA-Seq. *Bioinformatics.* **25**(9): 1105–11.
23. Ramensky, V., Bork, P., and Sunyaev, S. (2002) Human non-synonymous SNPs: server and survey. *Nucleic Acids Res.* **30**(17):3894–900.

Chapter 9

Comparative Study of Differential Gene Expression in Closely Related Bacterial Species by Comparative Hybridization

Ruisheng An and Parwinder S. Grewal

Abstract

The ability to profile bacterial gene expression has markedly advanced the capacity to understand the molecular mechanisms of pathogenesis, epidemiology, and therapeutics. This advance has been coupled with the development of techniques that enable investigators to identify bacterial specifically expressed genes and promise to open new avenues of functional genomics by allowing researchers to focus on the identified differentially expressed genes. During the past two decades, a number of approaches have been developed to investigate bacterial genes differentially expressed in response to the changing environment, particularly during interaction with their hosts. The most commonly used techniques include in vivo expression technology, signature-tagged mutagenesis, differential fluorescence induction, and cDNA microarrays, which fall into two broad classes: mutagenesis-based technologies and hybridization-based technologies. Selective capture of transcribed sequences, a recently emerging method, is a hybridization-based technique. This technique is powerful in analyzing differential gene expression of the bacteria, with the superb ability to investigate the bacterial species with unknown genomic information. Herein, we describe the application of this technique in a comparative study of the gene expression between two closely related bacteria induced or repressed under a variety of conditions.

Key words: Closely related bacteria, Differential gene expression, Genomic presence, Competitive hybridization, Comparative hybridization, Selective capture of transcribed sequences

1. Introduction

Today, lots of sequenced bacterial genomes, along with many more currently being sequenced, provide tremendous new opportunities for research into bacterial pathogenesis, epidemiology and therapeutics. At the same time, difficulties have also arisen concerning the efficient use of this accumulated genetic information. Facing thousands of genes, how to narrow down the number and decide

Michael Kaufmann and Claudia Klinger (eds.), *Functional Genomics: Methods and Protocols*,
Methods in Molecular Biology, vol. 815, DOI 10.1007/978-1-61779-424-7_9, © Springer Science+Business Media, LLC 2012

which one should be focused on appears critical. Profile of differentially expressed bacterial genes, especially in closely related bacterial species, provides excellent insight into how these organisms selectively employ their genome during contact with the host and other environments encountered in their life cycle.

During the past two decades, a number of techniques have been developed to study bacterial genes that are expressed specifically in the host during infection or that are required for survival in desired growth conditions. These methods are either based on protein or gene levels. The commonly used protein-based methods include two-dimensional gel electrophoresis and in vivo induced antigen technology (IVIAT). Two-dimensional gel electrophoresis relies on the separation of whole proteins by gel electrophoresis in two dimensions and the subsequent comparative identification of individual proteins from bacteria grown under different conditions through mass spectrometry. Several studies have been reported to analyze bacterial proteins produced during growth in vitro under conditions that mimic some aspects of infection. Theoretically, to identify in vivo induced gene products, two-dimensional gel electrophoresis can be performed to compare protein patterns present in vitro and in vivo, but up to now, no studies of global protein expression analysis of a bacterial pathogen within its natural host or in an animal model have been published. This is because of the technical hurdles associated with separating bacteria from the host tissues and obtaining enough material to perform serial statistical analysis. Thus, the potential of this technique for bacterial in vivo gene expression analysis is limited. IVIAT has been developed to identify proteins produced by pathogenic bacteria during an actual infectious process by probing antigens using pooled sera from infected animals (1). This method overcomes limitations of animal models, allowing direct identification of microbial proteins produced during infection (2–4). However, because this technique is based on the immune reactions, its application to bacteria–invertebrate interactions is limited (5). The gene-based methods are either mutagenesis based or hybridization based. Commonly used mutagenesis-based methods include in vivo expression technology (IVET), differential fluorescence induction (DFI), signature-tagged mutagenesis (STM), and genomic analysis and mapping by in vitro transposition (GAMBIT). IVET can be used to positively select promoters that are turned on in specific growth conditions (6). This system relies on the generation of transcriptional fusions of genomic sequences to a reporter gene (7). The major technical problems limiting the use of this technique include the need of a suitable animal model and the success of transformation and recombination (8). DFI is another promoter selection method using green fluorescent protein as a report marker (9). Like IVET, a common major disadvantage for DFI is the need of the animal model (5). STM is a negative selection approach in which tagged mutants unable to survive are identified (10). This technique is limited by the need for an adequate model that facilitates recovery of bacteria from an infected host. In addition, bacterial mutants that are slow-growing may be underrepresented (11–13). GAMBIT is introduced to identify the essential genes that are required for bacterial growth (14). Like STM, the use of GAMBIT in animal models constitutes a negative selection in which certain mutants are eliminated by selection in the animal. The major disadvantage is that it is necessary to design a large number of PCR primers to cover entire genomes and that it can only be applied to the naturally competent bacterial cells (15). As a hybridization-based method, cDNA microarray is generally used to determine the difference in mRNA levels among bacterial strains grown at different conditions (16). Although a cDNA microarray has the major advantage allowing to compare the same set of genes under many experimental conditions and to quickly analyze a large set of clones, it is limited in gene expression analysis due to the lack of a large set of clones carrying known genes for the array, the low numbers of bacteria in living tissues during infection but on the contrary the need for large amounts of mRNA to prepare probes, and the difficulty in purifying the bacteria from the eukaryotic tissue (16, 17). Therefore, a cDNA microarray can currently only be applied to bacterial infections that lead to high titers in host tissues (18–20).

While the development of these elegant techniques has remarkably contributed to the study of bacterial differentially expressed genes, each of them has advantages and disadvantages. The common limitations of these techniques include the isolation of considerable quantities of high quality starting materials such as proteins and mRNA, difficulties in differentiating between bacterial and host genes or gene products, and the requirement of proper animal models. For mutagenesis-based approaches, such as IVET and STM techniques, well-developed genetic manipulation systems of the pathogen are required including the ability to mutagenize the bacteria. An improved approach, the selective capture of transcribed sequences (SCOTS), overcomes most of these limitations noted above. The SCOTS procedure (Fig. 1) allows the selective capture of a great diversity of bacterial cDNAs that are induced or uniquely expressed by bacteria in a specific culture condition from total cDNA prepared from infected cells or tissue by hybridization to biotinylated bacterial genomic DNA. cDNA mixtures obtained are then enriched for sequences that are preferentially transcribed during growth in the desired condition by additional hybridizations to bacterial genomic DNA in the presence of cDNA similarly prepared from bacteria grown in the condition being compared. This approach was originally developed by Graham and Clark-Curtiss (21) for the identification of genes expressed by *Mycobacterium tuberculosis* upon growth in macrophages. It has subsequently shown the advance of analyzing differential gene expression in a great diversity of bacteria under a

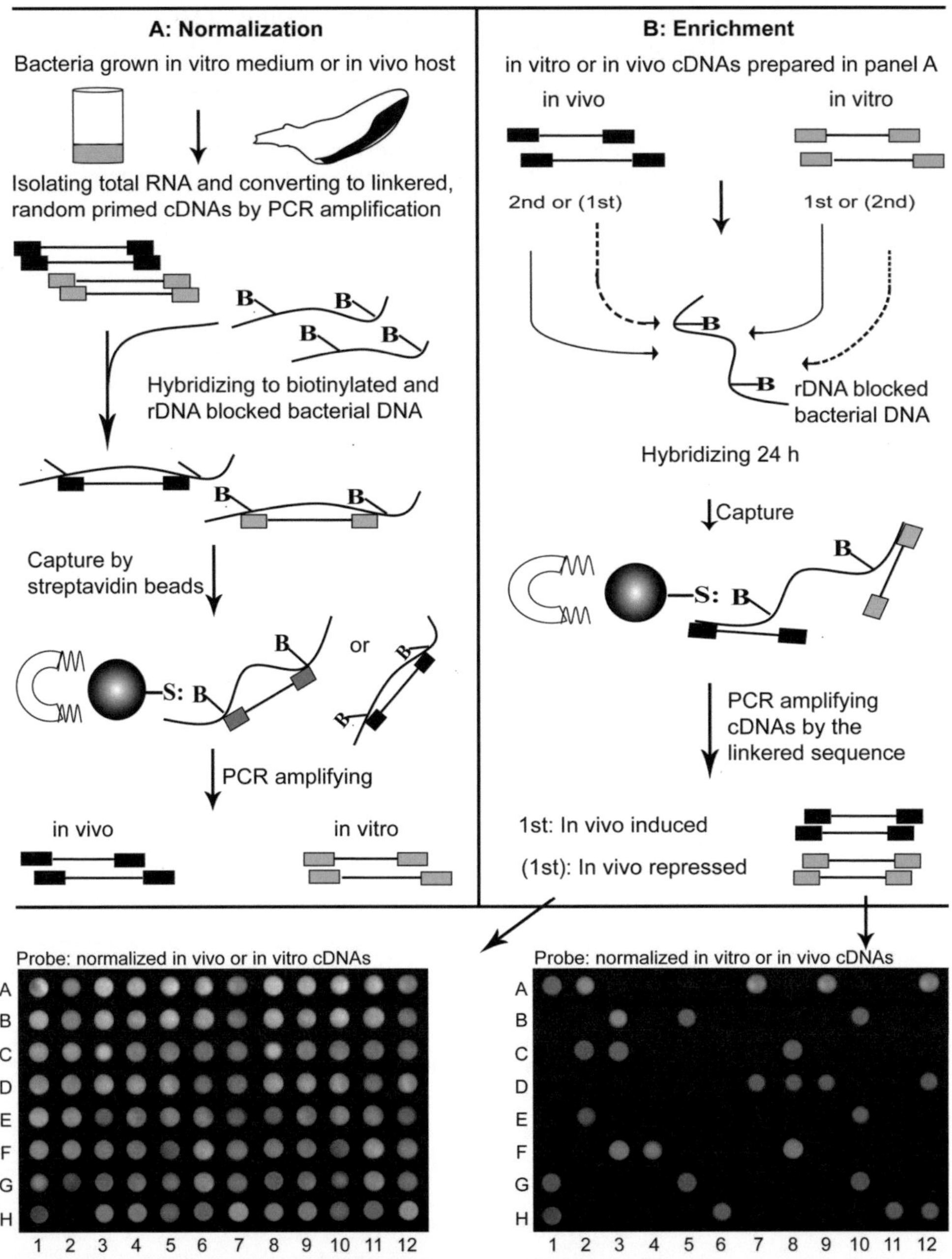

Fig. 1. Schematic presentation of the selective capture of transcribed sequences approach followed by southern blot analysis of the identified sequences. (**a**) Normalized bacterial cDNAs are obtained directly from bacteria grown in vitro medium or in vivo host tissues. (**b**) cDNAs corresponding to genes preferentially induced or repressed in the host relative to the medium are enriched by differential cDNA hybridization. The enriched cDNAs are transformed into a cloning vector to build the cDNA library. Cloned *inserts* are amplified by PCR, equally aliquoted to the same position of two nylon membranes, and probed with digoxigenin-labeled normalized in vivo or in vitro cDNAs.

variety of growth conditions (22–29). SCOTS can also be used to determine the expression of any given gene of interest by southern hybridization of the PCR-amplified gene with the SCOTS derived cDNA probe mixture from the desired growth condition (30). With the SCOTS technique, no specific genetic manipulation of the bacteria is required, and the host and bacterial cDNAs can be easily differentiated. Moreover, the ability to investigate differential gene expression in the bacterial species with unknown genomic information has become possible using the SCOTS approach. In addition, differences in gene transcription in host cells or tissues could be established by comparative blocking between different strains belonging to the same or similar species with high overall DNA homology (25). Such comparative hybridization can result in the identification of pathogen-specific and conserved bacterial genes that are expressed during infection and provide further insight into the mechanisms by which bacteria colonize host tissues, cope with, or circumvent host defenses and adjust to the nutrient limitations and other stresses that could occur in different host environments. In this article, we will describe the SCOTS approach which combines commonly used techniques in molecular biology such as nucleic acid isolation, cDNA synthesis, DNA hybridization, PCR amplification, cloning, southern blot hybridization and comparative hybridization.

2. Materials

2.1. General Consumables and Equipments

1. PCR system including thermal cycler (Bio-Rad), DNA polymerase and reaction buffer (Promega), and dNTP (10 mM of each).
2. Gel electrophoresis system including gel electrophoresis apparatus, gel staining chemicals, and gel imaging system with Chemiluminescent detection.
3. Isopropanol (100%).
4. 100% Ethanol, which can be diluted to the specified concentration whenever needed.
5. QIAquick PCR Purification Kit (QIAGEN).
6. NanoDrop (Thermo Scientific).

2.2. Genomic DNA Manipulation

1. Proper growth medium for the bacteria being studied.
2. Genomic-tip 100/G: Genomic DNA Buffer Set (QIAGEN) for bacterial genomic DNA isolation; alternatively, the bacterial genomic DNA can be isolated according to the method described by Aljanabi and Martinez (31).
3. Sterile salt homogenizing buffer: 0.4 M NaCl, 10 mM Tris–HCl pH 8.0, 2 mM EDTA pH 8.0.

4. Sodium dodecyl sulfate (SDS) solution (20% w/v) is prepared by dissolving 20 g SDS in distilled, deionized water to a final volume of 100 ml.
5. Proteinase K solution (20 mg/ml).
6. Saturated NaCl solution (6 M) is prepared by dissolving 36 g sodium chloride in distilled, autoclaved water to a final volume of 100 ml.
7. Pairs of primers which are designed for PCR amplification of ribosomal RNA operon for each bacterial species being studied.
8. Psoralen-PEO-Biotin (PIERCE) is prepared by dissolving it in distilled water to make a stock at 20 mM (about 13.8 μg/μl), stored in aliquots at −20°C, and added to DNA samples as required. The prepared Psoralen-PEO-Biotin solution has to be protected from light by storing in a dark colored tube.
9. UV Cross Linker FB-UVXL-1000 (Fisher Scientific).
10. Sonicator.
11. EPPS–EDTA buffer stored at room temperature: 10 mM EPPS [*N*-(2-hydroxyethyl) piperazine-N9-(3-propanesulfonic acid)] and 1 mM EDTA (ethylenediamine tetraacetic acid).

2.3. Total RNA Isolation and cDNA Synthesis

1. Liquid nitrogen.
2. Mortar and pestle for crushing the tissues and bacterial cells.
3. Trizol reagent (Invitrogen) for total RNA isolation.
4. Chloroform.
5. DEPC-treated water.
6. Random primers (see Note 1) with a defined sequence at the 5′ end and random nonamers at the 3′ end (PCR primer-dN9) are used for both first- and second-strand cDNA synthesis.
7. SuperScript II reverse transcriptase (50 U/μl, Invitrogen).
8. 10× Reverse transcription buffer: 200 mM Tris–HCI pH 8.4, 500 mM KCl, 50 mM $MgCl_2$, and 100 mM DTT.
9. RNase inhibitor (Invitrogen).
10. Klenow fragment (5 U/μl) and 10× Klenow buffer (BioLabs).
11. RNase-free DNase I (Ambion).

2.4. Identification of Bacterial Genes Specifically Expressed Under Desired Conditions

1. Streptavidin-coated magnetic Dynabeads M-280 (Dynal, Lake Success, NY).
2. Magna-Sep Magnetic Particle Separator (Invitrogen).
3. Mineral oil (Bio-Rad).
4. NaCl (1 M).
5. 20× SSC: 3 M NaCl and 0.3 M sodium citrate is prepared as a stock solution and can be diluted to the specified concentration whenever needed.

6. NaOH–NaCl elution buffer: 0.5 M NaOH in 0.1 M NaCl.
7. Original TA Cloning Kit with One Shot TOP ten chemically competent *E. coli* cells (Invitrogen).

2.5. Southern Blot Hybridization

1. 1× PBS buffer: 0.1 M NaCl, 7 mM Na_2HPO_4, and 3 mM NaH_2PO_4, pH 6.8.
2. NaCl–NaOH denaturation buffer: 3 M NaCl in 0.4 M NaOH.
3. Nylon membranes (Bio-Rad) and 3 mm Whatman filter paper.
4. Vacuum.
5. Southern dot-blot apparatus.
6. Amersham ECL Plus western blotting detection reagents (GE Healthcare Bio-Sciences Corp).
7. Anti-digoxigenin-HRP (Roche).
8. Dry milk powder.
9. 1× SSC: 0.1% SDS buffer.
10. Hybridization oven and hybridization bottle (Bio-Rad).
11. PCR DIG Probe Synthesis kit and Dig easy hyb granules (Roche).

3. Methods

From the very beginning, the investigator should determine the appropriate growth conditions under which bacterial gene expression will be analyzed. It is generally interesting to study gene expression by bacteria at some time points during growth under conditions relevant to the pathogenesis of these bacteria: in vivo growth within an animal host. Also, another growth condition needs to be carefully chosen by the investigator since transcripts from bacterial cells grown under this condition at some time points will be used to prepare cDNA mixtures to enrich for capture of bacterial cDNA molecules representing genes that are more specifically expressed under the above desired condition. For the best convenience of describing the methods, it is defined here that the growth conditions for the bacteria are in the proper artificial medium to mid-logarithmic phase (in vitro) and in the host specific tissue at 48 h postinfection (in vivo) which can be considered as a time point by that the bacteria have adapted to the host environment and are actively multiplying. Bacteria being studied are two closely related species (Bacterium #1 and #2), which means that there will be four pools: B1-vitro, B1-vivo, B2-vitro, and B2-vivo. In the procedures described below, we want to comparatively study differential gene expression of these two bacterial species grown in the host specific tissue relative to be grown within the in vitro medium.

Prior to the procedure description, it is assumed that the investigator has appropriately harvested the samples (the bacteria grown in vitro and the bacterial infected tissue) which can be stored in liquid nitrogen or at –70°C until required according to the requirement for each bacterial species being studied (see Note 2). In addition, the most commonly used molecular techniques such as PCR amplification, agarose gel electrophoresis, and cloning are either based on the standard protocols in Molecular Cloning (32) or according to the manufacturers' introductions unless otherwise stated.

3.1. Isolation of Bacterial Genomic DNA

1. The bacterial genomic DNA is isolated either using the QIAGEN kit or according to the modified procedures of Aljanabi and Martinez (31) as described below.
2. Overnight cultured bacterial cells in 2 ml broth are harvested in a microcentrifuge tube by centrifuging for 10 min at 5,000 × *g*.
3. After discarding the supernatant, the pellet is resuspended in 400 μl of salt homogenizing buffer and homogenized by vortexing, followed by adding 40 μl of 20% SDS and 8 μl of 20 mg/ml proteinase K.
4. The well mixed sample is then incubated at 60°C for 2 h or overnight, after which 300 μl of saturated NaCl solution is added and the sample is vortexed for 30 s at maximum speed, followed by centrifugation at 10,000 × *g* for 30 min.
5. The supernatant is transferred to a fresh tube and an equal volume of isopropanol is added and mixed well.
6. After incubation at –20°C for 2 h, the sample is centrifuged for 20 min at 10,000 × *g* at 4°C, and the pellet is washed with 70% ethanol, dried, and finally dissolved in 50 μl of sterile water.

3.2. Blocking of Ribosomal Genes and Biotinylating of Bacterial Genomic DNA

1. The ribosomal operon (see Note 3) is amplified from the bacterial genomic DNA using proper primers designed for each bacterial species and high fidelity DNA polymerase at a standard PCR condition.
2. The amplified ribosomal DNA is sonicated to obtain a smear of DNA with most fragments within a size range of 1 kb (see Note 4).
3. The sonicated ribosomal DNA is precipitated with ethanol and resuspended in EPPS-EDTA hybridization buffer at a concentration of 6 μg/2 μl.
4. The isolated genomic DNA is adjusted to a concentration of 0.5 μg/μl in sterile water.
5. The Psoralen-PEO-Biotin solution is mixed with the bacterial genomic DNA at a ratio of 1:99 (*v*/*v*) in a 1.5 ml microtube, and here 2 μl of biotin solution is mixed with 198 μl of genomic DNA solution (see Note 5).

6. The open microtube is incubated on ice and irradiated in a UV Cross-Linker FB-UVXL-1000 (Fisher Scientific) at C-L 125 mJ for 30 min.
7. After 30 min, additional 2 μl of biotin solution is added into reaction mixture which is irradiated for another 30 min, and the irradiation experiment is repeated three times in total, leading to a dense biotin labeled pattern.
8. The biotin labeled genomic DNA is purified from excess biotin reagent by ethanol precipitation. After adding 2 volumes of 100% ethanol, being incubated at –20°C for 1 h and following centrifugation at 10,000 × *g* for 20 min, the pellet is washed with 70% ethanol, dried, and dissolved in sterile water.
9. The biotin labeled genomic DNA is sonicated to obtain a smear of DNA with most fragments within a size range of 1–5 kb (see Note 4).
10. The sonicated, biotin labeled genomic DNA is then precipitated with ethanol and resuspended in EPPS-EDTA buffer at a concentration of 0.3 μg/2 μl.
11. The sonicated ribosomal DNA (2 μl) is mixed with the sonicated, biotin labeled genomic DNA (2 μl) at a ratio of 20:1 (6 μg of ribosomal DNA to 0.3 μg of genomic DNA per 4 μl mixture).
12. After adding 4 μl EPPS-EDTA buffer to 4 μl rDNA–genomic DNA mixture, the sample is denatured by boiling under mineral oil for 3 min, and followed by immediate incubation on ice for 2 min and addition of 2 μl 1 M NaCl.
13. The mixture is incubated at a temperature 20°C below the T_m of the bacterial DNA for 30 min.

3.3. Isolation of Total RNA and Preparation of cDNA Pools

1. The harvested tissues or bacterial cells are crushed in sterile mortar with liquid nitrogen, followed by adding approximately 1 ml Trizol reagent to 50–100 mg sample.
2. After incubation at room temperature for 5 min, 0.2 ml chloroform per 1 ml Trizol is added and mixed well by shaking for 15 s.
3. After incubation at room temperature for another 5 min, the sample is centrifuged at 4,000 rpm or (3000 × *g*), 4°C for 15 min.
4. The colorless upper aqueous phase (containing RNA) is transferred to a fresh tube and 0.5 ml isopropanol per 1 ml Trizol used in the initial homogenization step is added.
5. After incubation at room temperature for 10 min, the sample is centrifuged at 4,000 rpm or (3000 × *g*), 4°C for 10 min.
6. After removing the supernatant, the RNA pellet is washed in 1 ml of 75% (*v*/*v*) ethanol per 1 ml of Trizol used in the initial

homogenization step by vortexing to resuspend the pellet and centrifuging at 4,000 rpm or (3000×*g*), 4°C for 5 min.

7. The RNA pellet is dried in air, and dissolved in appropriate volume of DEPC-treated water.
8. The isolated total RNA is quantified by a NanoDrop and treated with RNase-free DNase I according to the manufacturer's instructions.
9. The first-strand cDNA for each bacterial species and each growth condition is independently prepared from 5 μg of isolated total RNA which is suspended in 8 μl of DEPC-treated water, heated to 65°C for 5 min and rapidly cooled on ice for 1 min by adding 1 μl dNTP (10 mM), 1 μl random primer (50 ng/μl) (see Note 1), 2 μl 10× reverse transcription buffer, and 1 μl RNase inhibitor.
10. After incubation at 42°C for 2 min, 1 μl SuperScript II reverse transcriptase is added and incubation is continued at 42°C for 1 h, after which the reaction is boiled at 95°C for 5 min and rapidly cooled on ice for 1 min.
11. 21 μl Distilled water, 2 μl dNTP (10 mM), 5 μl 10× Klenow buffer, and 2 μl Klenow fragment, are then added for second-strand cDNA synthesis at 37°C for 40 min.
12. After 40 min incubation at 37°C, the reaction is inactivated at 75°C for 5 min, and the synthesized cDNA is purified on a QIAquick Spin column to eliminate the excess of the random primers.
13. The synthesized double-stranded cDNA is then amplified by PCR using a defined terminal sequence (see Note 1) and the product is precipitated and resolved in EPPS-EDTA buffer at a concentration of 6 μg/8 μl.

3.4. Selective Capture of Bacterial Genes Specifically Expressed Under Desired Conditions

1. PCR amplified cDNA (8 μl) is denatured by boiling under mineral oil for 3 min and rapidly cooled on ice for 1 min, after which 2 μl NaCl (1 M) is added and the sample is incubated at a temperature 20°C below the T_m of the bacterial genomic DNA for 30 min.
2. The 10 μl self-hybridized cDNA is then added to the 10 μl mixture of ribosomal and genomic DNA, and hybridization is allowed to proceed for 24 h at 20°C below the T_m of the bacterial DNA.
3. After hybridization, 20 μl DNA mixtures are diluted to 500 μl with 0.5× SSC and added to the tube containing 120 μg streptavidin-coated magnetic beads resuspended in 200 μl 0.5× SSC.
4. The mixture is incubated at room temperature for 10 min with gently inverting the tube every 2–3 min.

5. The magnetic beads binding with the bacterial cDNA–genomic DNA hybrids are captured using a magnetic stand, and the supernatant is removed carefully without disturbing the magnetic beads.
6. The magnetic beads are washed three times with 0.1× SSC (500 μl per wash) by gently flicking the bottom of the tube to resuspend the beads and recapturing the beads using the magnetic stand.
7. After the final wash, the washing buffer is removed as much as possible without disturbing the magnetic beads.
8. The magnetic beads are eluted with 200 μl NaOH–NaCl elution buffer for 10 min at room temperature.
9. After capturing the beads with the magnetic stand, the supernatant (containing the cDNA–genomic DNA hybrids) is transferred to a new tube.
10. The cDNA–genomic DNA hybrids are ethanol precipitated and dissolved in 100 μl sterile water.
11. Captured cDNA–genomic DNA hybrids are amplified with the defined primers at the standard PCR condition (see Note 1).
12. The PCR product is ethanol precipitated and resolved in EPPS-EDTA buffer at a concentration of 6 μg/8 μl for additional rounds of hybridization by repeating the above steps.
13. After three rounds of hybridization (see Note 6) amplified cDNA samples representing the normalized bacterial genes are pooled (see Note 7), precipitated with ethanol and resuspended in EPPS-EDTA buffer at a concentration of 6 μg/4 μl.
14. To isolate bacterial genes preferentially induced in the host tissue compared to the culture, normalized in vivo cDNAs (6 μg) are enriched by subtractive hybridization to the biotinylated bacterial genomic DNA (0.3 μg) that has been prehybridized with the rRNA operon (6 μg) and normalized in vitro cDNAs (6 μg) as described above. The hybrids are removed from the hybridization solution by binding to streptavidin-coated magnetic beads, and bacterial cDNAs are eluted, PCR-amplified using defined sequences specific for the cDNAs from the in vivo grown bacteria and precipitated for the next round of enrichment, as described above. A total of three rounds of enrichment can be performed empirically (see Note 8).
15. Similarly, to isolate bacterial genes preferentially repressed in the host tissue compared to the culture, normalized in vitro cDNAs (6 μg) are enriched (three rounds of enrichment) by subtractive hybridization to the biotinylated bacterial genomic DNA (0.3 μg) that has been prehybridized with the rRNA operon (6 μg) and normalized in vivo cDNAs (6 μg). The hybrids are removed from the hybridization solution by binding

to streptavidin-coated magnetic beads, and bacterial cDNAs are eluted, PCR-amplified using defined sequences specific for the cDNAs from the in vitro grown bacteria and precipitated for the next round of enrichment.

16. The enriched bacterial cDNAs from steps 14 and 15 are independently cloned into an original TA cloning vector to construct libraries representing bacterial genes induced or repressed in the host relative to the in vitro culture.

3.5. Southern Blot Screening of Enriched cDNAs

1. Individual clones from each enriched cDNA library are randomly picked and amplified by PCR using universal M13 primers.
2. Positively charged nylon membrane and 3 mm Whatman filter paper are cut to fit the support plate of the multiwell southern dot-blot apparatus. After soaking in distilled water for 5 min, the filter paper and the nylon membrane are fixed on the support plate of the dot-blot apparatus with the nylon membrane on the top. For each screening, two membranes are prepared: one for in vitro cDNA probe and another for in vivo cDNA probe which are synthesized in step 7 as described below.
3. Twenty microliter of each amplified product in step 1 is mixed with 140 μl 20× SSC, and an equal amount of the mixture is transferred to each of the two nylon membranes (each PCR product per dot) by filtrating at low vacuum.
4. The nylon membrane with samples is denatured with NaCl–NaOH denaturation buffer for 10 min at room temperature, followed by neutralization in 1× PBS buffer for 10 min at room temperature.
5. The membrane is dried by baking at 80°C for 2 h, and then soaked in 2× SSC for 5 min.
6. The membrane is transferred into a hybridization bottle for hybridization using Dig easy hyb granules according to the manufacturer's instruction.
7. Normalized, in vitro and in vivo cDNA pools are digoxigenin-labeled using the PCR DIG Probe Synthesis kit for the use as probes.
8. The probes are denatured and added to the hybridization bottles containing the membrane and the hybridization buffer, and hybridization continues at 65°C for approximately 24 h.
9. The membrane is washed briefly with 2× SSC at room temperature and then twice with 1× SSC – 0.1% SDS for 15 min at 65°C.
10. A final brief rinse with 0.1× SSC at room temperature completes the washing process.
11. The membrane is incubated at room temperature with 4 ml 1× SSC with 8% (w/v) dry milk for 30 min, followed by adding 4 μl dilution of anti-digoxigenin-HRP conjugate (1:800).

12. After incubation at room temperature for 1 h, the membrane is washed as describe above.
13. The successful hybridization can be detected by Amersham ECL Plus western blotting detection reagents using chemiluminescent detection.
14. The individual clones (from the cDNA library representing induced genes) that only hybridized to the probe made from normalized in vivo cDNAs and the individual clones (from the cDNA library representing repressed genes) that only hybridized to the probe made from normalized in vitro cDNAs are chosen for sequence analysis.

3.6. Analysis of cDNA Clones

1. The selected clones are sequenced at a sequencing facility using the universal primers M13.
2. Similar sequences are identified using BLAST algorithms available from the National Center for Biotechnology Information (NCBI) (see Note 9).
3. The functions of the identified sequences are assigned by searching databases of NCBI and BioCyc (http://biocyc.org).
4. The sequences identified from one bacterium are individually screened by hybridization to the sonicated biotinylated genomic DNA of the other, followed by PCR detection of the streptavidin beads captured hybrids using the defined primers as described above to determine if an identified gene in bacterium #1 is present in the genome of bacterium #2, or vice versa.
5. The genes with sequences presented in both bacterial genomes are singled out to further evaluate their induction or repression specificity to bacterium #1 or #2. The selected genes differentially expressed in one bacterium are individually screened by southern blot hybridization to the digoxigenin-labeled enriched cDNAs of the other as described above. If the individual genes from one bacterial species can also hybridize to the digoxigenin-labeled probe made from enriched cDNAs of another bacterial species, this gene is similarly regulated in both bacterial species.
6. The genes identified to be differentially expressed under one specific condition (it is the in vivo host tissue in this article) can be screened to determine whether these genes are also expressed when the bacteria are growing under other conditions by southern blot hybridization to the digoxigenin-labeled enriched bacterial cDNAs prepared from different growth conditions (such as low nutrition, high pH, low iron, and other hosts).
7. Quantitative real-time PCR can be performed to further validate and quantify the expression changes profiled by the selective capture of transcribed sequences. It can be conducted in an IQ5 system (Bio-Rad) using QuantiTect SybrGreen PCR Kit (Qiagen) according to the manufacturer's instructions.

8. The potential function of the identified genes can be determined by inactivating these genes through insertion–deletion mutation and then comparing the phenotypes of the strains carrying the mutated genes to the phenotypes of their wild-type parent strains.

4. Notes

1. The random primer comprises two parts: the defined terminal sequence and a random nonamer at its 3′ end. For example, the random primer SeqB1vitro-dN9 (5′-*ATC CAC CTA TCC CAG TAG GAG* NNN NNN NNN) being used to synthesize cDNA from total RNA isolated from bacterium #1 grown in vitro contains the defined terminal sequence SeqB1vitro (*ATC CAC CTA TCC CAG TAG GAG*) and the random nonamer dN9 (NNN NNN NNN). The synthesized cDNAs are then amplified using the corresponding primer SeqB1vitro to generate the cDNA pool for bacterium #1 grown in vitro. The defined terminal sequence used for each bacterial species and each growth condition should be derived from different linkers or adaptors that will not hybridize with the genome of the bacterium being studied.
2. The bacterial growth conditions should be adjusted by the investigator depending on the objective of the studies. Empirically, 10^9 broth-grown bacteria can give good yields of high quality total RNA and a minimum of 10^6 bacterial cells together with the host tissue should be used for isolation of total RNA.
3. Compared to ribosomal RNA (rRNA), which constitutes a large portion (>82%) of total prokaryotic RNA, messenger RNA (mRNA) only counts up to about 4%. To effectively capture the mRNA transcripts by chromosomal DNA during hybridization, it is necessary to mask the loci representing abundant rRNA sequences on chromosomal DNA to significantly expose the loci representing mRNA sequences. For this purpose, the ribosomal operon is prepared.
4. The DNA mixture can be sonicated by using a sonicator's mini tip to pulse for 5 s each time at 20% maximum intensity. After each pulsing, the size range of the DNA fragments should be determined by agarose gel electrophoresis (0.7%).
5. As 0.3 μg of chromosomal DNA is needed for each round of normalization (ten individual reactions for the first round) and each round of enrichment reaction, it is suggested that a sufficient amount of biotinylated chromosomal DNA is prepared so that all successive rounds of normalization and enrichment can be done with the same preparation of biotinylated chromosomal DNA.

6. A eukaryotic housekeeping 18S rRNA gene can be used as a control to ensure that bacterial cDNAs are purified apart from the host cDNAs after normalization. For this purpose, the presence of 18S rRNA gene in the in vivo cDNA populations before and after normalization is measured by PCR using 50 ng cDNA samples and primers 18SF (5′-GGA ATT GAC GGA AGG GCA CCA) and 18SR (5′-CCA GAC AAA TCG CTC CAC CAA C).
7. In the beginning, ten individual cDNA samples synthesized independently can be used for the first round of normalization to guarantee the complexity of the cDNA mixtures. After three rounds of normalization, they can be pooled for subtractive enrichment.
8. A prokaryotic housekeeping gene gyrase A (*gyrA*) can be used as another control to ensure that only differentially expressed genes are captured by rounds of enrichment. The presence of *gyrA* in enriched cDNAs, and cDNAs before and after normalization is evaluated by PCR using primers gyrAF (5′-ACG CGA CGG TGT ACC GGC TT) and gyrAR (5′-GCC AGA GAA ATC ACC CCG GTC).
9. The average number of clones for each identical gene should be expected to occur at least twice in the sequenced samples. Rarefaction analysis (33) can be used to estimate coverage of the enriched cDNA libraries for the identified genes as described previously (34–37). In case of low coverage, more clones can be selected from the enriched libraries for southern blot screening and sequencing.

Acknowledgments

This work was supported by a competitive grant from the Ohio Agricultural Research and Development Center, The Ohio State University, Wooster, OH, USA.

References

1. Deb, D.K., Dahiya, P., Srivastava, K.K., Srivastava, R. and Srivastava, B.S. (2002) Selective identification of new therapeutic targets of *Mycobacterium tuberculosis* by IVIAT approach. *Tuberculosis* **82**, 175–182.
2. Handfield, M., Brady, L.J., Progulske-Fox, A. and Hillman, J.D. (2000) IVIAT: a novel method to identify microbial genes expressed specifically during human infections. *Trends Microbiol.* **8**, 336–339.
3. Handfield, M., Seifert, T. and Hillman, J.D. (2002) *In vivo* expression of bacterial genes during human infections. *Methods Mol. Med.* **71**, 225–242.
4. Hang, L.M., John, M., Asaduzzaman, E.A., Bridges, C., Vanderspurt, T.J., Kirn, R.K., Taylor, R.K., Hillman, J.D., Progulske-Fox, A., Handfield, M., Ryan, E.T. and Calderwood, S.B. (2003) Use of *in vivo*-induced antigen technology (IVIAT) to identify genes uniquely

expressed during human infection with *Vibrio cholerae*. *Proc. Natl. Acad. Sci. USA* **100**, 8508–8513.

5. Rollins, S.M., Peppercorn, A., Hang, L., Hillman, J.D., Calderwood, S.B., Handfield, M. and Ryan, E.T. (2005) Technoreview: *in vivo* induced antigen technology (IVIAT). *Cull. Microbiol.* 7, 1–9.
6. Mahan, M.J., Slauch, J.M. and Mekalanos, J.J. (1993) Selection of bacterial virulence genes that are specifically induced in host tissues. *Science* **259**, 686–688.
7. Veal-Carr, W.L. and Stibitz, S. (2005) Demonstration of differential virulence gene promoter activation *in vivo* in *Bordetella pertussis* using RIVET. *Mol. Microbiol.* **55**, 788–798.
8. Angelichio, M.J. and Camilli, A. (2002) *In vivo* expression technology. *Infect. Immun.* **70**, 6518–6523.
9. Valdivia, R.H. and Falkow, S. (1996) Bacterial genetics by flow cytometry: rapid isolation of *Salmonella typhimurium* acid-inducible promoters by differential fluorescence induction. *Mol. Microbiol.* **22**, 367–378.
10. Hensel, M., Shea, J.E., Gleeson, C., Jones, M.D., Dalton, E. and Holden, D.W. (1995) Simultaneous identification of bacterial virulence genes by negative selection. *Science* **269**, 400–403.
11. Shea, J.E., Santangelo, J.D. and Feldman, R.G. (2000) Signature-tagged mutagenesis in the identification of virulence genes in pathogens. *Curr. Opin. Microbiol.* **3**, 451–458.
12. Lehoux, D.E. and Levesque, R.C. (2000) Detection of genes essential in specific niches by signature-tagged mutagenesis. *Curr. Opin. Biotech.* **11**, 434–439.
13. Mecsas, J. (2002) Use of signature-tagged mutagenesis in pathogenesis studies. *Curr. Opin. Microbiol.* **5**, 33–37.
14. Judson, N. and Mekalanos, J.J. (2000) TnAraOut, a transposon-based approach to identify and characterize essential bacterial genes. *Nat. Biotech.* **18**, 740–745.
15. Akerley, B.J., Rubin, E.J., Lampe, D.J. and Mekalanos, J.J. (1998) PCR-mediated detection of growth-attenuated mutants in large pools generated by *in vitro* transposon mutagenesis. *Am. Soc. Microbiol. Gen. Meet.* 98th, Atlanta.
16. Shelburne, S.A. and Musser, J.M. (2004) Virulence gene expression *in vivo*. *Curr. Opin. Microbiol.* 7, 283–289.
17. To, K.Y. (2000) Identification of differential gene expression by high throughput analysis. *Comb Chem High Throughput Screen* **3**, 235–241.
18. Boyce, J.D., Cullen, P.A. and Adler, B. (2004) Genomic-scale analysis of bacterial gene and protein expression in the host. *Emerg. Infect. Dis.* **10**, 1357–1362.
19. Hinton, J.C., Hautefort, I., Eriksson, S., Thompson, A. and Rhen, M. (2004) Benefits and pitfalls of using microarrays to monitor bacterial gene expression during infection. *Curr. Opin. Microbiol.* 7, 277–282.
20. Jansen, A. and Yu, J. (2006) Differential gene expression of pathogens inside infected hosts. *Curr. Opin. Microbiol.* **9**, 138–142.
21. Graham, J.E. and Clark-Curtiss, J.E. (1999) Identification of *Mycobacterium tuberculosis* RNAs synthesized in response to phagocytosis by human macrophages by selective capture of transcribed sequences (SCOTS). *Proc. Natl. Acad. Sci. USA* **96**, 11554–11559.
22. Liu, S., Graham, J.E., Bigelow, L., Morse, P.D., 2nd and Wilkinson, B.J. (2002) Identification of *Listeria monocytogenes* genes expressed in response to growth at low temperature. *Appl. Environ. Microbiol.* **68**, 1697–1705.
23. Hou, J.Y., Graham, J.E. and Clark-Curtiss, J.E. (2002) *Mycobacterium avium* genes expressed during growth in human macrophages detected by selective capture of transcribed sequences (SCOTS). *Infect. Immun.* **70**, 3714–3726.
24. Daigle, F., Graham, J.E. and Curtiss, R., 3 rd (2001) Identification of *Salmonella typhi* genes expressed within macrophages by selective capture of transcribed sequences (SCOTS). *Mol. Microbiol.* **41**, 1211–1222.
25. Dozois, C.M., Daigle, F. and Curtiss, R., 3 rd (2003) Identification of pathogen-specific and conserved genes expressed *in vivo* by an avian pathogenic *Escherichia coli* strain. *Proc. Natl. Acad. Sci. USA* **100**, 247–252.
26. Baltes, N., Buettner, F.F. and Gerlach, G.F. (2007) Selective capture of transcribed sequences (SCOTS) of *Actinobacillus pleuropneumoniae* in the chronic stage of disease reveals an HlyX-regulated autotransporter protein. *Vet. Microbiol.* **123**, 110–121.
27. Graham, J.E., Peek, R.M., Jr., Krishna, U. and Cover, T.L. (2002) Global analysis of *Helicobacter pylori* gene expression in human gastric mucosa. *Gastroenterology* **123**, 1637–1648.
28. An, R., Sreevatsan, S. and Grewal, P.S. (2008) *Moraxella osloensis* gene expression in the slug host *Deroceras reticulatum*. *BMC Microbiol.* **8**, 19.
29. An, R., Sreevatsan, S. and Grewal, P.S. (2009) Comparative *in vivo* gene expression of the closely related bacteria *Photorhabdus temperata* and *Xenorhabdus koppenhoeferi* upon infection of the same insect host, *Rhizotrogus majalis*. *BMC Genomics* **10**, 433.
30. Haydel, S.E. and Clark-Curtiss, J.E. (2004) Global expression analysis of two-component

system regulator genes during *Mycobacterium tuberculosis* growth in human macrophages. *FEMS Microbiol. Lett.* **236**, 341–347.

31. Aljanabi, S.M. and Martinez, I. (1997) Universal and rapid salt-extraction of high quality genomic DNA for PCR-based techniques. *Nucleic Acids Res.* **25**, 4692–4693.
32. Sambrook, J., Russell, D.W. and Russell, D. (2000) *Molecular Cloning: A Laboratory Manual.* Cold Spring Harbor Laboratory Press, New York.
33. Heck, K.L., Jr., Belle, G.V. and Simberloff, D. (1975) Explicit calculation of the rarefaction diversity measurement and the determination of sufficient sample size. *Ecology* **56**, 1459–1461.
34. Wang, Y.L. and Morse, D. (2006) Rampant polyuridylylation of plastid gene transcripts in the dinoflagellate *Lingulodinium. Nucleic Acids Res.* **34**, 613–619.
35. Suga, K., Mark Welch, D., Tanaka, Y., Sakakura, Y. and Hagiwara, A. (2007) Analysis of expressed sequence tags of the cyclically parthenogenetic rotifer *Brachionus plicatilis. PLoS ONE* **2**, e671.
36. Zhu, X.C., Tu, Z.J., Coussens, P.M., Kapur, V., Janagama, H., Naser, S. and Sreevatsan, S. (2008) Transcriptional analysis of diverse strains *Mycobacterium avium* subspecies paratuberculosis in primary bovine monocyte derived macrophages. *Microb. Infect.* **10**, 1274–1282.
37. Frias-Lopez, J., Shi, Y., Tyson, G.W., Coleman, M.L., Schuster, S.C., Chisholm, S.W. and Delong, E.F. (2008) Microbial community gene expression in ocean surface waters. *PNAS* **105**, 3805–3810.

Chapter 10

Whole-Genome RT-qPCR MicroRNA Expression Profiling

Pieter Mestdagh, Stefaan Derveaux, and Jo Vandesompele

Abstract

MicroRNAs (miRNAs) are small noncoding RNA molecules that function as negative regulators of gene expression. They are essential components of virtually every biological process and deregulated miRNA expression has been reported in a multitude of human diseases including cancer. Owing to their small size (20–22 nucleotides), accurate quantification of miRNA expression is particularly challenging. In this chapter, we present different RT-qPCR technologies that enable whole genome miRNA expression quantification.

Key words: microRNA, Stem-loop, RT-qPCR, Global mean normalization

1. Introduction

miRNAs represent one of the largest classes of gene regulators. Currently, the miRbase sequence database (Release 16, http://www.mirbase.org) contains over 17,000 entries of mature miRNAs in 142 species including 1,223 mature human miRNAs. Their involvement in human disease has important implications for translational research, as miRNA expression signatures have been correlated to diagnosis and prognosis, and are eligible as excellent targets for therapy. Unfortunately, accurate quantification of miRNA expression levels is a major challenge in the field. Several hybridization-based methods, such as microarray and bead-based flow cytometry, have been introduced to quantify the expression of hundreds of miRNAs in a single experiment. However, these approaches require substantial amounts of input RNA, which precludes the use of small biopsies, single cells or body fluids such as serum, plasma, urine, or sputum. While the reverse transcription quantitative PCR (RT-qPCR) in principle has a much higher sensitivity, down to a single molecule, the RT reaction requires

Michael Kaufmann and Claudia Klinger (eds.), *Functional Genomics: Methods and Protocols*,
Methods in Molecular Biology, vol. 815, DOI 10.1007/978-1-61779-424-7_10, © Springer Science+Business Media, LLC 2012

modification to enable the detection of small RNA molecules such as miRNAs (see ref. 1 for a review on all available RT-qPCR platforms for miRNA detection). One approach relies on the use of stem-loop RT primers (2, 3), while another is based on polyadenylation of the mature miRNA prior to oligo-dT primed cDNA synthesis (4). Next to sensitivity, RT-qPCR based approaches have a superior specificity and a high level of flexibility, allowing additional assays to be readily included in the workflow.

1.1. Stem-Loop Reverse Transcription miRNA Profiling

Stem-loop reverse transcription is based on the use of a looped miRNA specific RT-primer that will hybridise to the 3′ end of the mature miRNA to initiate cDNA synthesis (2). Upon denaturation, the loop unfolds, providing a longer template for detection in a qPCR reaction (Fig. 1a). Since this process is miRNA specific, multiplex pooling of individual stem-loop primers is necessary to produce cDNA template for multiple miRNAs.

The stem-loop RT-qPCR miRNA profiling platform is provided by Applied Biosystems and uses a miRNA specific forward primer and hydrolysis probe together with a universal reverse primer to measure miRNA expression. Stem-loop primers for more than 700 mature human miRNAs are pooled in two Megaplex primer pools (pool A and pool B) to allow whole genome miRNA expression profiling. An optional limited-cycle preamplification step is introduced to increase the sensitivity of the reaction, enabling miRNA profiling studies of single cells and body fluids. The preamplification procedure uses the same miRNA specific forward and universal reverse primers to amplify the cDNA template in a 12-cycle PCR reaction. As is the case for the stem-loop primers, the forward and reverse preamplification primers are pooled in two pools that match the Megaplex RT primer pools. In order to assess whether the use of a preamplification step introduced a bias in miRNA expression values we compared two workflows, by including or excluding preamplification. We profiled the expression of 430 miRNAs in different neuroblastoma cell lines and evaluated the differential miRNA expression between different cell lines for both procedures. If no bias is introduced, the ΔCq (Cq or quantification cycle according to MIQE-guidelines) (5) for any given miRNA should be similar

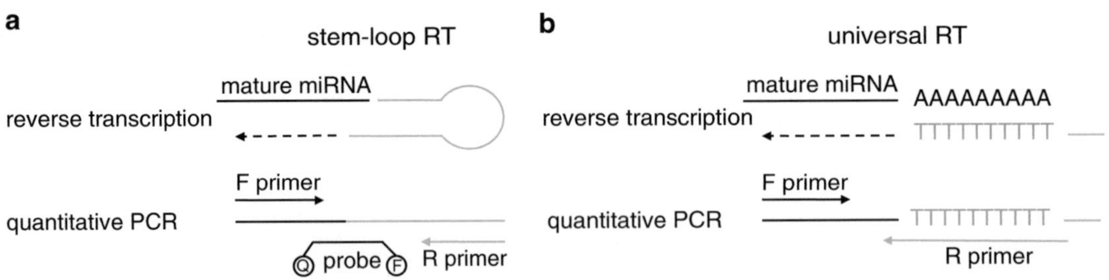

Fig. 1. Schematic overview of the stem-loop RT-qPCR (**a**) and universal RT-qPCR (**b**) miRNA profiling platforms.

for both approaches. Therefore, the difference in ΔCq (ΔΔCq) as measured by both approaches should approach zero. We found that 80% of all detected miRNAs had a ΔΔCq < 1 and 75% had a ΔΔCq < 0.5. Following analysis of only the most abundant miRNAs (Cq < 30), 94% had a ΔΔCq < 1, suggesting that both procedures give similar results. For low-abundant miRNAs, results should be interpreted with caution. The higher variation observed for the low-abundant miRNAs is partly attributable to increased variation in the RT-reaction, which is typically observed for low copy templates (3).

1.2. Universal Reverse Transcription miRNA Profiling

This approach is based on polyadenylation of the mature miRNA (4). Reverse transcription is initiated using a polyT primer that can be tagged (Fig. 1b). This reaction is universal, providing cDNA template for quantification of any miRNA. Several suppliers provide such a platform, including Exiqon that uses LNA-modified miRNA specific forward and reverse primers to measure miRNA expression. The use of LNA-modified primers precludes the need for a preamplification step and enables the study of miRNA expression when limited amounts of RNA are available.

1.3. Normalizing Whole Genome RT-qPCR miRNA Expression Data

The accuracy of the results obtained through RT-qPCR miRNA expression profiling is largely dependent on proper normalization of the data (Fig. 2). Several parameters inherent to the RT-qPCR reaction need to be controlled for to distinguish technical variation from true biological changes. For normalization of RT-qPCR data, the use of multiple stable reference genes is accepted as the gold standard method (6). As there is no such thing as a set of universal stable reference genes, each individual experiment requires careful selection of the most stable candidates. Typically, a set of ten

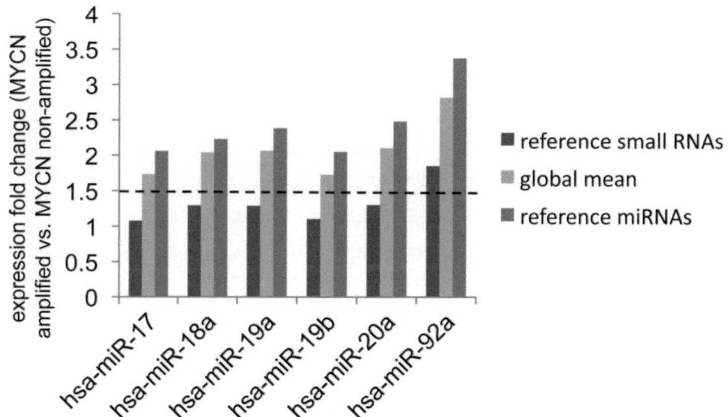

Fig. 2. Expression fold change of miR-17-92 miRNAs in MYCN amplified versus MYCN nonamplified neuroblastoma tumor samples for three different normalization methods: reference small RNAs (i.e., a selection of stable small nuclear/nucleolar RNAs), global mean, and reference miRNAs (i.e., miRNAs that resemble the global mean expression). The dashed line indicates a 1.5-fold change in expression.

candidate reference genes is evaluated in a pilot experiment with representative samples from the different experimental conditions under investigation. The most stable reference genes are subsequently identified using well established algorithms such as geNorm or Normfinder (6, 7). While candidate reference mRNA genes are well established, candidate reference miRNA genes are not. Only few candidate reference miRNA genes have been reported in the scientific literature and all too often, small nuclear or nucleolar RNAs (such as U6, U24, U26) are used instead.

For whole genome miRNA expression profiling, we have successfully introduced the global mean miRNA expression value as a virtual reference gene representing the best normalization factor (8). The global mean expression value is calculated as the average Cq of all expressed miRNAs per sample, where miRNAs with a Cq-value < 32 are considered expressed. Compared to small nuclear and nucleolar RNAs, global mean normalization is by far better in reducing the technical variation and consequently allows a more accurate interpretation of the biological changes (8). This was illustrated by evaluating the differential expression of miRNAs from the miR-17-92 cluster in primary neuroblastoma tumor samples with and without overexpression of the MYCN transcription factor, known to activate the miR-17-92 cluster by binding to its promoter. Neuroblastoma tumor samples with overexpression of the *MYCN* gene should therefore have increased expression of miR-17-92 miRNAs. Surprisingly, only one of the miRNAs from the miR-17-92 cluster was found to be differentially expressed when normalizing with small nuclear or nucleolar RNAs. Upon normalization using the global mean expression value, all miR-17-92 miRNA were found to be differentially expressed.

1.4. Identification of Stably Expressed Reference miRNAs

Typically, whole-genome miRNA expression studies are followed by focused validation studies for a selection of miRNAs. In this case, the global mean expression can no longer be used for normalization. Our group demonstrated that it is possible to identify miRNAs that resemble the global mean expression value and that the geometric mean of their expression levels can be successfully used to mimic global mean expression value normalization (8). As with the global mean expression, normalization using these miRNAs results in a higher reduction of technical variation and a more accurate interpretation of the biological changes. Alternatively, selection of miRNAs that resemble the global mean expression value can be performed using the miRNA body map Web tool (http://www.mirnabodymap.org) (9). The miRNA body map contains whole genome RT-qPCR miRNA expression data for over 700 samples from varying tissue and disease origin (see Note 1). For normalization of experiments in which only a few miRNAs are measured, we recommend to consult the miRNA body map to evaluate whether it contains samples of similar tissue or disease origin and use the integrated tool to identify stable reference miRNAs. Candidate reference miRNAs for a subset of normal and disease tissues were also identified by Peltier and Latham (10). The authors report that miR-191 and miR-103, among others, were found to be stably expressed across 13 normal tissues and five pair of distinct tumor/normal adjacent tissues (see Note 2). Ultimately, a selection of small nuclear and/or nucleolar RNAs could be applied for miRNA expression normalization, given that these are stably expressed across the samples under investigation. Of note, the use of small nuclear/nucleolar RNAs can, at least in some cases, lead to a misinterpretation of the biological changes (8).

2. Materials

2.1. Stem-Loop RT-qPCR

1. TaqMan microRNA reverse transcription kit (Applied Biosystems) containing dNTPs (100 nM), MultiScribe reverse transcriptase (50 U/μl), reverse transcription buffer (10×), RNase inhibitor (20 U/μl).
2. Human Megaplex primer pools A and B (Applied Biosystems).
3. Human Megaplex PreAmp primer pools A and B (Applied Biosystems). The use of PreAmp primers is optional and depends on the amount of available input RNA (see Methods for details on minimal amounts of input RNA).
4. TaqMan PreAmp master mix (Applied Biosystems). Optional, use in combination with PreAmp primers.
5. TaqMan universal PCR master mix II (2×) (Applied Biosystems).
6. TaqMan miRNA assays, either as single tube assays or predispensed in TaqMan array miRNA cards matching Megaplex pool A and B (Applied Biosystems).
7. $MgCl_2$ (50 mM).
8. Nuclease-free water.

2.2. Universal RT-qPCR

1. Universal cDNA synthesis kit (Exiqon) containing reaction buffer (5×) and enzyme mix. The universal reverse transcription primer is included in the reaction buffer.
2. SYBR Green master mix (2×) (Exiqon).
3. Forward and reverse LNA primers, either as single tube assays or predispensed (Human panel I and II) in 384-well plates (Exiqon).
4. Nuclease-free water.

3. Methods

3.1. Stem-Loop RT-qPCR

1. Dilute RNA sample to a concentration of 10 ng/μl (total RNA) for the workflow with preamplification and 500 ng/μl for a workflow without preamplification (see Notes 3 and 4). Sensitivity can be improved by increasing the amount of input RNA. Similarly, decreasing the amount of input RNA will result in a lower sensitivity. Keep RNA on ice at all times to prevent degradation (see Note 3 and Fig. 3).
2. Prepare the reverse transcription reaction mix by combining 0.8 μl of Megaplex primers pool, 0.2 μl of dNTPs, 1.5 μl of Multiscribe reverse transcriptase, 0.8 μl of reverse transcription buffer, 0.45 μl of $MgCl_2$, 0.1 μl of RNase inhibitor, 0.65 μl of nuclease-free water, and 3 μl of the diluted RNA sample. To avoid pipetting volumes below 1 μl, scale up the individual volumes to process at least 10 samples. Prepare an individual RT reaction for each Megaplex primer pool (pool A and B). Mix reagents by pipetting and spin down (see Note 5).
3. Incubate the reverse transcription mix on ice for 5 min.
4. Run the reverse transcription reaction as follows: (16°C for 2 min, 42°C for 1 min, 50°C for 1 s) × 40 cycles, 85°C for 5 min, and cooling down to 4°C.
5. In case a preamplification reaction is performed, spin down each sample and add 2.5 μl of reverse transcription product to 12.5 μl of TaqMan PreAmp master mix, 2.5 μl of the matching PreAmp primer pool (pool A or B), and 7.5 μl of nuclease-free water. Pipette to mix and spin down.

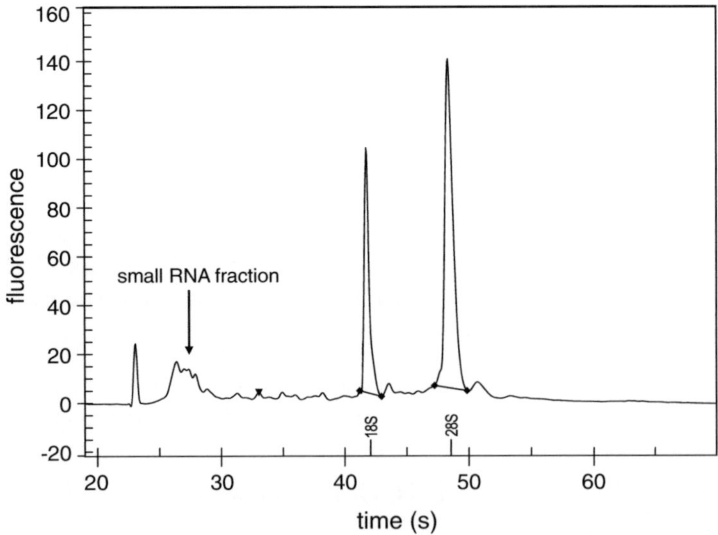

Fig. 3. Experion profile of a good-quality total RNA sample. The small peak represents the small RNA fraction.

6. Run a 12-cycle preamplification reaction as follows: 95°C for 10 min, 55°C for 2 min, 72°C for 2 min, (95°C for 15 s, 60°C for 4 min) × 12 cycles, 99°C for 10 min, and cooling down to 4°C.
7. Dilute the preamplification product 4× by adding 75 µl of nuclease-free water to each sample. Pipette to mix and spin down.
8. For each TaqMan miRNA array, combine 450 µl of TaqMan universal PCR master mix, 441 µl of nuclease-free water, and 9 µl of diluted preamplification product. Pipette to mix and spin down. Increase the fraction of preamplification product to increase sensitivity.
9. For a workflow without preamplification, combine 450 µl of TaqMan universal PCR master mix, 444 µl of nuclease-free water, and 6 µl of Megaplex reverse transcription product.
10. Pipette 100 µl of the PCR reaction mix into each port of the TaqMan miRNA array, centrifuge, and seal.
11. When profiling individual assays, dilute the Megaplex reverse transcription product or the 4× diluted preamplification product 50× and add 2.25 µl of this dilution to 0.25 µl of TaqMan miRNA assay and 2.5 µl of TaqMan universal PCR master mix in a 384-well plate.
12. Run the PCR reaction as follows: 95°C for 10 min, (95°C for 15 s, 60°C for 1 min, optical read) × 40 cycles.

3.2. Universal RT-qPCR

1. Dilute RNA sample to a concentration of 5 ng/µl (total RNA). Sensitivity can be improved by increasing the amount of input RNA (see Notes 3 and 4). Similarly, decreasing the amount of input RNA will result in a lower sensitivity. Keep RNA on ice at all times to prevent degradation. When using the predispensed panels, dilute RNA sample to 5.5 ng/µl to get a 10% excess when preparing the PCR reaction mix.
2. Prepare the reverse transcription reaction mix by combining 4 µl of reaction buffer, 2 µl of enzyme mix, 10 µl of nuclease-free water, and 4 µl of the RNA sample (5 ng/µl). Mix by pipetting and spin down. Prepare two reverse transcription mixes to analyze both human panels.
3. Incubate the reverse transcription mix at 42°C for 60 min, followed by reverse transcriptase heat inactivation at 95°C for 5 min.
4. When profiling individual assays, dilute the reverse transcription product 80× in nuclease-free water and prepare PCR reaction mix by combining 2.5 µl of SYBR Green master mix with 0.5 µl of LNA primer mix and 2 µl of diluted cDNA in a 384-well plate.

5. When using predispensed panels, combine both 20 μl transcription reactions per sample and dilute the reverse transcription product 110× by adding 4,360 μl of nuclease-free water to 40 μl of reverse transcription product. Prepare the PCR reaction mix for each sample by combining 4,360 μl of SYBR Green master mix with 4,360 μl of diluted reverse transcription product and pipette 10 μl in each well of the predispensed panels.
6. Run the PCR reaction as follows: 95°C for 10 min, (95°C for 10 s, 60°C for 1 min, optical read) × 40 cycles, melting curve analysis.

3.3. Data Normalization

1. When profiling all (or a substantial subset of) miRNAs, normalize miRNA expression using the global mean expression value (μ) (see Note 6). Given *k* expressed miRNAs, the normalized relative quantity (in log scale) for miRNA *i* in sample *j* is defined as:

$$\mathrm{NRQ}_{i,j} = Cq_{i,j} - \mu_j$$

$$\mu_j = \frac{\sum_{i=1}^{k} Cq_{i,j}}{k}$$

Alternatively, it is also possible to calculate the normalized relative quantity in linear space (see Note 7). Given *k* expressed miRNAs in sample *j*, the normalized miRNA *i* is defined as:

$$\mathrm{NRQ}_{i,j} = \frac{\mathrm{RQ}_{i,j}}{\sqrt[k]{\prod_{i=1}^{k} \mathrm{RQ}_{i,j}}}$$

2. To identify a set of reference miRNAs resembling the global mean, calculate the geNorm pairwise variation *V* value to determine robust similarity in expression of a given miRNA with the global mean expression value. For each miRNA, calculate the difference between its *Cq*-value and the global mean expression value in each sample. Next, determine the standard deviation of these differences for each miRNA. The miRNAs with the lowest standard deviation most closely resemble the global mean expression value. The optimal number of miRNAs for normalization should be determined through geNorm analysis of the ten best ranked miRNAs. To avoid including miRNAs that are putatively coregulated, exclude those miRNAs that are located within 2 kb of each other. Coregulated miRNAs are replaced by the next best ranked miRNA.

4. Notes

1. The miRNA body map Web tool is available at http://www.mirnabodymap.org. To identify stably expressed miRNAs, navigate to the "data analysis" section by clicking the "data analysis" icon in the top left icon bar. Next, choose your species of interest and select a dataset. Under the "miRNA centric analysis" option, choose "Select most stably expressed miRNAs." Finally, select your samples of interest and click "next" to view stable reference miRNAs for your sample subset.
2. The stability of candidate reference miRNAs depends on the tissue or disease type but also on the experimental conditions (e.g., treatment of the cells with siRNA or compound). When changing experimental conditions, verify the stability of the reference miRNAs by measuring their expression on a representative selection of samples followed by geNorm or Normfinder analysis.
3. miRNA expression profiling will only be successful if the small RNA fraction is retained after RNA isolation. Several commercial kits are available that enable the extraction of total RNA including the small RNA fraction. The presence of the small RNA fraction can be evaluated using mircofluidics-based electrophoresis systems such as the Bioanalyser (Agilent) or the Experion (Bio-Rad) (Fig. 3). We strongly encourage to include only RNA samples of sufficient quality. In addition, enrichment of the small RNA fraction is not advised.
4. There is no need to perform a DNase-treatment prior to miRNA expression profiling when using the stem-loop RT-qPCR platform. When using the universal RT-qPCR platform, DNA contamination can be an issue. This can be evaluated by profiling a sample for which the reverse transcription reaction was performed without reverse transcriptase. qPCR signals that are detected in this sample typically indicate a contamination with genomic DNA.
5. The stem-loop RT-qPCR miRNA expression profiling protocol can be adjusted to a multiplex format (both with and without preamplification of the reverse transcription product), which allows to perform reverse transcription (and preamplification) for a limited number of miRNAs as compared to the classical Megaplex format where reverse transcription for all miRNAs is performed. Consult Applied Biosystems for further information on the "Protocol for Custom RT and Preamplification Pools with TaqMan MicroRNA Assays."
6. Baseline and threshold settings should be carefully evaluated when determining Cq-values. Typically, the baseline should be

set to the cycle interval where no amplification takes place. The threshold is set, with the Y-axis in log-scale, where all assays are in log linear phase.

7. Biogazelle's qbase^{PLUS} software ((11); http://www.qbaseplus.com) employs an improved version of the global mean normalization method based on geometric averaging of all expressed miRNAs, as well as an improved version of the geNorm method (enabling identification of the single best reference gene).

References

1. Benes, V and Castoldi, M (2010) Expression profiling of microRNA using real-time quantitative PCR, how to use it and what is available, *Methods 50*, 244–249.
2. Chen, C Ridzon, D. A Broomer, A. J Zhou, Z Lee, D. H Nguyen, J. T Barbisin, M Xu, N. L Mahuvakar, V. R Andersen, M. R Lao, K. Q Livak, K. J and Guegler, K. J. (2005) Real-time quantification of microRNAs by stem-loop RT-PCR, *Nucleic Acids Res 33*, e179.
3. Mestdagh, P Feys, T Bernard, N Guenther, S Chen, C Speleman, F and Vandesompele, J. (2008) High-throughput stem-loop RT-qPCR miRNA expression profiling using minute amounts of input RNA, *Nucleic Acids Res 36*, e143.
4. Shi, R and Chiang, V. L. (2005) Facile means for quantifying microRNA expression by real-time PCR, *Biotechniques 39*, 519–525.
5. Bustin, S. A Benes, V Garson, J. A Hellemans, J Huggett, J Kubista, M Mueller, R Nolan, T Pfaffl, M. W Shipley, G. L Vandesompele, J and Wittwer, C. T. (2009) The MIQE guidelines: minimum information for publication of quantitative real-time PCR experiments, *Clin Chem 55*, 611–622.
6. Vandesompele, J De Preter, K Pattyn, F Poppe, B Van Roy, N De Paepe, A and Speleman, F. (2002) Accurate normalization of real-time quantitative RT-PCR data by geometric averaging of multiple internal control genes, *Genome Biol 3*, RESEARCH0034.
7. Andersen, C. L Jensen, J. L and Orntoft, T. F. (2004) Normalization of real-time quantitative reverse transcription-PCR data: a model-based variance estimation approach to identify genes suited for normalization, applied to bladder and colon cancer data sets, *Cancer Res 64*, 5245–5250.
8. Mestdagh, P Van Vlierberghe, P De Weer, A Muth, D Westermann, F Speleman, F and Vandesompele, J. (2009) A novel and universal method for microRNA RT-qPCR data normalization, *Genome Biol 10*, R64.
9. Mestdagh, P Lefever, S Pattyn, F Ridzon, D. A Fredlund, E Fieuw, A Vermeulen, J De Paepe, A Wong, L Speleman, F Chen, C and Vandesompele, J. (in preparation) The microRNA body map: dissecting microRNA function through integrative genomics.
10. Peltier, H. J and Latham, G. J. (2008) Normalization of microRNA expression levels in quantitative RT-PCR assays: identification of suitable reference RNA targets in normal and cancerous human solid tissues, *RNA 14*, 844–852.
11. Hellemans, J Mortier, G De Paepe, A Speleman, F and Vandesompele, J. (2007) qBase relative quantification framework and software for management and automated analysis of real-time quantitative PCR data, *Genome Biol 8*, R19.

Chapter 11

Using Quantitative Real-Time Reverse Transcriptase Polymerase Chain Reaction to Validate Gene Regulation by PTTG

Siva Kumar Panguluri and Sham S. Kakar

Abstract

Pituitary tumor transforming gene is an important gene which is involved in many cellular functions including cell division, DNA repair, organ development, expression, and secretion of various angiogenic and metastatic factors. Overexpression of this gene has also been reported in many cancers. Understanding the molecular pathways induced by this oncogene is therefore important not only to understand the development of the disease but also for proper diagnosis and treatment. Gene profiling is an excellent tool to identify the genetic mechanisms, networks, and pathways associated with a particular disease. Oligonucleotide microarrays can be everybody's choice as a first step to identify the global expression of genes involved in the study of interest. Each technique has its own limitation. Therefore, further confirmation of the results with a different technique is always necessary. Quantitative real-time reverse-transcriptase polymerase chain reaction (qRT-PCR) is one of the widely used and best described techniques to confirm the microarray data. Here, we describe the qRT-PCR techniques for gene profiling studies and the methods used for the analysis of the output data for further studies.

Key words: PTTG, Reverse transcriptase, Polymerase chain reaction, Complementary DNA, Oncogene, Securin, Relative expression

1. Introduction

Pituitary tumor transforming gene (PTTG), also known as securin, is highly expressed in a variety of human primary tumors as well as tumor cell lines, including carcinoma of the ovary, testis, kidney, colon, thyroid, pituitary, liver, adrenal gland, breast as well as melanoma, leukemia, and lymphoma (1). In addition to its role in cell cycle during sister chromatid separation, it is also reported to be involved in the expression and secretion of various growth and

Michael Kaufmann and Claudia Klinger (eds.), *Functional Genomics: Methods and Protocols*,
Methods in Molecular Biology, vol. 815, DOI 10.1007/978-1-61779-424-7_11, © Springer Science+Business Media, LLC 2012

angiogenic factors including bFGF, VEGF, and IL-8 (1). Although PTTG has been reported to stimulate vascular endothelial growth factor (VEGF) (2, 3), basic fibroblast growth factor (2, 4), IL-8 (2), and matrix metalloproteinase (MMP)-2 (3, 5, 6), the precise mechanism by which PTTG contributes to angiogenesis and metastasis is still poorly understood.

Identification of genes differentially expressed by PTTG treatment using microarrays will be the best choice in order to understand networks and pathways involved in different stages of tumor development (7). From our previous studies on cDNA-microarray analysis of HEK 293 cells infected with adenovirus overexpressing PTTG, a large number of genes (~67%) were found to be down-regulated including c-Jun, v-Maf, and Dicer1 or up-regulated including many of histones (7). The reliability and quality of microarray results depend on several factors such as array production, RNA extraction, probe labeling, hybridization conditions, and image analysis. Therefore, the genes identified as differentially expressed by this method warrants validation using another independent technique such as quantitative real-time PCR, which is quantitative, rapid, sensitive, and requires 1,000-fold less RNA than conventional techniques (8).

In this chapter, we are discussing the details on the methodology of quantitative real-time reverse-transcriptase PCR analysis. Though most of the procedures given in this chapter are similar across all types of real-time PCR machines, the analysis of final data (either C_t/cycle threshold or absolute quantities or relative quantities) varies from the final output data. Here, we are describing the analysis of qRT-PCR data with the C_t values (Applied Biosystems) and relative quantities (Bio-Rad).

2. Materials

Prepare all solutions using nuclease-free water and conditions. Prepare and store all cDNA synthesis and qRT-PCR reagents at −20°C (unless indicated otherwise). Diligently follow all waste disposal regulations (especially phenol-based reagents used during RNA isolation) according to the material safety data sheet provided by the manufacturer. Store all other general reagents such as agarose gel-electrophoresis reagents at room temperature.

2.1. Cell Culture and Adenovirus

Perform all the following steps aseptically under a hood.

All reagents are from Invitrogen unless specified otherwise.

1. Growth medium: Dulbecco's Modified Eagle's Medium (DMEM) with high Glucose 1× with 1% HEPES (4-(2-hydroxyethyl)-1-

piperazineethanesulfonic acid), 1% penicillin G (sodium salt) and streptomycin sulfate solution (1,000 units of each per ml), and 10% fetal bovine serum (FBS).

2. 0.25% Trypsin–EDTA: 0.25% Trypsin in 0.1% EDTA solution.
3. The full-length PTTG cDNA was subcloned into the adenovirus shuttle vector pShuttle. Positive clones were sequenced to confirm the sequence and orientation of the cDNA. The adenovirus expression system was generated and purified in association with the Gene Therapy Center, Virus Vector Core Facility, University of North Carolina at Chapel Hill.

2.2. Total RNA Isolation

Prepare all the solutions at nuclease-free conditions. Always use gloves and eye protection. Avoid contact with skin or clothing. Work in a chemical hood. Avoid breathing vapor.

1. TRIzol Reagent (Invitrogen): Store the reagent at 4°C in an amber-colored bottle. TRIzol reagent may be corrosive and cause irritation, so avoid contacting the skin directly. Use gloves and lab coats while working with TRIzol reagent.
2. DEPC-treated water: Add 1 ml of 0.1% diethyl pyrocarbonate (DEPC) from Sigma-Aldrich to sterile water, mix well. Alternatively, water can be autoclaved after adding DEPC (see Note 1).
3. Isopropyl alcohol of molecular biology grade.
4. 70% ethanol: Add 30 ml of absolute ethanol to 70 ml of DEPC-treated water. Refrigerate until use.

2.3. Formaldehyde Agarose Gel

Prepare all reagents in this section with nuclease-free or DEPC-treated water.

1. Ethidium bromide: Add 2 mg of ethidium bromide to 10 ml of water, mix well, and store in a dark bottle. The final concentration of the solution will be 200 μg/ml (see Note 2).
2. Formaldehyde: Prepare a 37–40% w/v (12.3 M) solution, which may contain a stabilizer such as methanol (10–15%).
3. Formamide: aliquot in small quantities and store at −20°C.
4. 10× RNA gel-loading buffer: Prepare 0.1 M EDTA (pH 8), 50% glycerol, 0.25% of bromophenol blue (w/v), and 0.25% of xylene cyanol FF (w/v) in DEPC-treated water.
5. 10× MOPS buffer: Prepare 0.5 M MOPS (3-[*N*-morpholino] propanesulfonic acid), 10 mM EDTA (pH 8.0) in sterile DEPC-treated water and adjust the pH to 7.0 with NaOH. Sterilize the solution by filtering through a 0.45-μm filter and store at room temperature protected from light (see Note 3).

2.4. cDNA Synthesis

All reagents are from Applied Biosystems unless specified otherwise.

cDNA can be synthesized using High Capacity cDNA Reverse Transcription Kit. The kit consists of the following components:

1. 10× RT Buffer: Hydroxylated Organoamine 1–10%, Halide Salt 1–10%.
2. 100 mM dNTP mix: 25 mM each dATP, dGTP, dCTP, and dTTP.
3. 10× RT Random Primers.
4. Multiscribe Reverse Transcriptase 50 U/μl.

In addition to the above components from the kit we also require DNase/RNase-free water.

Composition: 1 ml diethyl pyrocarbonate in 1 l water.

2.5. Syber-Green qRT-PCR Components

The following reagents are required for performing real time PCR using Syber-green.

1. cDNA made previously.
2. DNase/RNase-free water: diethyl pyrocarbonate (0.1%) and water.
3. 20 μM primers, forward and reverse.
4. Power SYBR Green PCR Master Mix (Bio-Rad): Prepare 10–30% dimethyl sulfoxide (DMSO), 1–10% Tris Base.
5. 96-Well optical reaction plate.
6. MicroAmp optical adhesive film, PCR compatible, DNA/RNA/RNase-free.

2.6. Primers

The primers can be designed using the Vector NTI software.

3 Methods

Carry out all procedures for RNA extraction, cDNA synthesis, and qRT-PCR analysis at 4°C (or in ice) unless specified otherwise. All the cell culture work should be carried out under sterile conditions using a Biosafety Level-II hood.

3.1. Treatment of Cells with Ad-PTTG

1. HEK293 cells or another cell line of interest should be trypsinized when they are at log phase and plated on T-75 flasks. Cells will be infected with adenovirus vector (blank control), adenovirus vector expressing PTTG shRNA (AdPTTG shRNA) at variable multiplicities of infection (MOI) (see Note 4).

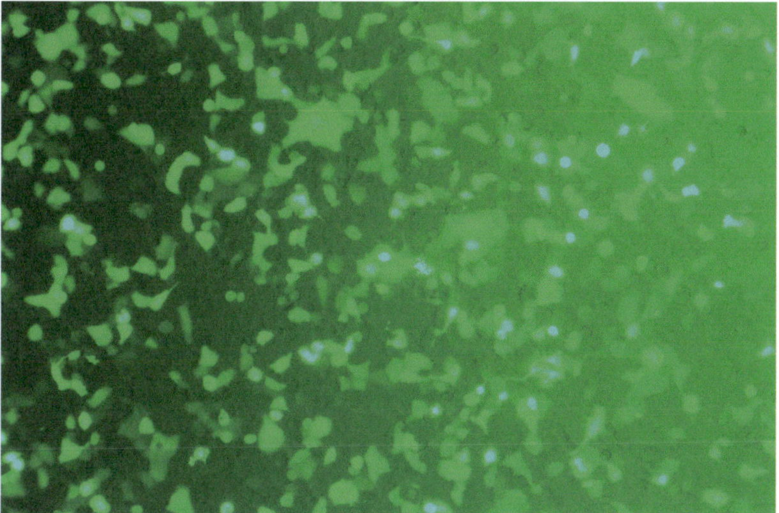

Fig. 1. Knocking-out of PTTG in lung carcinoma cells. A549 lung carcinoma cells were plated in each well of six-well plates and incubated overnight in a CO_2 incubator. Next day 5 µl of 1:100 diluted Adenovirus (from the main stock of ~ 10^{11} MOI) carrying PTTG shRNA to knock out the PTTG gene was added and incubated for 4 h in serum-free medium. Then the cells were incubated in growth medium and collected for total RNA isolation 48 h of postinfection.

2. Infection of cells should be carried out in serum-free DMEM medium for 2 h. After 2 h, the medium will be replaced with regular growth medium and incubated at 37°C under 5% CO_2. The optimum MOI should provide a 90–95% level of infection.
3. All the experiments using adenoviruses should be performed in accordance with instructions from the University Institutional Biosafety Committee (see Note 5).
4. The infection of the virus can be monitored after 12–24 h under a florescent microscope equipped with a FITC filter. Maximum expression levels can be achieved 48 h postinfection. Figure 1 shows that over 90% of expression can be achieved when A549 cells were treated with Adenovirus serotype 5 carrying scrambled PTTG shRNA at 48 h postinfection. The cells were treated with 5 µl/well of 1:100 diluted virus from the main stock (~10^{11} MOI) in a six-well plate.

3.2. Extraction of Total RNA from Myotubes/Skeletal Muscle Tissue by TRIzol

1. Trypsinize adenovirus infected cells 48 h after infection with 0.25% Trypsin and pellet the cells by centrifugation at 2,000 × *g* for 2 min at 4°C.
2. Discard the supernatant and add 1 ml TRIzol Reagent. Resuspend the cell pellet in TRIzol many times to ensure proper lysis of the cells.

3. To the homogenate add 200 μl chloroform and mix well by repeatedly inverting the tube gently for 1 min. Incubate the mixture on ice for 3 min and centrifuge at 12,000 × *g* for 15 min at 4°C. This step will separate proteins in the phenol phase, DNA in the inter-phase leaving RNA in the aqueous phase.
4. Transfer the aqueous phase into a fresh Eppendorf tube slowly with a 200-μl pipette tip without disturbing the interphase. Then add an equal volume of isopropanol (approximately 600 μl), mix well, and centrifuge at 12,000 × *g* for 10 min at 4°C.
5. Discard the supernatant slowly, without disturbing the pellet (see Note 6), and wash the pellet with 500 μl 70% ethanol prepared in DEPC-treated water. Then centrifuge at 12,000 × *g* for 5 min at 4°C and discard the supernatant without disturbing the pellet.
6. Dry the pellet on air or in a laminar-flow hood for 10 min (see Note 7) and dissolve the pellet in 50–100 μl DEPC-treated water. Incubate for 10 min at 60°C to ensure the total RNA is dissolved properly and measure the quantity using a NanoDrop photometer at 260 nm.

3.3. Quantification of Total RNA by Formamide Agarose Gel Electrophoresis

1. For the analysis of total RNA in a formamide agarose gel, firstly all the electrophoresis apparatus needs to be washed and rinsed with 70% ethanol to ensure RNase-free conditions. To make 1.5% agarose gel, add 1.5 g of agarose (regular use) to 72 ml of sterile water, dissolve the agarose by boiling, and cool the solution down to 55°C. Add 10 ml of 10× MOPS electrophoresis buffer and 18 ml of deionized formaldehyde. Cast the gel in a chemical fume hood and allow the gel to set for at least 1 h at room temperature (9).
2. When the gel is ready set up the RNA sample mixture with the loading dye. For this purpose, take 2 μl of RNA sample (up to 2 μg) in a clean RNAse-free tube, add 2 μl of 10× MOPS buffer, 4 μl of deionized formaldehyde, 10 μl formamide, and 1 μl ethidium bromide (200 μg/ml). Close the tube and incubate the mixture at 55°C for 1 h and then chill the samples in ice for 10 min. Quick spin the tube to collect the mixture and add 2 μl of 10× formaldehyde gel-loading buffer to each sample and keep the samples on ice until ready for loading.
3. Add sufficient 1× MOPS buffer to cover the gel and remove the comb. Prerun the gel for 5 min at 5 V/cm. Then load the RNA samples into the wells leaving the first well for the RNA ladder (RNA size standard). Run the samples at 5 V/cm until the bromophenol blue dye migrates to 3/4th of the gel. Once the gel run is finished then visualize the RNA using a UV transilluminator and photograph the gel (see Note 8).

3.4. cDNA Synthesis

1. Use 2 μg of RNA in a final volume of 10 μl with DNase/RNase-free water in a 0.2-ml PCR tube (see Note 9).
2. Make PCR master-mix using the High-Capacity cDNA Reverse Transcription Kit as follows (see Note 10):

Reagents	Volume (μl)
10× RT Buffer	2.0
20× dNTP mix (100 mM)	0.8
10× RT Random primers	2.0
Multiscribe reverse transcriptase	1.0
Nuclease-free water	4.2
Total	10.0

3. Add 10 μl of PCR master-mix to the RNA sample (i.e., from step 1 above). Total volume is now 20 μl.
4. Briefly vortex and centrifuge the tubes.
5. Program a PCR machine as follows:

	Step 1	Step 2	Step 3	Step 4
Temperature	25°C	37°C	85°C	4°C
Time	10 min	120 min	5 s	Hold

6. Place the tubes on the PCR heating block and run the PCR reaction. Store the tubes at −20°C after the PCR reaction is completed.

3.5. Quantitative Real-Time RT-PCR

3.5.1. Designing Primers

We design primers using the Vector NTI software (see Note 11). Basic steps for designing good quality primers using the Vector NTI software are as follows:

1. Find the cDNA or mRNA sequence for the gene of interest in the NCBI nucleotide database and copy the gene's unique gene ID number.
2. Open the Vector NTI software (Invitrogen) followed by clicking on Tools in the main menu, and then click on open link GID.
3. Paste the gene ID in the dialog box and click OK. The program will download the nucleotide sequence on your computer.
4. Save this file on your computer under an appropriate name.
5. Select a region of 300–400 bp within the cDNA sequence.
6. Go to Analyses in main menu and click on Primer Design. A dialog box will appear.

7. Make only the entries specified below, all others leave to their default values.

Product length: Min: 100 bp, Max: 200 bp

Maximum number of output options: 50

T_m (C): ≥55 and ≤60

%GC: ≥55 and ≤60

Length: ≥20 and ≤25

8. Click "Apply." A window will open at the upper left corner containing sequence of 50 primer sets.
9. Look at each forward and reverse primer sequence individually. The GC difference should be 0°C. T_m difference should not be more than ±1°C between two primers.
10. Click on the first primer meeting the required parameters, and then by right clicking the mouse, select Analyze. A new window will open containing the selected primer information.
11. Check for palindromes and repeats. Palindromes and Repeats should be 0. If there is a Palindrome or Repeat in either the forward or the reverse primers, do not use this pair. Perform the same on another set of primers.
12. Next click on "Dimers and Hairpin Loops" icon. A new window will open providing separately the number of hairpin loops and dimers in the selected primer. The ideal situation is that we should have no hairpin loop and no dimer. However, this is rather rare for most of the primers. The following criteria can be used to pick the good primers even with those having hairpin loops and primer dimers.
13. Make sure that the primer does not have more than 8 hairpin loops or dimers. Less than 8 is better but up to 8 are still acceptable.
14. Check the dimer dG and hairpin dG for each dimer and hairpin loop, respectively by clicking the >> button on the window.
15. The best value for dG should be 0 kcal/mol. However, dG values between −1.8 and +1.8 kcal/mol can be accepted. If any of the two primers in the pair has a dG value outside this range then do not use this pair and analyze other primer sets (see Note 12).
16. Once a right primer set is found, copy the primer sequences and send them for primer synthesis (see Note 13).
17. Test run qRT-PCR using these primers with a few samples. The primer set which shows a good dissociation curve should be used for qRT-PCR. Figure 2 shows the example of good and bad primers. A good primer will always give a single peak in the melting curve and a bad primer shows two peaks. Primers with a very low peak or a flat melting curve represent very low or no amplification of product.

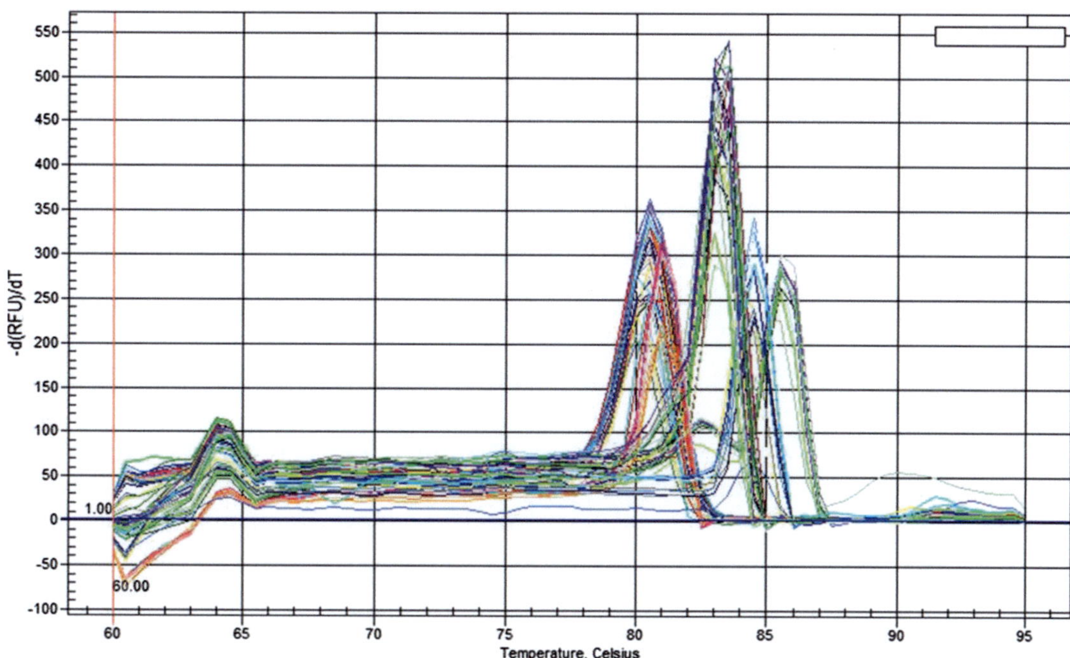

Fig. 2. Melting curves in qRT-PCR analysis. Here, 1 μl of cDNA was amplified with different primers (different GC contents and T_m) by qRT-PCR and analysis was done using Syber-green in a Bio-Rad iQ Cycler. The melt peaks showed here have different colors for each samples. The presence of two or more peeks represents primer dimmer formation.

3.5.2. Quantitative Real-Time PCR Reaction

After synthesizing the cDNA and obtaining primers, the real-time PCR reaction is set up as follows:

Components	Volume (μl)
cDNA	1
20 μm Stock Primer 1	1
20 μM Stock primer 2	1
RNase-free water	7
2× SYBER-Green Master Mix	10
Total	20

1. Prepare the master-mix with all the ingredients except the cDNA.
2. Dispense 1 μl of cDNA in the individual wells of 96-well Optical Reaction Plate with barcode 128 and then add 19 μl of master mix to each well. All reactions should be carried out in duplicate or triplicate to reduce the variation.
3. Data normalization is accomplished using the endogenous control such as glyceraldehyde-3-phosphate dehydrogenase (GAPDH) or β-actin (see Note 14).

4. Seal the plate using MicroAmp optical Adhesive Film and spin the plate in a PCR plate centrifuge.
5. Insert the plate in to the 7300 Sequence Detection system or iCycler iQ system.
6. Set the thermal conditions for qRT-PCR using 7300 system SDS software as follows:
   (a) Denaturation at 95°C for 10 min
   (b) 40 cycles of denaturation at 95°C for 15 s, annealing and extension at 60°C for 1 min
   (c) Finally, a melting curve of 95°C for 15 s, 60°C for 15 s, and 95°C for 15 s.
7. For the Bio-Rad iQ system the cycle conditions are initial denaturation at 95°C for 3 min followed by 45 cycles of denaturation at 95°C for 10 s, annealing at 60°C for 20 s, and extension at 72°C for 30 s.
8. Click on the 7300 system SDS software icon on the computer attached with the 7300 Sequence Detection system. Click on the Create New Document tab.
9. A new window will appear, select $\Delta\Delta C_t$ (Relative Quantitation) Plate in the Assay pull down menu.
10. Click "Next" and enter the name of the primes to be used in the left side window. Then select the primer in the left window and click "Add button." After adding all primer names, click the Next tab.
11. Enter the sample information for each well and save the file as .sds document.
12. In the same window, click the "Instrument" tab and then click on the "Add Dissociation Stage" tab.
13. Finally, click on the "Start" tab. This will start the program. Do not disturb the program until the run is finished.

3.6. Analysis of qRT-PCR Data

3.6.1. For Applied Biosystem

1. Open the 7300 system SDS software on the computer.
2. Click on "Create New Document."
3. Select ddC_t (Relative Quantitation) Study from the pull-down menu of the Assay tab.
4. Click on the "Next" tab. A new window will appear. Click on the "Add plates" tab.
5. Select the desired.sds file saved at the time of setting-up the qRT-PCR assay. Click "open." The file will appear. Click the "Finish tab" in the dialog box.
6. A new window will open. Select all the fields in the upper left box and click the green arrow in the main menu.

7. Again select all the fields in upper left window. The corresponding C_t values will appear in the bottom left window.
8. Go in the main menu and save the file as an .sdm (SDS Multiplate documents).
9. Click on the "main menu File" tab, and sequentially click on "Export," "Results," and "Both." Save the file as a .csv file.
10. Close the application and proceed for the analysis part using the .csv file.
11. Open the .csv file using Microsoft Excel, calculate the averages for the duplicates/triplicates of each sample and normalizing gene. This gives us average CT values.
12. Subtract the average CT values of the normalizing gene from the corresponding average CT values of the sample. This is ΔCT of that sample (ΔCT = average CT of sample – average CT of normalizing gene).
13. Calculate the final average by taking the average of all control ΔCT values.
14. Subtractinδ the CT values from the final average gives us the $\Delta\Delta C_t$ values.
15. The corresponding fold change is calculated as two to the power of $\Delta\Delta C_t$ values. This gives us the fold change in the samples as compared to the control which can be plotted on a graph (10).

3.6.2. For Bio-Rad

1. If the qRT-PCR reaction was performed using Bio-Rad equipment, the raw data will be either an absolute quantification or a relative quantification. To calculate an absolute quantification, a standard graph with the known concentration of a gene of choice will be taken at five different dilutions (10^{-1} dilutions each) in duplicates or triplicates.
2. Most of the researchers prefer to use a housekeeping gene for the standard curve. But it is always preferred to have one standard curve with a housekeeping gene and one with the gene to be tested. When the known quantities (in ng or μg) of standards were given, then the unknown/test samples will be calculated based on the standard curve and their absolute quantities will be given in the results data. The absolute quantities of each test sample triplicates will be used to generate mean quantities and standard errors in Microsoft Excel.
3. To identify the significance of the gene expression, a student *t*-test can be calculated in an Excel file. If the absolute quantity of the gene is not required, alternatively, all the cDNA samples to be run will be pooled and can be taken as a standard curve at five dilutions (10^{-1}).

4. Any housekeeping gene like GAPDH or β-actin or 18 S rRNA primers can be used to generate a standard curve. The starting quantity of the samples in the standard curve can be give as 100% and the relative quantities of the unknown or test samples will be calculated in the final results (which are displayed under starting quantity for each sample including the standards).
5. In this case, the relative quantities of the control and the treatments were taken in an Excel file and the mean, standard error and *p*-values (student *t*-test) can be calculated from each sample triplicates to generate final bar diagrams.
6. The final graphs from both absolute quantities and relative quantities will have normalized quantity or normalized relative expressions on its *y*-axis, respectively (7). Here, the normalized values will be obtained by the division of absolute quantities or relative quantities of test by its corresponding absolute quantities or relative quantities of the housekeeping enzyme.
7. The best standard curve will have efficiency values from 90 to 100% (e.g., $E = 100.6\%$ in the standard curve reaction shown in Fig. 3a and 99.4% in Fig. 3b). Also the R^2 value given below the standard graph (Fig. 3) should be from 0.9 to 1. If multiple PCRs with multiple plates are to be done, a separate standard curve should be in every plate set.
8. For normalization, the relative quantities or absolute quantities of the housekeeping gene for all the samples can be done once and can be used across all different primers to be used for the same samples. Sometimes, the PCR data will have a very bad standard curve. Even in this case, the data will have threshold cycle values (C_t values) for all samples. If there is a good standard

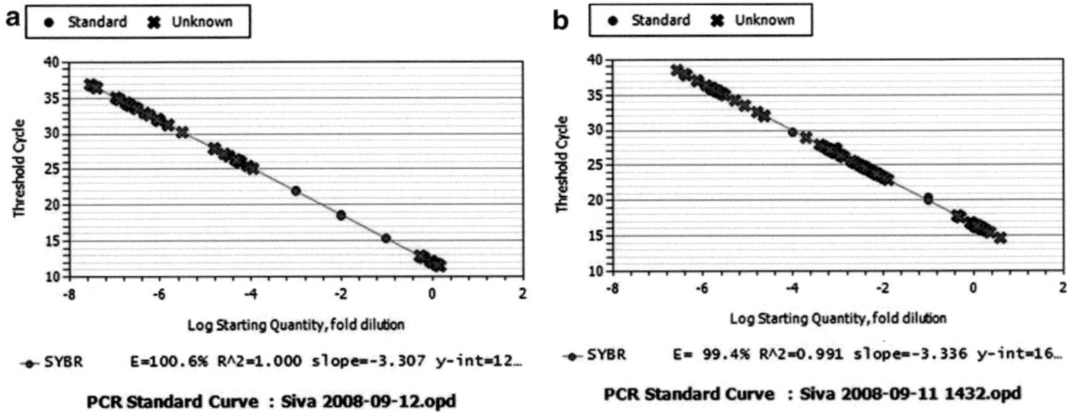

Fig. 3. Good and bad standard curves. Here, 1 μl of cDNA was taken for each dilution and a total of five different dilutions were used to plot standard curve by using the housekeeping enzyme GAPDH. The qRT-PCR analysis was done using Sybergreen in Bio-Rad iQ Cycler and standard curves were analyzed. The standard curved with efficiency (*E*) more than 99% was considered as best standard curves. All the samples were run in triplicates.

curve obtained for the same cDNA in the previous run, the C_t values of each sample from the current run can be used to calculate the relative quantities from the previous standard graph using an equation given below the standard graph.

9. Alternatively, by using the formula bellow if the efficiency is known (e.g., efficiency (*E*) in the standard graph in Fig. 3a is 100.6%).

$$\text{Relative quantity} = \text{efficiency}\ (\text{control } C_t - \text{unknown } C_t)$$

4. Notes

1. DEPC is not miscible with water; shake vigorously after adding DEPC to water for proper incubation. Overnight incubation with continuous mixing works very good.
2. Ethidium bromide is mutagenic and forms fume. Avoid direct contact to the body and store it in a dark colored bottle. Always store it at room temperature in a fume hood.
3. The MOPS buffer yellows with age if it is exposed to light or autoclaved. Straw colored buffer works well, but dark colored buffers are no longer recommended.
4. Adenovirus stock should be stored at −80°C in small aliquots. There will be a 50% loss of virus efficiency after every freeze–thaw circuit; therefore, the main stock should be stored in aliquots to avoid repeated freezing–thawing.
5. Working with Adenovirus needs a Bio-safety level-II hood. All the safety and precautions should be taken while working with Adenoviruses. Though these viruses are replication deficient, there is always a potential risk for pregnant or sick or injured persons. Replication deficiency should be examined periodically.
6. Generally, RNA pellets can be seen as small white pellets. Sometimes the pellet may not be visible. Therefore, mark the bottom corner of the Eppendorf tube with a marker before centrifugation and keep the tube facing the marked corner outside in the rotor. In this way, even if the pellet is not visible, the RNA will be supposed to be pelleted at the marked corner of the tube which will help to aspirate the supernatant slowly without loss of RNA.
7. Drying the RNA pellet after the ethanol wash is very important. Especially for down-stream applications such as cDNA synthesis, probe labeling, and other hybridization experiments. Any remains of ethanol will interfere with the enzymes in the down-stream applications. Too much ethanol traces will also give trouble while loading the gel (sample will float away from the wells).

8. The quality of RNA can be estimated by the presence of two bands, 28S and 18S species of rRNA. If the RNA is degraded, smeary appearance of either 28S or 18S rRNA or both will be visible. The un-degraded RNA samples should have the 28S rRNA approximately twice intense than the 18S rRNA. In some cases, RNA may be of high quality but may appear to be degraded due to contamination of the running buffer or the electrophoresis apparatus. Therefore, examine RNA quality using another technique such as RNA Bioanalyzer (Agilent Technologies, Valer, Kratzmeier).
9. To avoid multiple pipetting, it is better to make a master mix and then add master mix to the tubes containing cDNA.
10. Having a good primer set is critical for the success of the qRT-PCR assay.
11. It is quite possible that you may not get any good quality primers in a selected region of 300–400 bp, therefore, move across the sequence (by shifting the starting point 200 bp downstream) and perform the same search. Sometimes, it may takes more than an hour to get a really good set of primers.
12. If the gene of interest is not giving a good primer sets, we order two to three best possible sets of primers and test them separately in the qRT-PCR assays.
13. After finishing the run, it is a good idea to run the PCR products on agarose gel electrophoresis to examine the amplified product(s). This gel should show a single PCR product without primer dimers.
14. Since GAPDH, 18 S rRNA, tubulin, and β-actin are expressed by all cell types, any of these genes can be used as a housekeeping internal control for normalization. But in many cases there could be a differential regulation of these genes caused by the treatment in your experiments. Therefore, it is always good to use all these genes initially to identify the best housekeeping gene which does not have any differential regulation in both controls and the experimental groups.

References

1. Panguluri, S.K., C. Yeakel, and S.S. Kakar, *PTTG: an important target gene for ovarian cancer therapy.* J Ovarian Res, 2008. **1**(1): p. 6.
2. Hamid, T., M.T. Malik, and S.S. Kakar, *Ectopic expression of PTTG1/securin promotes tumorigenesis in human embryonic kidney cells.* Mol Cancer, 2005. **4**(1): p. 3.
3. McCabe, C.J., et al., *Vascular endothelial growth factor, its receptor KDR/Flk-1, and pituitary tumor transforming gene in pituitary tumors.* J Clin Endocrinol Metab, 2002. **87**(9): p. 4238–44.
4. Ishikawa, H., et al., *Human pituitary tumor-transforming gene induces angiogenesis.* J Clin Endocrinol Metab, 2001. **86**(2): p. 867–74.
5. Pei, L. and S. Melmed, *Isolation and characterization of a pituitary tumor-transforming gene (PTTG).* Mol Endocrinol, 1997. **11**(4): p. 433–41.
6. Malik, M.T. and S.S. Kakar, *Regulation of angiogenesis and invasion by human Pituitary tumor transforming gene (PTTG) through increased expression and secretion of matrix

metalloproteinase-2 (MMP-2). Mol Cancer, 2006. **5**: p. 61.

7. Panguluri, S.K. and S.S. Kakar, *Effect of PTTG on endogenous gene expression in HEK 293 cells.* BMC Genomics, 2009. **10**: p. 577.
8. Rajeevan, M.S., et al., *Validation of array-based gene expression profiles by real-time (kinetic) RT-PCR*. J Mol Diagn, 2001. **3**(1): p. 26–31.
9. Sambrook, J. and Russell, D.W. *Molecular Cloning: A Laboratory Manual.* Cold Spring Harbor Laboratory Press, Cold Spring Harbor, New York, 2001.
10. Panguluri, S.K., et al., *Genomic profiling of messenger RNAs and microRNAs reveals potential mechanisms of TWEAK-induced skeletal muscle wasting in mice.* PLoS ONE, 2010. **5**(1): p. e8760.

Chapter 12

FRET-Based Real-Time DNA Microarrays

Arjang Hassibi, Haris Vikalo, José Luis Riechmann, and Babak Hassibi

Abstract

We present a quantification method for affinity-based DNA microarrays which is based on the real-time measurements of hybridization kinetics. This method, i.e., real-time DNA microarrays, enhances the detection dynamic range of conventional systems by being impervious to probe saturation, washing artifacts, microarray spot-to-spot variations, and other intensity-affecting impediments. We demonstrate in both theory and practice that the time-constant of target capturing is inversely proportional to the concentration of the target analyte, which we take advantage of as the fundamental parameter to estimate the concentration of the analytes. Furthermore, to experimentally validate the capabilities of this method in practical applications, we present a FRET-based assay which enables the real-time detection in gene expression DNA microarrays.

Key words: Microarray, Real-time, Gene expression, Time-constant, FRET

1. Introduction

Massively parallel affinity-based detection of nucleic acid fragments, i.e., the DNA microarray technology, has become a key assaying technique in molecular biology (1, 2). Although relatively new, DNA microarrays have enabled a variety of important high-throughput applications, for example, genome-wide quantitative analysis of gene expression and large-scale single nucleotide polymorphism (SNP) discovery and genotyping (3–5).

In DNA microarrays, the minimum detection level (MDL) is generally limited by the nonspecific capturing events and the inherent uncertainty of the analyte–probe interactions. However, the highest detection level (HDL) is a function of the capturing probe concentration in individual spots and its associated saturation level. Today, the achievable MDL and HDL of microarray systems do not

Michael Kaufmann and Claudia Klinger (eds.), *Functional Genomics: Methods and Protocols*,
Methods in Molecular Biology, vol. 815, DOI 10.1007/978-1-61779-424-7_12, © Springer Science+Business Media, LLC 2012

satisfy the stringent requirements of many biotechnology applications; microarrays are considered semiquantitative platforms and are best suited for applications where parallelism is the most imperative criteria (6–8). Accordingly, improving both the MDL and HDL of microarrays is extremely critical. The motivation is to not only enhance the quality of the data in the existing high-throughput applications (e.g., gene expression profiling), but also facilitate the adoption of microarrays in emerging high-performance applications such as in vitro diagnostics and forensics.

1.1. Binding Kinetics

The capturing process (for both specific and non-specific analytes) is a dynamic process that occurs over time. What we detect in a microarray is the total number of captured analytes at each spot, denoted by $n_c(t)$. When we first introduce the sample to array, we have $n_c(0)=0$, but this value monotonically increases until it reaches the biochemical steady-state, i.e., where the capturing and release processes have equal rate, making $n_c(\infty)$ constant.

Molecular binding, like any other biochemical process, is a stochastic process, making $n_c(t)$ a random variable (6). Yet, $\langle n_c(t)\rangle$, the expected value (ensemble average) of $n_c(t)$, can be approximated by the well-known rate equation in the form of

$$\frac{\mathrm{d}\langle n_c(t)\rangle}{\mathrm{d}t}=k_1(n_t-\langle n_c(t)\rangle)(n_p-\langle n_c(t)\rangle)-k_{-1}\langle n_c(t)\rangle, \qquad (1)$$

where k_1 and k_{-1} are the association and dissociation rate constants of the capturing (hybridization for DNA molecules), respectively, n_p is the total number of DNA capturing probe molecules immobilized on the surface, and n_t is the number of existing analyte molecules in the sample.

For solid-phase reactions (e.g., hybridization in DNA microarrays), the rate equation is different than it is the case for homogeneous reactions. It can be shown that in microarrays we have

$$\frac{\mathrm{d}\langle n_c(t)\rangle}{\mathrm{d}t}=k_1^*\left(\frac{n_p-\langle n_c(t)\rangle}{n_p}\right)(n_t-\langle n_c(t)\rangle)-k_{-1}\langle n_c(t)\rangle, \qquad (2)$$

where k_1^* is the association rate constant when there is unlimited abundance of capturing probes and the term $(n_p-\langle n_c(t)\rangle)/n_t$ represents the availability of the probes, i.e., the probability of finding an unoccupied probe. Assuming that there is negligible analyte depletion in the system due to hybridization (i.e., n_t remains relatively constant during experiments), eq. 2 can be rewritten as

$$\frac{\mathrm{d}\langle n_c(t)\rangle}{\mathrm{d}t}=k_1^*\left(\frac{n_p-\langle n_c(t)\rangle}{n_p}\right)n_t-k_{-1}\langle n_c(t)\rangle. \qquad (3)$$

The solution for eq. 3 with the assumption of $\langle n_c(0)\rangle=0$ is

$$\langle n_c(t)\rangle = \frac{k_1^* n_t n_p}{k_1^* n_t + k_{-1} n_p}\left(1 - e^{-(\frac{k_1 n_t}{n_p} + k_{-1})t}\right), \tag{4}$$

which is essentially the approximation for the capturing kinetics.

1.2. Capturing Time-Constant

Microarray protocols generally allocate a fixed (and consistent) amount of time for the incubation step (e.g., 5–24 h for the hybridization step for gene expression DNA microarrays). At the end of this step with a duration of t_0, the solution containing the sample is carefully removed (washing step), and the intensity of the fluorescent signal is measured, which is an indication of the amount of captured analytes at different capturing stops. According to eq. 4, this procedure creates a nonlinear relationship between n_t and $\langle n_c(t_0)\rangle$ which is an exponential function of t_0. Although this is the approach typically used in many systems, there are two fundamental challenges associated with it. The first is that the relationship between the measured $n_c(t)$ and the desired n_t is nonlinear.

The alternative approach that we discuss here in this chapter is to estimate n_t, not based on a single measurement of eqs. 4 and 5 but, by looking at the kinetics of $n_c(t)$. Using this full trajectory, one may estimate, τ_c, the time-constant (or, its inverse, the rate constant) of the capturing for a specific analyte at each capturing spot. Using eq. 4, it is easy to see that

$$\tau_c = \frac{1}{\frac{k_1^* n_t}{n_p} + k_{-1}} = \frac{n_p}{k_1^* n_t + k_{-1} n_p}, \tag{5}$$

and for high-affinity probe–analyte moieties, we have

$$\tau_c \approx \frac{n_p}{k_1^* n_t}, \tag{6}$$

which shows that the time constant of capturing is proportional to the number of probe molecules and inversely proportional to the analyte concentration.

Now if we want to quantify the concentrations of the analytes, we should noninvasively measure the capturing kinetics and evaluate eq. 4 from the change in the signal, rather than stop the reaction to measure the signal from the captured analytes via end point estimation, as done in conventional DNA microarray platforms. In other words, this is a paradigm shift in terms of detection in microarrays. In this chapter, we discuss this in further detail (9).

2. Materials

2.1. Reagents and Consumables

1. Modified DNA oligonucleotides synthesized by TriLink BioTechnologies, USA (sequences shown in Table 1).
2. Printing buffer (100 mM Na-phosphate pH 8.5; 0.005% w/v SDS).
3. ArrayControl RNA Spikes (Ambion Inc., USA).
4. MessageAmpII aRNA Kit (Ambion, USA).
5. RNA Fragmentation Reagent Kit (Ambion, USA).
6. QSY9 carboxylic acid, succinimidyl ester quencher (Molecular Probes, USA).
7. Sephadex Spin-50 Mini Columns (USA Scientific, USA).
8. DNA printing buffer (100 mM Na-phosphate pH 8.5; 0.005% w/v SDS).
9. CodeLink activated slides (GE Healthcare, USA).
10. SlideHyb Glass Array Hybridization Buffer #1 (Ambion, USA).
11. Mouse total RNA.

2.2. Instrument

1. 24-Well hybridization cassette (TeleChem International, Inc., USA).
2. 3A peltier thermoelectric heating/cooling modules (Velleman, The Netherlands).
3. 5C7-195 benchtop temperature controller (McShane Inc., USA).
4. MicroGrid II microarrayer (Biorobotics/Genomics Solutions, USA).
5. Zeiss LSM Pascal Inverted Laser Scanning Microscope (Zeiss, Germany).

Table 1
Oligonucleotide sequences

Oligonucleotide name	Sequence (5′–3′)
Probe A	[Cy3]-TACTTTCTCAGTACCATTAG-GGCAA-[Amin]
Probe B	[Cy3]-CCCGGTTTCCCGGGTAAACACCACC-[Amin]
Control Probe	[Cy3]-GTTGCCAAGTGCAGCAGGCGAAAGT-[Amin]
Target A	ACTTTCGCCTGCTGCACTTGGCAAC-[BHQ2]

2.3. Software

1. Matlab (Mathworks, USA).

3. Methods

3.1. Overview of the FRET-Based Technique

To enable real-time detection in microarrays with little background interference, we need to ensure that only the captured analytes in intimate proximity of the capturing probes contribute to the measured signal. Since short distances in a molecular scale is critical, here we have used fluoresce resonance energy transfer (FRET) moieties to create binding-specific signals (10–12). In this approach, in each capturing spot we attach radiating donor molecules (e.g., fluorescent molecules) to the capturing probes (method "A" in Fig. 1) or to a "dummy" probe near the capturing probes (method "B" in Fig. 1). This can be done prior to array spotting and during

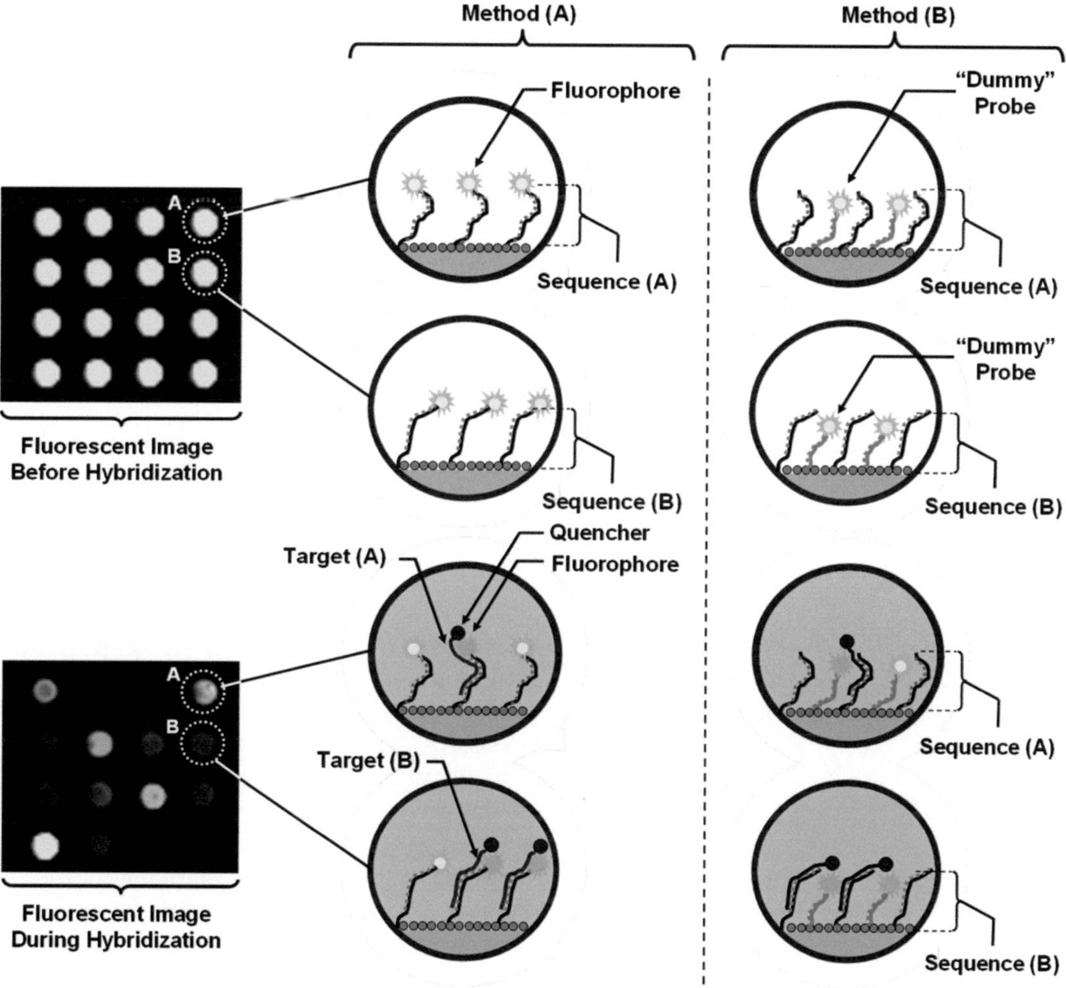

Fig. 1. Two FRET-based real-time DNA microarray assaying alternative methods where the analyte and probe layer comprise of the quencher (acceptor) and fluorophores (donor), respectively. In both Method (A) and Method (B), the bindings of the analytes quench the fluorescent signal of the capturing spot and hence can be used in the measurements as a quantitative signal indicating hybridization.

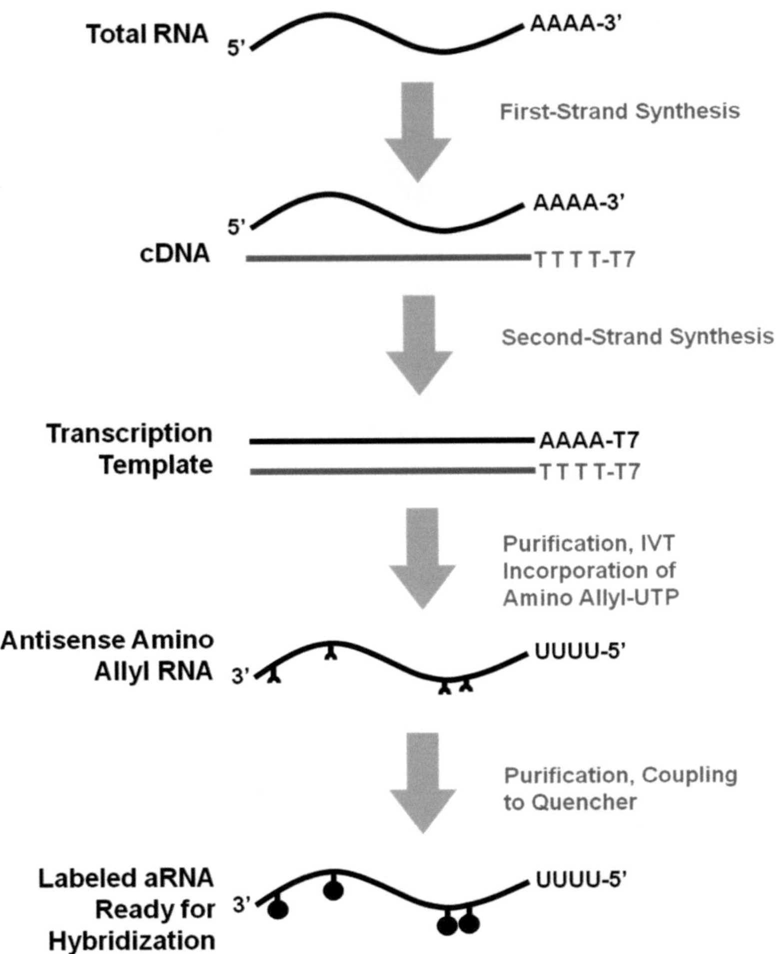

Fig. 2. mRNA quencher labeling protocol.

the synthesis of the probes. For instance, in the case of DNA microarrays, as shown in Fig. 3, the DNA oligonucleotides that act as the capturing probes are end-labeled with Cyanine (Cy) fluorophores. Subsequently, in the sample preparation process, we attach the acceptor molecules of the FRET system to the analytes. Now when the sample containing the analytes is applied to the array, which consists of capturing spots with donors, hybridization events bring the donor and acceptor into intimate proximity resulting in a molecular FRET system. In this particular implementation, we use nonradiating acceptors (i.e., quenchers), such that hybridization "turns off" the fluorophore of the capturing probe or the "dummy" probe, and hence reduces the overall emitted fluorescent signal of the spot as. From an imaging point of view, this method requires identical instrumentation compared to other fluorescence-based assays, while the hybridization solution containing the sample

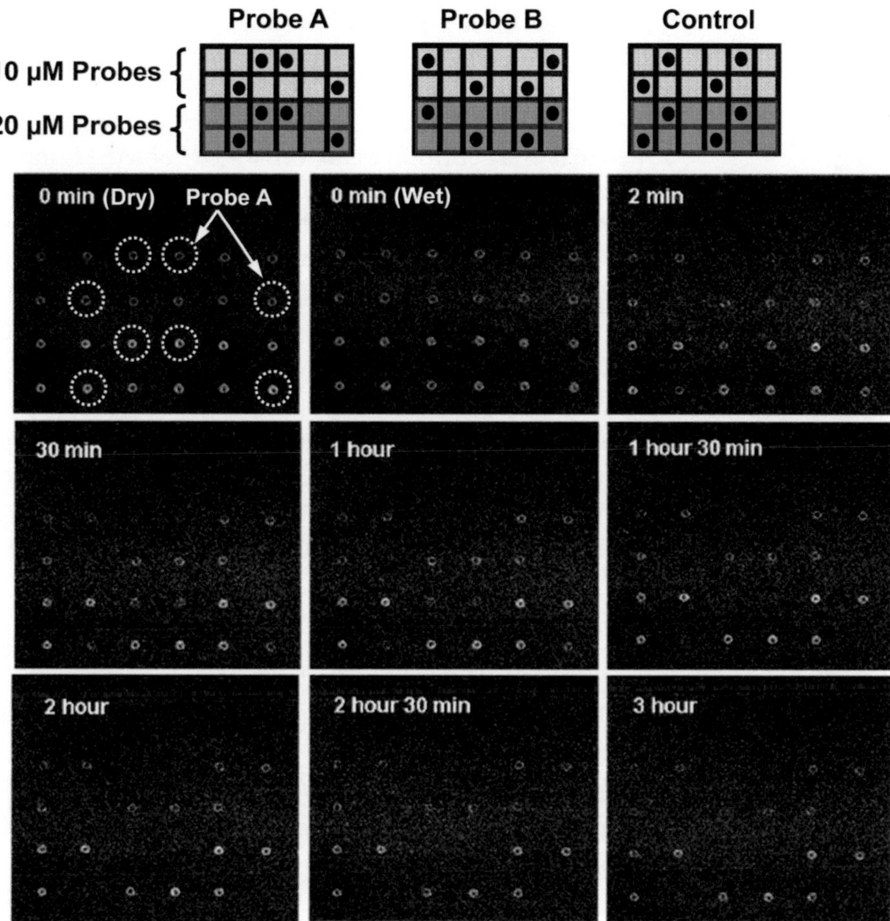

Fig. 3. The experimental results of this real-time method where the donor and acceptor moieties in the FRET assay are Cy3 and BHQ2, respectively, with Förster distance of approximately 6 nm. The sequences of the three different sequences of printed oligonucleotides are listed in Table 12.1 and they all are printed in four replicates from a solution at 10 and 20 μM concentration. The concentration of Target A was 20 ng/100 μl and the hybridization temperature was 44°C. The selected fluorescent images (every 30 min) are shown during the first 3 h of the hybridization step.

introduces little fluorescence background. In addition, parallel measurements can be carried out and the method is scalable to large size arrays. It is also important to recognize that the low background signal in this method enables the effective detection of the capturing events with a high signal-to-noise ratio (SNR) during the DNA hybridization phase.

3.2. Target Labeling

Real-time microarray experiments were performed using either target DNA oligonucleotides or target in vitro transcribed RNAs that were labeled with quencher residues.

1. Target oligonucleotides were 3′ modified during synthesis with QSY9.
2. To prepare in vitro transcribed, QSY9-labeled target RNA, the Amino Allyl MessageAmpII aRNA Kit and QSY9 carboxylic

acid, succinimidyl ester were used, as shown in Fig. 2. Manufacturers' protocols were used, with the following modifications: 50 ng of each spike RNA was used per cDNA synthesis reaction; the amount of amino allyl UTP used per in vitro transcription reaction was doubled; 5 mg of QSY9 were dissolved in 220 μl of DMSO, and 11 μl of the dissolved succinimidyl ester were used per labeling reaction.

3. In vitro transcribed RNA was cleaved using Ambion's Fragmentation Reagent (following the manufacturer's instructions) and purified using Sephadex Spin-50 Mini Columns. Approximately, one QSY9 residue was incorporated for every 20 nucleotides of the target IVT RNA.

3.3. Microarray Manufacturing

1. Probes for the real-time microarrays were designed against the ArrayControl RNA Spikes. These RNA Spikes are a collection of eight individual RNA transcripts (Spikes 1 through 8) that range in size from 750 to 2,000 bases, and each transcript has a 30-base 3′ poly(A) tail.
2. Probes were custom synthetic DNA oligonucleotides modified during synthesis with a Cy3 fluorophore at the 5′ end and an amine residue at the 3′ end. The control probes were designed such that they would not specifically hybridize to any of the targets (RNA Spikes) used.
3. Dilutions of the labeled DNA oligonucleotides were prepared in printing buffer at the appropriate concentration (usually, 0.8, 4, and 20 μM DNA).
4. Dilutions were dispensed in 384-well plates (15 μl of dilution per well), and DNA was spotted onto slides using a microarrayer (see Note 1).
5. After printing, the slides were processed (DNA coupling and slide blocking) following the manufacturer's protocols with minor modifications: DTT was added to the wash buffer (1 mM DTT, final concentration), and slide exposure to light was minimized to protect the fluorophores.
6. The printed microarrays were packed individually under vacuum with a flush of nitrogen gas, and stored at room temperature until use.

3.4. Microarray Hybridization and Data Acquisition

1. Labeled target oligonucleotides or IVT RNAs were diluted at the indicated concentration in 50 μl of hybridization buffer.
2. To initiate the hybridization the microarray was first put in contact with 50 μl of hybridization buffer (without labeled target), and then (at $t = 0$) the labeled target(s) were added in a volume of 50 μl of hybridization buffer (i.e., the final hybridization volume was 100 μl). The labeled targets were preheated for 5 min at 70–80°C.
3. Hybridization temperature was controlled (with an accuracy of 1°C) using the temperature controller. The Peltier thermoelectric

heating and cooling modules were placed on top of the hybridization cassette and the temperature sensor of the temperature controller was placed in the well adjacent to the sample well and in contact with the cassette.

4. The fluorescence imaging was done using a Zeiss LSM Pascal Inverted Laser Scanning Microscope from below the microarray slide which was mounted in the hybridization cassette. The time-series images (see Fig. 3) were analyzed using our own software, developed in Matlab. The software initially uses the $t = 0$ image to find the coordinate and area of individual capturing spots (see Note 2).

5. The signal degradation was measured at each capturing spot using the time-series images. The signal of the control spot of each image is used to compensate for possible fluorophore bleaching (see Notes 3 and 4).

6. Signal changes as a function of time were analyzed to find the dynamics of capturing, as shown in Fig. 4. This data were then used to evaluate the analyte concentrations (see Tables 2 and 3 example data sets).

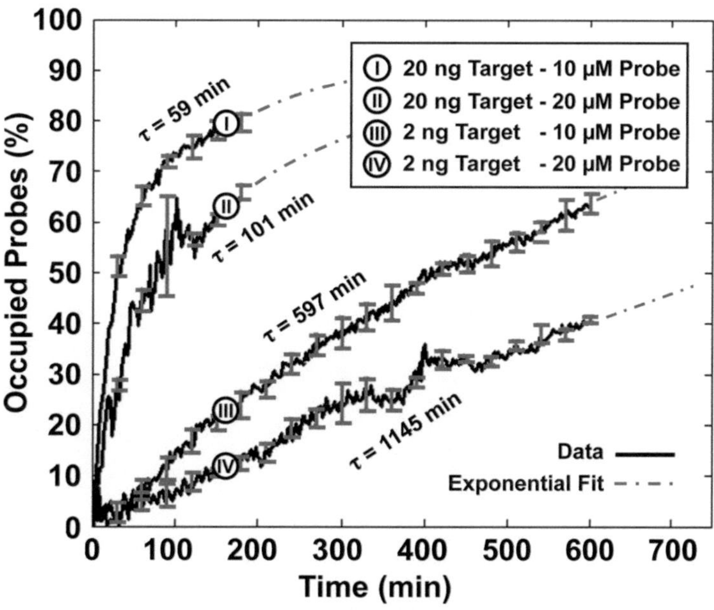

Fig. 4. The real-time capturing data acquired from four microarray experiments, for different concentrations of Probe A and Target A. Each curve is generated using the results of eight independent spots on the array. According to eq. 6, τ should be proportional to n_p and inversely proportional to n_t. As evident, this is indeed the case, i.e., the computed τ changes due to target and probe concentrations behave as predicted.

Table 2
Real-time microarray experiment results for different concentrations of 720 bp mRNA target and oligonucleotide probes with no biological background

n_t(ng)	n_p(μM)	τ_A(min)	τ_B(min)	$\frac{\tau_A}{\tau_B}$	$\frac{n_p}{n_t}$	$\left(\frac{n_p}{n_t}\right)_n$	$(\tau_A)_n$	$(\tau_B)_n$	$\log_2 \frac{(\tau_A)_n}{\left(\frac{n_p}{n_t}\right)_n}$	$\log_2 \frac{(\tau_B)_n}{\left(\frac{n_p}{n_t}\right)_n}$
400	20	120	188.6	0.636	0.05	0.08	0.097	0.082	0.285	0.033
80	20	442.5	979.4	0.451	0.25	0.4	0.354	0.425	-0.154	0.088
16	20	2058.8	4040.9	0.509	1.25	2	1.672	1.754	-0.258	-0.190
400	10	68.8	161.9	0.425	0.025	0.04	0.056	0.070	0.482	0.8130
80	10	214.7	472.4	0.454	0.125	0.2	0.174	0.205	-0.197	0.036
16*	10	1230.8	2308.8	0.534	0.625	1	1	1	0	0
3.2	10	7073.5	11149.9	0.634	3.125	5	5.747	4.840	0.201	-0.047
0.64	10	15940.5	29373.4	0.543	15.625	25	12.951	12.750	-0.949	-0.971
400	2	51.5	158.3	0.325	0.005	0.008	0.0418	0.0687	2.387	3.102
80	2	272.3	631.1	0.431	0.025	0.04	0.221	0.274	2.468	2.776
16	2	1603.7	3815.1	0.420	0.125	0.2	1.303	1.656	2.704	3.050
3.2†	2	13686	71913.9	0.190	0.625	1	11.120	31.215	3.475	4.964
0.64†	2	15790.2	76395.2	0.207	3.125	5	12.829	33.16	1.360	2.729

τ_A and τ_B are the measured hybridization time-constants for Method (A) and Method (B) as shown in Fig. 1 at a 50:50 ratio of dummy to capturing probe of the FRET-based assay, respectively. The (*) indicates the reference capturing experiment where the time-constant of other spots are compared to. The index (*n*) shows the normalized data. The (†) indicate the experiments where the light intensity SNR was unacceptable (i.e., background variation was higher than the signal value). The $\log_2$ function is to create a figure-of-merit for the quality of quantification of the data

Table 3
Real-time microarray experiment results for different concentrations of 720 bp mRNA target and oligonucleotide probes with 7.5 μg/50 μl aRNA prepared from total mouse RNA as the complex biological background

n_t (ng)	n_p (μM)	τ_A (min)	τ_B (min)	$\frac{\tau_A}{\tau_B}$	$\frac{n_p}{n_t}$	$\left(\frac{n_p}{n_t}\right)_n$	$(\tau_A)_n$	$(\tau_B)_n$	$\log_2 \frac{(\tau_A)_n}{\left(\frac{n_p}{n_t}\right)_n}$	$\log_2 \frac{(\tau_B)_n}{\left(\frac{n_p}{n_t}\right)_n}$
400	20	251	562.3	0.446	0.05	0.08	0.159	0.225	0.987	1.4935
80	20	703.4	1420.1	0.495	0.25	0.4	0.444	0.569	0.152	0.508
16	20	2726.3	4519.4	0.603	1.25	2	1.722	1.810	-0.215	-0.143
3.2	20	9253.3	37270.7	0.248	6.25	10	5.847	14.931	-0.774	0.578
400	10	119.9	228.1	0.526	0.025	0.04	0.076	0.0913	0.921	1.192
80	10	394.1	944	0.417	0.125	0.2	0.249	0.378	0.316	0.919
16*	10	1582.5	2496.1	0.633	0.625	1	1	1	0	0
3.2	10	5359.8	11052.4	0.485	3.125	5	3.387	4.427	-0.561	-0.17

τ_A and τ_B are the measured hybridization time-constants for Method (A) and Method (B) as shown in Fig. .1 at a 50:50 ratio of dummy to capturing probe of the FRET-based assay, respectively. The (*) indicates the reference capturing experiment where the time-constant of other spots are compared to. The index (n) shows the normalized data. The $\log_2$ function is to create a figure-of-merit for the quality of quantification of the data. The low SNR measurements are not reported in this table in contrast to Table 2

4. Notes

1. Dye-modified oligonucleotides change the contact angle of the printing buffer and can result in doughnut-shaped printed spots. To mitigate this, we added detergents (e.g., 0.005% w/v SDS) to the printing buffer.
2. Self-quenching of fluorophores was observed in the capturing spot when the printing concentrations of the dye-modified oligonucleotides were higher than 5 μM. This affects the amplitude of the signal, but the time-constant remains untouched.
3. During the initial 5–10 min of hybridization after the sample is introduced to the array, the signal was unpredictable and somewhat noisy. At this point, we have no conclusive explanation for this phenomenon. Regardless, this does not affect the time-constant estimation, since we do not include the images of this initial phase to the calculations and rely on the rest.
4. We were able to heat the array for 5 min to 90°C resulting in a release of the targets and repeated the experiment. Overall the signal amplitude was degraded; however, the time-constants remained the same.

Acknowledgments

We are grateful to Vijaya Kumar for experimental assistance with microarray manufacture and target labeling. We also want to thank Professor Scott Fraser at Caltech for technical feedback in the imaging aspects of this project.

References

1. Monitoring of Gene Expression Patterns with a Complementary DNA Microarray, *Science*, **270**, 467–470.
2. Chee, M., Yang, R., Hubbell, E., Berno, A., Huang, X.C., Stern, D., Winkler, J., Lockhart, D.J., Morris, M.S. and Fodor, S.P. (1996) Accessing Genetic Information with High-Density DNA Arrays, *Science*, **274**, 610–614.
3. Lockhart, D.J. and Winzeler, E.A., (2000) Genomics, Gene Expression and DNA Arrays, *Nature*, **405**, 827–836.
4. Wang, D.G., Fan, J.–B., Siao, C.–J., Berno, A., Young, P., Sapolsky, R., Ghandour, G., Perkins, N., Winchester, E., Spencer, J., et al. (1998) Large-Scale Identification, Mapping, and Genotyping of Single-Nucleotide Polymorphisms in the Human Genome, *Science*, **280**, 1077–1082.
5. Hardenbol, P., Baner, J., Jain, M., Nilsson, M., Namsaraev, E.A., Karlin-Neumann, G.A., Fakhrai-Rad, H., Ronaghi, M., Willis, T.D., Landegren, U. *et. al.*, (2003) Multiplexed Genotyping with Sequence-Tagged Molecular Inversion Probes, *Nat. Biotechnol.*, **21**, 673–678.
6. Hassibi, A., Zahedi, S., Navid, R., Dutton, R.W. and Lee, T.H., (2005) Biological Shot-Noise and Quantum-Limited Signal-to-Noise Ratio in Affinity-Based Biosensors, *J. Appl. Phys.*, **97**–084701:1–10.

7. Hassibi, A., Vikalo, H. and Hajimiri, A., (2007) On Noise Processes and Limits of Performance in Biosensors, *J. Appl. Phys.*, **102**–014909:1–12.
8. Tu, Y., Stolovitzky, G. and Klein, U., (2002) Quantitative Noise Analysis for Gene Expression Microarray Experiments, *Proc. Natl. Acad. Sci.*, **99**–22:14031–14036.
9. Hassibi, A., Vikalo, H., Riechmann, J.L. and Hassibi, B. (2009) Real-time DNA Microarray Analysis, *Nucleic Acids Research*, doi:10.1093/nar/gkp **675**, 1–12.
10. Okamura, Y., Kondo, S., Sase, I., Suga, T., Mise, K., Furusawa, I., Kawakami, S. and Watanabe, Y. (2000) Double-Labeled Donor Probe can Enhance the Signal of Fluorescence Resonance Energy Transfer (FRET) in Detection of Nucleic Acid Hybridization, *Nucleic Acids Research*, **28**, e107.
11. Rajendran, M. and Ellington, A.D. (2003) In vitro Selection of Molecular Beacons, *Nucleic Acids Research*, **31**–19, 5700–5713.
12. Marras, S.A.E., Tyagi, S. and Kramer, F.R. (2006) Real-time Assays with Molecular Beacons and other Fluorescent Nucleic Acid Hybridization Orobes, *Clin Chim Acta*, **363**, 48–60.

Part IV

Protein Analysis I: Quantification and Identification

Chapter 13

2-D Gel Electrophoresis: Constructing 2D-Gel Proteome Reference Maps

Maria Paola Simula, Agata Notarpietro, Giuseppe Toffoli, and Valli De Re

Abstract

Two-dimensional gel electrophoresis (2-DE) is the most popular and versatile method of protein separation among a rapidly growing array of proteomic technologies. Based on two independent biochemical characteristics of proteins, it combines isoelectric focusing, which separates proteins according to their isoelectric point (pI), and SDS-PAGE, which separates them further according to their molecular mass. An evolution of conventional 2-DE is represented by the 2D-Difference in Gel Electrophoresis (2D-DIGE) that allows sample multiplexing and achieving more accurate and sensitive quantitative proteomic determinations. The 2-DE separation permits the generation of protein maps of different cells or tissues and the study, by differential proteomics, of protein expression changes associated to the different states of a biological system. In order to identify the molecular bases of pathological processes, it is also useful to characterize the physiological protein homeostasis in healthy cells or tissues. On these grounds, the availability of detailed 2D reference maps could be very useful for proteomic studies. The protocol described in this chapter is based on the 2D-DIGE technology and has been applied to obtain the first 2-DE reference map of the human small intestine.

Key words: Reference map, 2D-DIGE, 2D-electrophoresis, MALDI-TOF, Proteome

1. Introduction

In the post-genome era, the major efforts of the scientific community are focused on proteome definition, which describes the complete set of proteins expressed in the lifetime of cells and tissues, and on shedding light on the complex relationship between known genome sequences, cell function, and organization.

In proteome analysis, two-dimensional gel electrophoresis (2-DE), introduced by O'Farrell and Klose in 1975 (1, 2), is the milestone of proteomic tools. Thanks to its high resolving power

Michael Kaufmann and Claudia Klinger (eds.), *Functional Genomics: Methods and Protocols*,
Methods in Molecular Biology, vol. 815, DOI 10.1007/978-1-61779-424-7_13, © Springer Science+Business Media, LLC 2012

and its large sample loading capacity, it allows several hundred proteins to be displayed simultaneously on a single gel, producing a direct and global view of a sample proteome at a given time.

The 2-DE separation permits the generation of protein maps of different cells or tissues and the study, by differential proteomics, of protein expression changes associated to the different states of a biological system.

To identify the molecular bases of pathological processes, it is also useful to characterize the physiological protein homeostasis in healthy cells or tissues. The availability of detailed reference maps could be very useful for comparative proteomic studies, for possible biomarkers identification, for the study of proteomic modulation associated with disease progression, and for new developments in the field of pharmacological treatments.

Specific 2-DE maps, generated for several tissues and cell lines of human and other species, are accessible via public networks (e.g., http://www.expasy.ch, http://www.ludwig.edu.au, http://www.gelbank.anl.gov). In these maps, various proteins, separated by 2-DE, are classified by specific landmarks which provide the connection to protein databases of different tissues or organisms (3).

Moreover, several human protein maps of different biological fluids and tissues have been published (4–14) in an effort to study proteins and produce more global information regarding normal protein expression and alterations in several diseases.

Two-DE can be applied to almost any type of protein-containing sample, including eukaryotic tissue and derived extracts, cells and organelles, biological fluids, prokaryotic organisms, seeds, vegetal matter, and plants. However, adequate sample preparation, tailored to the specific aim of the proposed study, is of utmost importance (15). In general, samples contain hundreds, if not thousands, of different polypeptides spanning several orders of magnitude in concentration, with a few well-known abundant proteins overshadowing many more relevant, but lower expressed ones. Enrichment of such low-abundance proteins is often a necessary pre-analytical step that should be considered seriously, in particular when analyzing complex body fluids such as plasma or urine (13, 16, 17). To achieve the desired level of sample purification, various pre-fractionation methods have been developed, including specific cell or microorganism cultures, laser capture microdissection (18), fluorescence-activated cell sorting of antibody-bound cells, differential centrifugation of organelles, and reversed-phase high-performance liquid chromatography or affinity chromatography (19), among others (20). In general, sample pre-fractionation markedly increases the number of detected proteins as compared with crude extracts, provided that the total amount of proteins loaded onto the gel is adequate.

The apparently unlimited diversity of 2-DE-related pre-analytical and analytical methods, the combination of which differs from one laboratory to another, makes the establishment of an ultimate

2-DE method a difficult task. Moreover, 2-DE has some limitations that must be taken into account. Gel to gel variation due to differences in electrophoretic conditions, different first dimension strips and second dimension gels, gel distortions and user to user variations, represent important limits for the use of a generalized protein map. The normalization of the spot amount is also questionable as the individual expression profile could be very changeable and, in addition, extractive and technical procedures can introduce further quantitative variability. This problem can be largely circumvented by using 2D-Difference in Gel Electrophoresis (2D-DIGE) that allows more accurate and sensitive quantitative proteomic studies (21). The 2D-DIGE technology relies on direct labelling of the lysine groups on proteins with cyanine (Cy) dyes before isoelectric focusing (IEF). As a consequence, it enables the analysis of up to three different samples (labelled with the three different cyanine dyes Cy2, Cy3, and Cy5) in the same gel. The introduction of an internal standard in every gel greatly decreases the system variation and hinders the need of technical replicates. The internal standard, which is a pool of all the samples within the experiment, and therefore contains every protein from every sample, better increases the certainty of data. It is used to match the protein patterns across gels thereby excluding the problem of inter-gel variation, a common problem with standard 2-D assays. Moreover, it allows the most reliable quantitation of any 2-DE method. Normally, the Cy2 dye is assigned to internal standard labelling.

The protocol described below has been applied to obtain the first 2-DE reference map of the human small intestine (10). The sample consisted of small tissue intestine biopsies. Total protein extracts have been labelled with cyanine dyes (2D-DIGE approach) and then separated on 11 cm IPG Strips pH 3–10 NL (first dimension), followed by a second dimension separation on criterion 8–16% precast gels.

The rationale of this protocol is summarized in Fig. 1.

2. Materials

2.1. Equipment

1. IPG Dry Strip reswelling and focusing trays.
2. Protean IEF Cell (Bio-Rad).
3. Criterion 8–16% precast gels (Bio-Rad).
4. Criterion cell (Bio-Rad).
5. Microwave oven, laboratory vortex, and stirrer.
6. Typhoon Trio scanner and DeCyder software (GE HealthCare).
7. ZipTips (Millipore).
8. Maldi-Tof Voyager De-Pro mass spectrometer with Data explorer version 5.1 (Applied Biosystems).

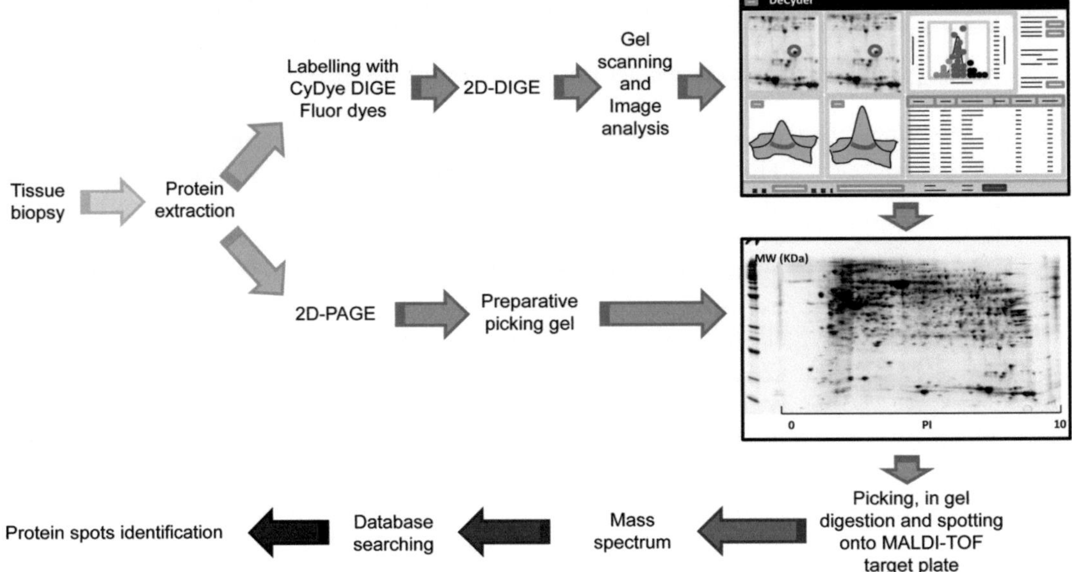

Fig. 1. The creation of reference maps of different cells or tissues is possible with two-dimensional gel electrophoresis. The proteins extracted from tissue biopsies are labelled with CyDye DIGE Fluor minimal dyes and run on a 2D-DIGE experiment; the analytical gels are scanned using a Typhoon Trio scanner and the images obtained are analyzed with the DeCyder software to obtain mean normal volumes for each spot. In parallel, a pool comprising equal amounts of all the samples analyzed is used to create a preparative picking gel that will be stained with Coomassie G-250. When constructing reference maps, all visible spots, in the preparative picking gel, should be picked, in gel trypsin digested and spotted onto MALDI-TOF target plate. MALDI mass spectra are then acquired and processed and the peak lists obtained are used to perform database searching for the identification of protein spots.

2.2. Reagents

1. Sample grinding kit and 2D clean-up kit (GE HealthCare).
2. Sample lysis buffer: 30 mM Tris–HCl, 2 M thiourea, 7 M urea, and 4% CHAPS, pH 8.5.
3. Tris–HCl 1 M pH 9.
4. Bradford protein assay kit.
5. CyDye DIGE Fluor minimal dyes 2, 3, and 5 (GE HealthCare).
6. 99.8% pure dimethylformamide.
7. 10 mM lysine.
8. Rehydration buffer: 2 M thiourea, 7 M urea, 40 mM dithiothreitol (DTT), 4% CHAPS, 0.5% IPG buffer pH range 3–10 and traces of bromophenol blue (BBF).
9. 11 cm IPG Strips pH 3–10 NL.
10. Mineral oil.
11. Reducing buffer: 4 M urea, 2 M thiourea, 50 mM Tris–HCl, pH 8.8, 30% (vol/vol) glycerol, 2% (wt/vol) SDS, and 1% (wt/vol) DTT.

12. Alkylating buffer: 4 M urea, 2 M thiourea, 50 mM Tris–HCl, pH 8.8, 30% (vol/vol) glycerol, 2% (wt/vol) SDS, and 2.5% (wt/vol) iodoacetamide.
13. Agarose 1% (wt/vol) in SDS-PAGE running buffer with traces of BBF.
14. SDS-PAGE running buffer: 0.025 M Trizma base, 0.192 M glycine, 0.1% (wt/vol) SDS.
15. Fixing solution: 50% (vol/vol) ethanol and 2% (vol/vol) orthophosphoric acid.
16. Staining solution: 34% (vol/vol) methanol, 2% (vol/vol) orthophosphoric acid, 17% (wt/vol) ammonium sulphate, and 0.065% (wt/vol) Coomassie G-250.
17. Spot destaining solution: 25 mM ammonium bicarbonate in 50% acetonitrile.
18. 100% acetonitrile.
19. Sequencing grade lyophilized trypsin.
20. 10% (vol/vol) trifluoroacetic acid (TFA).
21. MALDI-TOF MS matrix solution (10 g/l α-cyano-4-hydroxy-cinnamic acid in 50% (vol/vol) acetonitrile/0.3% (vol/vol) TFA).

3. Methods

1. Extract proteins from tissue biopsies using sample grinding kit and then precipitate the extracts using the 2D clean-up kit according to the manufacturer's recommendations. Pellet the sample by centrifugation at 13,000 × *g* for 5 min at 4°C and air-dry the pellet (see Note 1).
2. Re-suspend the pellet in adequate rehydration buffer. The optimal amount of rehydration buffer must be determined experimentally. As an indication, start by adding 40 μl of sample buffer and then add additional 10 μl volumes until the pellet dissolves (see Note 2).
3. Check the pH of all the samples by adding a very small amount of sample (3 μl) to a pH indicator strip. Make sure that the sample pH is between 8.0 and 9.0. You can increase the pH by carefully adding 50 mM NaOH in rehydration buffer or a final 30 mM Tris–HCl.
4. Determine the protein concentration using the Bradford protein quantitation kit following the manufacturer's instructions. Make sure that the protein concentration is around 5 μg/μl by adequately concentrating or diluting the samples.

5. Prepare a normalization pool (standard pool) comprising equal amounts from all the samples to be analyzed.
6. Reconstitute the CyDye DIGE Fluor minimal dyes in 5 μl of DMF by centrifuging at 12,000 × *g* for 30 s to make a stock of 500 pmol/μl. Create a working solution of 100 pmol/μl of CyDye by adding 9 μl of DMF to 1 μl of stock solution. Store the tubes at −20°C until needed.
7. Label 50 μg of each sample with 2 μl (200 pmol) of working solution of either Cy3 or Cy5 dye in such a way that Cy3 and Cy5 are swapped equally among the samples from the different conditions; in this case, for the construction of a reference map, label an equal number of samples with each dye. At the same time, label an aliquot of 50 μg of internal standard sample with 2 μl of Cy2 for each gel. Mix the sample and dye by vortexing vigorously, centrifuge at 13,000 × *g* for 10 s and keep on ice for 30 min in the dark (see Note 3).
8. Add 1 μl of 10 mM lysine to each sample to quench the labelling reaction. Vortex, centrifuge at 13,000 × *g* for 10 s, and keep on ice for 10 min.
9. At this stage, matched samples labelled with Cy3 or Cy5 along with an aliquot of internal standard (labelled with Cy2) should be pooled together.
10. Create a sample for the "pick gel," by mixing equal amounts of all the samples to make up 400 μg (final sample load depends on the strip length).
11. Add adequate rehydration buffer to make the volume of the samples prepared up to 200 μl (final rehydration volume depends on the strip length). The samples are now ready for IEF (see Note 4).
12. Pipette the sample as a line along the edge of a channel in the IEF rehydration tray. With utmost care, remove the protective cover from the IPG Dry Strip and position it with the gel side down into the IPG tray and, finally, overlay mineral oil over the IPG Strip to prevent evaporation and urea crystallization. Passive strip rehydration must be carried out for at least 12 h.
13. After rehydration, wet two paper wicks (for each strip) with deionized water and position them over the electrodes of the focusing tray. Transfer the strips onto the focusing tray, cover them with mineral oil, and run the appropriate IEF protocol (see Note 5).
14. When the electrophoresis has been completed, remove the strips from the focusing tray and transfer them, gel side up, into a clean tray. Store the focused IPG Strips at −70°C.
15. Thaw the strips and leave them onto the lab bench for no longer than 15 min. Equilibrate the IPG Strips to reduce the

disulphide bonds by gently rocking them in 2–5 ml of reducing buffer/strip for 15 min. Immediately after this, alkylate the –SH groups of proteins by gently rocking the strips in 2–5 ml of alkylating buffer/strip for 15 min. The SDS in the buffers also helps the proteins to acquire a negative charge, which drives their migration under the electrical current (see Note 6).

16. During the incubation, melt the 1% agarose overlay solution in a microwave oven.
17. Fill a 100-ml graduate cylinder with SDS-PAGE running buffer. Remove the IPG Strip from the rehydration tray and dip it briefly into the graduated cylinder containing the running buffer. Lay the strip, with the gel side towards you, onto the gel being careful not to trap any bubble between the gel and the strip.
18. Slowly pipette the agarose solution, making sure that no bubbles are introduced, up to the top of the IPG well of the pre-cast 8–16% gradient gel. Ensure that the agarose has solidified before starting the second dimension run.
19. Mount the gel(s) into the electrophoresis cell following the instructions provided with the apparatus.
20. Fill the reservoirs with SDS-PAGE running buffer and begin the electrophoresis.
21. When the electrophoresis is stopped, the analytical gels must be scanned with a Typhoon Trio scanner, while the preparative picking gel must be stained with Coomassie G-250 (see below).
22. Scan the analytical gels using a Typhoon Trio scanner at 100 μm resolution. The first step is to perform a quick pre-scan at 500–1,000 μm resolution to figure out an optimal photomultiplier tube (PMT) value. Once a PMT value is noted for each channel, which gives the desired pixel intensity, the gels should be scanned at 100 μm resolution (see Note 7).
23. Analyze the images using the DeCyder™ software. The DIA should be used for intra-gel analysis for protein spot detection as well as for normalization of Cy3 and Cy5 gel images with respect to the Cy2 image. After spot detection, the abundance changes are represented by the normalized volume ratio (Cy3:Cy2 and Cy5:Cy2) (see Note 8).
24. Use the BVA for inter-gel analysis. After manually landmarking all the gels, the remaining protein spots can be matched in the automatic mode. Alternatively, both DIA and BVA can be run automatically by using the Batch Processor module of DeCyder software. Spot data (in this case the normal volume) can be exported from the DeCyder software using the XML toolbox. This step allows to import the data on an Excel spreadsheet to

perform the required calculations (mean and CV for each spot).

25. As regards the picking gel it is stained with Coomassie G-250:
    (a) Rinse the gel(s) in deionized water for 10 min.
    (b) Cover the gel(s) with fixing solution and incubate for 45 min with gentle agitation.
    (c) Wash the gel(s) two times for 10 min in deionized water.
    (d) Cover the gel(s) with staining solution and incubate for 45 min with gentle agitation (see Note 9).
    (e) The optimum staining is achieved after 24–48 h.
26. When constructing reference maps, all visible spots, in the preparative picking gel, should be picked manually or with a Spot Handling Workstation. The following steps describe the manual protocol.
27. Transfer the spots to 500 or 1,500 μl tube and wash them with 200–400 μl of destaining solution until they turn transparent.
28. Soak in 200–400 μl of 100% acetonitrile twice for 10 min, the spots will turn opaque white.
29. Discard the acetonitrile and dry spots in a Speed-Vac for 20–30 min.
30. Rehydrate the spots with the trypsin containing solution following the manufacturer's instructions; incubate for 30–60 min in ice.
31. Incubate overnight at 37°C.
32. Soak the gel slice in 25–50 μl of 1% (vol/vol) TFA for 30–60 min with gentle agitation.
33. Transfer the supernatant to a second clean tube.
34. Extract the gel again with 25–50 μl of 1% (vol/vol) TFA for 30–60 min with gentle agitation.
35. Combine the two extracts and Speed-Vac to complete dryness; heat to no more than 30°C (see Note 10).
36. Reconstitute the dried sample(s) with 0.1% (vol/vol) TFA; desalt with ZipTips following manufacturer's instructions.
37. Following ZipTip clean-up, directly elute peptides with 2 μl of freshly prepared matrix solution onto the MALDI-TOF target plate.
38. Peptide mass fingerprinting (PMF) can be performed on a Voyager-DE PRO Biospectrometry Workstation mass spectrometer (Applied Biosystems). MALDI mass spectra are acquired in 700–4,000 Da molecular weight range, in reflector and in positive ion mode, with 150 ns delay time and an ion

acceleration voltage of 20 kV. Spectra are externally calibrated using Peptide calibration Mix 4, 500–3,500 Da (Laser Bio Labs).

39. Mass spectra, obtained by collecting 1,000–2,000 laser shots, are then processed using Data Explorer version 5.1 software (Applied Biosystems).
40. Peak lists can be obtained from the raw data following advanced baseline correction (peak width 32, flexibility 0.5, degree 0.1), noise filtering (noise filter correlation factor 0.7), and monoisotopic peak selection.
41. Database searching can be done with the online MASCOT search engine (http://www.matrixscience.com), Aldente (http://www.expasy.org/tools/aldente), and ProFound (http://www.prowl.rockefeller.edu/prowl-cgi/profound.exe), among others, PMF tools, against the NCBInr and Swiss Prot databases, limiting the search to the appropriate taxonomy, allowing for one trypsin missed cleavage and with a 50 ppm mass tolerance error. The fixed modification to be selected is cysteine carbamidomethylation, while the variable modification is the methionine oxidation.

4. Notes

1. Keep proteins as much as possible on ice; do not over-dry the pellet (max. 5 min) or it can became difficult to re-solubilize. As an alternative to 2D-clean up kit, chloroform/methanol or acetone precipitation can be used.
2. Before determining protein concentration, allow the complete protein solubilization by keeping the samples at 4°C for at least 1 h to overnight. We usually determine protein concentration with the Bradford assay but this is not mandatory and there are several valid alternatives on the market.
3. A range of ratios for protein concentration: CyDye amount (50 μg: 100 pmol to 50 μg: 400 pmol) should be tested to ensure the optimal ratio for the sample of interest.
4. The rehydration buffer added in this step has to be implemented with 40 mM DTT, 0.5% (vol/vol) IPG buffer and traces of BBF. You can prepare a stock of rehydration buffer, make 1 ml aliquots and store at –20°C until use.
5. The length of the pH strip, its pH range as well as the IEF running protocol should be empirically determined to provide the best possible resolution for your sample. Following IEF, the IPG Strips can be stored at –70°C for up to 3 months.

6. You can prepare equilibration buffer, without DTT or iodoacetamide, and store it at room temperature for up to 1 week. Before starting the equilibration step add fresh DDT/iodoacetamide in the reducing and alkylating buffers, respectively. Do not exceed the stipulated times of alkylation and reduction, as there is a possibility of protein loss. Perform this step as close as possible to run the second dimension.
7. It is essential that the maximum pixel intensities of all the three images do not differ significantly. When acquiring gel images it is important to empirically find the PMT values, for the three channels, that allow to evidence the low abundant proteins and do not produce an excessive saturation for the high abundant ones. If most of the protein spots are saturated, the gels should be re-scanned using a lower PMT value to bring all the protein spots in the linear dynamic range. This is crucial to obtain a meaningful quantitative comparison between the gel images.
8. Make sure that extraneous protein spots are removed and that all true protein spots are included by manually examining all the protein spots detected. The eventual quantitative data are very robust if this is adhered to strictly, even though it is a time-consuming process. The spot filtering parameters can be optimized for a particular system based on the distribution of protein spots and their intensities.
9. The Coomassie G-250 must be added after 45 min of incubation in staining solution.
10. Spot handling must always be done wearing gloves to avoid keratin contamination; do not allow anyone to use Speed-Vac with ungloved hands during these steps as sample tubes will be uncapped.

References

1. Klose J (1975). Protein mapping by combined isoelectric focusing and electrophoresis of mouse tissues. A novel approach to testing for induced point mutations in mammals. Humangenetik 26:231–243.
2. O'Farrell PH (1975). High resolution two-dimensional electrophoresis of proteins. J.Biol. Chem. 250:4007–4021.
3. Edvardsson U, Alexandersson M, Brockenhuus von Lowenhielm H, et al (1999). A proteome analysis of livers from obese (ob/ob) mice treated with the peroxisome proliferator WY14,643. Electrophoresis 20:935–942.
4. Ahmed M, Forsberg J, Bergsten P (2005). Protein profiling of human pancreatic islets by two-dimensional gel electrophoresis and mass spectrometry. J Proteome Res 4:931–940.
5. Finehout EJ, Franck Z, Lee KH (2004). Towards two-dimensional electrophoresis mapping of the cerebrospinal fluid proteome from a single individual. Electrophoresis 25:2564–2575.
6. Guo X, Zhao C, Wang F, et al (2010). Investigation of Human Testis Protein Heterogeneity Using Two-Dimensional Electrophoresis. J Androl 31:419–429.
7. Jin M, Diaz PT, Bourgeois T, et al (2006). Two-dimensional gel proteome reference map of blood monocytes. Proteome Sci 4:16.
8. Kim J, Kim SH, Lee SU, et al (2002). Proteome analysis of human liver tumor tissue by two-dimensional gel electrophoresis and matrix assisted laser desorption/ionization-mass

spectrometry for identification of disease-related proteins. Electrophoresis 23:4142–4156.

9. O' Neill EE, Brock CJ, von Kriegsheim AF, et al (2002). Towards complete analysis of the platelet proteome. Proteomics 2:288–305.
10. Simula MP, Cannizzaro R, Marin MD, et al (2009). Two-dimensional gel proteome reference map of human small intestine. Proteome Sci 7:10.
11. Tomazella GG, Da Silva I, Laure HJ, et al (2009). Proteomic analysis of total cellular proteins of human neutrophils. Proteome Sci 7:32.
12. Magni F, Sarto C, Valsecchi C, et al (2005). Expanding the proteome two-dimensional gel electrophoresis reference map of human renal cortex by peptide mass fingerprinting. Proteomics 5:816–825.
13. Candiano G, Santucci L, Petretto A, et al (2010). 2D-electrophoresis and the urine proteome map: where do we stand? J Proteomics 73:829–844.
14. Wang L, Zhu YF, Guo XJ, et al (2005). A two-dimensional electrophoresis reference map of human ovary. J Mol Med 83:812–821.
15. Shaw MM and Riederer BM (2003). Sample preparation for two-dimensional gel electrophoresis. Proteomics 3:1408–1417.
16. Boschetti E and Righetti PG (2008). The ProteoMiner in the proteomic arena: a non-depleting tool for discovering low-abundance species. J Proteomics 71:255–264.
17. Ahmed N and Rice GE (2005). Strategies for revealing lower abundance proteins in two-dimensional protein maps. J Chromatogr B Analyt Technol Biomed Life Sci 815:39–50.
18. Banks RE, Dunn MJ, Forbes MA, et al (1999). The potential use of laser capture microdissection to selectively obtain distinct populations of cells for proteomic analysis--preliminary findings. Electrophoresis 20:689–700.
19. Lescuyer P, Hochstrasser DF, Sanchez JC (2004). Comprehensive proteome analysis by chromatographic protein prefractionation. Electrophoresis 25:1125–1135.
20. Righetti PG, Castagna A, Antonioli P, Boschetti E (2005). Prefractionation techniques in proteome analysis: the mining tools of the third millennium. Electrophoresis 26:297–319.
21. Yan JX, Devenish AT, Wait R, et al (2002). Fluorescence two-dimensional difference gel electrophoresis and mass spectrometry based proteomic analysis of Escherichia coli. Proteomics 2:1682–1698.

Chapter 14

The Use of Antigen Microarrays in Antibody Profiling

Krisztián Papp and József Prechl

Abstract

Technological advances in the field of microarray production and analysis lead to the development of protein microarrays. Of these, antigen microarrays are one particular format that allows the study of antigen–antibody interactions in a miniaturized and highly multiplexed fashion. Here, we describe the parallel detection of antibodies with different specificities in human serum, a procedure also called antibody profiling. Autoantigens printed on microarray slides are reacted with test sera and the bound antibodies are identified by fluorescently labeled secondary reagents. Reactivity patterns generated this way characterize individuals and can help design novel diagnostic tools.

Key words: Protein microarray, Antibody, Profiling, Immunoglobulin, Immunity, Immunoassay, Antigen

1. Introduction

Antibodies are glycoproteins secreted by B lymphocytes as part of the adaptive immune responses. They constitute a significant portion of the protein mass in blood plasma, which are actively transported into mucosal secretions and can appear in varying quantities in practically all body fluids. Generally, an antibody molecule unit consists of two heavy and two light chains, each containing constant and variable regions, forming a symmetric tetrameric structure. The peculiar genetic organization of the loci responsible for immunoglobulin production (1) and an enzymatic machinery controlling somatic mutations (2) allows B cells to [1] generate highly diverse variable domains and [2] couple these domains to several different constant regions. The first phenomenon promotes the generation of antigen-binding molecules with adequate affinity, while the second event – known as isotype switching – modulates the biological functions of these molecules.

Michael Kaufmann and Claudia Klinger (eds.), *Functional Genomics: Methods and Protocols*, Methods in Molecular Biology, vol. 815, DOI 10.1007/978-1-61779-424-7_14, © Springer Science+Business Media, LLC 2012

One can classify antibodies based on properties of the constant regions, such as isotype (heavy chain constant region) and glycosylation, or based on properties of the variable domains, such as specificity and affinity. While the peptide sequence adequately identifies a constant region, sequencing of the variable regions usually will not help in defining the most important aspect of the antibody: target specificity. Monoclonal antibodies, derived basically from a single immortalized B cell, recognize their target antigen with high specificity and affinity and are simply defined by their specificity. Polyclonal antibodies, raised against a given antigen by immunization, are clonally heterogenous and show varying degrees of specificity and affinity. Still more complex, the variety of serum antibodies circulating in an individual reflects immunological experiences of a lifetime. Classical serological assays sample from this diversity by looking at the presence of antibodies against a single particular target antigen. In contrast, antibody profiling is an attempt to assess at least a piece of the diversity of antigen binding molecules (3). Technically, a selected panel of antigens is immobilized on a relatively small surface in an addressable fashion, a device called microarray. This is followed by the treatment of the array of antigens with the tested serum and the detection of bound antibodies. From the proteomics point of view this approach is biased; we cannot identify and define all antibodies, only those whose targets are printed on the array will bind with adequately high affinity (4). From the immunological point of view, this is a multiplexed immunoassay with hundreds, or more, of antigen–antibody interactions taking place in parallel (5).

In fact, theoretical advantages of carrying out immunological reactions in microspot format have long been described (6). It was the machinery that was lacking, until the genomics era when robotics, laser scanning equipment and bioinformatics became available for proteomic approaches. Microarray production requires the reproducible "printing" of submillimeter antigen spots on a small surface, usually a microscope slide, with a positioning precision in the micrometer range. The procedure is carried out in a temperature- and humidity-controlled environment, to conserve structure and function of the printed material and also to ensure consistency from lot-to-lot. The phase of antibody binding and detection is comparable to immunoassays and does not require special equipment. Fluorescent detection has the advantage of high sensitivity and wide dynamic range; in addition, fluorescent laser scanners possess the resolution and precision necessary for quantitative results. Finally, image analysis software is used to extract the binding information from the images.

Generation of the antigen microarrays is the priciest part of the technology, especially if you use factory-made slides and purchase commercially available antigens. To avoid the generation of a huge amount of microarray data with little biological information, always spend ample time with designing the content of your arrays and the sample collection criteria and storage conditions before starting printing. Do pilots to confirm that all the standards and reference materials are properly working and to get used to the workflow. These procedures are explained in Subheading 4, as they do not form part of the streamlined protocol. Here, we describe the detection of bound IgG molecules; by modest modifications of this protocol other immunoglobulins or even other serum proteins can also be detected.

2. Materials

2.1. Microarray Production

1. Phosphate-buffered saline stock solution (10× PBS): 1.37 M NaCl, 27 mM KCl, 80 mM Na_2HPO_4, 14.6 mM KH_2PO_4 (adjust to pH 7.4 with HCl if necessary); pass through a 0.22-μm filter. Store at room temperature. Prepare working solution by dilution of one part with nine parts of water (see Note 1).
2. Spotting buffer: PBS containing 0.05% sodium azide. Store at room temperature.
3. Orientation spot solution: 0.05% Bromophenol blue and 0.05% sodium azide in PBS. Store at room temperature (see Note 2).
4. Pin washing solution for passive bath: 10% ethanol, 0.1% Tween 20 in water.
5. Nitrocellulose membrane covered 2-pad FAST Slides™ (GE Healthcare, UK) (see Note 3).
6. Theoretically, any soluble material can be used as antigen, as long as its solvents are compatible with nitrocellulose. Most proteins, carbohydrates, and nucleic acids meet this criterion; lipids can be difficult to keep in emulsion or in micelles and liposomes (see Note 4).
7. Standards and reference materials: Human IgG purified from human serum, protein G.
8. Source plate: 384-well plate, U-bottom, polypropylene.
9. Spotting pins: MCP310S solid pin (BioRad, Hercules, CA, USA) (see Note 5).
10. Protein microarray printer: BioOdyssey Calligrapher MiniArrayer (BioRad, Hercules, CA, USA). Alternatively, use a microarray printer capable of cooling the printing area and adjusting humidity of the printing chamber, with standard pin washing accessories.

2.2. Serum Treatment and Staining of Slides

1. Serum diluent buffer: 5% (w/v) bovine serum albumin, 0.05% sodium azide in PBS. Filtrate though 0.45 μm filter (see Note 6).
2. Washing buffer: 0.05% Tween 20 in PBS.
3. Reagent buffer: 5% (w/v) BSA, 0.05% Tween 20, 0.05% sodium azide in PBS. Filtrate though 0.45 μm filter.
4. Detecting antibody: $F(ab)_2$ fragment of goat anti-human IgG antibody DyLight 649-conjugated (Jacksons Immuno-Research, Suffolk, UK) (see Note 7).
5. Slide Spinner for microarray slide drying (Labnet International, Oakham, Rutland, UK).
6. FAST Frame and dual well incubation chamber (GE Health-care, UK).
7. Serum samples (see Note 8).

2.3. Scanning and Analysis

1. Fluorescent scanner: Axon GenePix 4300A (Molecular Devices, Sunnyvale, CA, USA). Alternatively, use a laser scanner with the following specifications: laser excitation at 635 nm, standard red filter (655–695 nm), resolution 20 μm or less per pixel, accepts regular 3" by 1" microscope slide format and scans the surface. For analyzing pictures use GenePix Pro 7 software (Molecular Devices, Sunnyvale, CA, USA) (see Note 9).

3. Methods

The steps of antibody profiling can be grouped into four stages: experimental design and sample collection, antigen microarray production, immunological reactions, scanning and analysis. Technical aspects of protein microarray experimental design are found in Subheading 4. Strategies for profiling antibodies against various antigens have been described; the protocol below can be adapted to characterize immune response to autoantigens (7–10), allergens (11, 12), microbes or microbial epitopes (13–15), or tumor antigens (16, 17).

3.1. Microarray Production

1. These instructions assume the use of BioOdyssey Calligrapher miniarrayer.
2. Prepare 1 and 0.2 mg/ml solution from each antigen and human IgG in spotting buffer and load 11 μl solution into wells of source plate (see Note 10) according to the design of your plate and microarray layout (Fig. 1).
3. Cool down arrayer to 15°C and set humidity to 50%. Insert source plate, FAST slides and suitable number of solid pins to

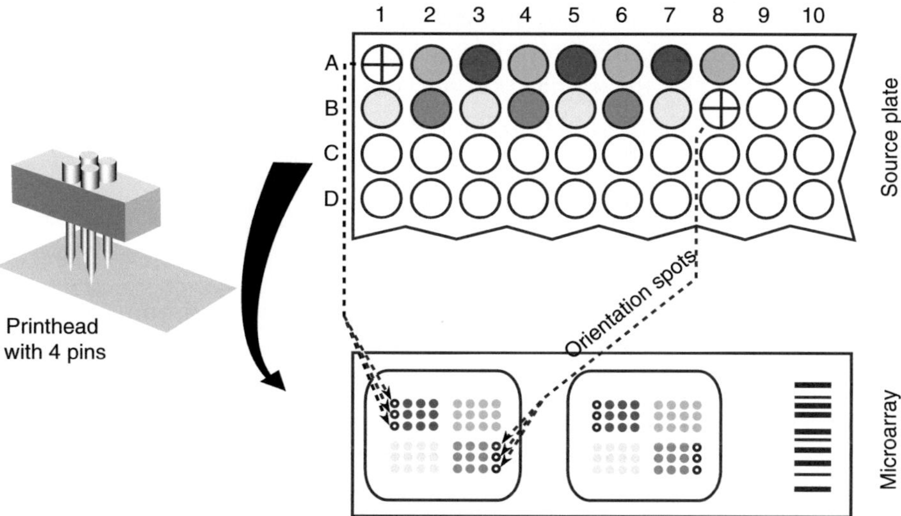

Fig. 1. Designing array layout. Printing pins transfer antigen solutions from the source plate to the microarray surface. Depending on the number of pins used, grids corresponding to each pin are generated on the slide. Using a colored solution, such as bromophenol-blue (wells marked with a *cross*), one can easily visualize the corners of the printed array. Please note that the layout of antigen solutions on the source plate is different from that of the printed array; the printer software generates a "map" of the layout (GAL file), which is used for the analysis.

their position. Fill up passive bath with pin washing buffer and the flow through bath with Milli Q water.

4. Set arrayer to print triplicates from each material, spot distance should be at least 750 μm for using MCP 310 S pins. In general, use a spot distance twice the diameter of the spots to ensure proper signal separation and thereby acquisition of reliable background signals around each spot.
5. Following printing, store slides in a cool, low humidity environment protected from light; slides can be used up to 6–12 months. Sealing individually in an aluminum package is a good way to store slides. Otherwise repack them in the factory box and then wrap it in parafilm to exclude moisture. Avoid touching the surface of the slide when packaging.

3.2. Serum Treatment and Staining of Slides

1. These instructions assume the use of FAST slide system (see Note 11). Place the double-well chamber onto a FAST Slide and insert it into a FAST Frame.
2. Thaw serum samples and keep them on ice until use.
3. Wash microarray pads four times for 5 min with 800 μl PBS. Incubate slides on orbital shaker at room temperature. Discard buffer using an aspirator (see Note 12).
4. Dilute serum sample 125 times in serum diluent buffer and centrifuge at 15,000 × *g* for 6 min for removing aggregates. Please note that from this point on you are handling potentially

infectious human serum; accordingly follow the biohazard guidelines of your institution regarding waste treatment and the disinfection of reusable materials.

5. Load 350 μl diluted serum into chambers and shake slides for 60 min at 37°C in a humidified chamber (see Note 13).
6. Discard serum sample and fill up chambers with 800 μl washing buffer for a quick wash. Wash chambers three more times for 5 min with 800 μl washing buffer.
7. Dilute DyLight 649 conjugated anti-human IgG antibody 5,000-fold in reagent buffer, add 700 μl to the chamber after removing washing buffer.
8. Cover slides with an aluminum foil as dye-conjugates are light sensitive. Shake microarray for 30 min at room temperature.
9. Discard detecting antibody and fill up chambers with 800 μl washing buffer for a quick wash. Wash chambers two more times for 5 min with 800 μl washing buffer.
10. Dismount FAST Frame and wash the whole slide in 7 ml washing buffer for a quick wash. Wash whole slides two more times for 10 min with 7 ml washing buffer.
11. Dry slides by spinning them for 2 min with Slide Spinner (see Note 14).

3.3. Scanning

1. These instructions assume the use of Axon GenePix 4300A scanner and GenePix Pro 7 software.
2. Let the scanner warm up for 20 min then insert dry slides upside down.
3. Set 635 nm laser excitation and 20 μm scanning resolution (see Note 15).
4. Set standard red filter (655–695 nm) and adjust laser power and PMT settings to get appropriate signal intensities (see Note 16).
5. Scan slide and save picture.
6. Quality control. The golden standard of quality control is visual inspection: a properly developed antigen microarray should possess a homogenous low background with sharp signals (Fig. 2). Though methods exist for correcting uneven background, generally it is more reliable to repeat the experiment if the slide does not pass this control step.

3.4. Analysis

1. Analyze the grayscale picture of microarray by GenePix Pro 7 software.
2. Open the image of microarray and load the array list (GAL file) (see Note 17).

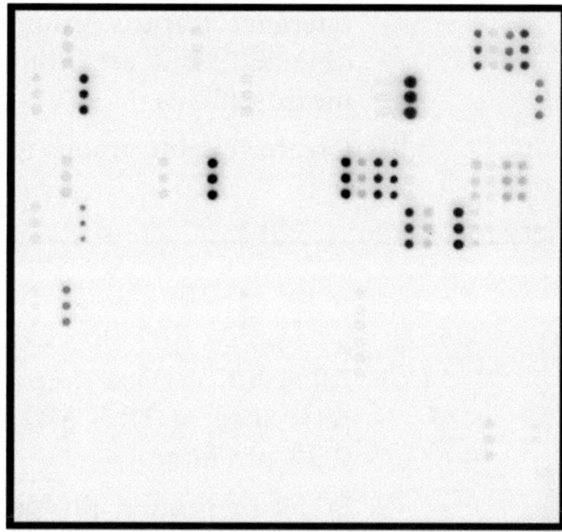

Fig. 2. Image of a scanned fluorescently labeled antigen microarray. Triplicates of antigens were printed; bound IgG was detected as described in the text.

3. Adjust the brightness and contrast of picture to visualize the spots (see Note 18).
4. Align the GAL file generated circles onto spots manually or let it do the software automatically.
5. Select the Analyze option to analyze image and generate GPR file after the spot finding has finished (see Note 19).
6. Extract feature specific information ("Name," "B635 SD," "F635 Median – B635") from GPR file by Microsoft Excel software (see Note 20).
7. Calculate the background limit that is the double of the average of "B635 SD."
8. Set the feature intensity to one if it does not reach the background limit. Negative values are not interpreted and all signal intensities below the background are set to an arbitrary value. Using one as this arbitrary value has the advantage that corresponds to zero on a logarithmic scale, which is often used to visualize and analyze data.
9. Calculate the median of feature replicates.
10. Normalize data to the signal intensity of the control feature. This is achieved by adjusting signal intensities of the control feature (printed IgG) to the same value in all the datasets to be compared. Calculate the average of the IgG signals in all microarray slides and multiply or divide signal intensities of all features with a value – this is the normalization factor – to obtain equal IgG signals. We prefer to use normalization factors that are less than 2 or greater than 0.5. Practically, this means that a linear normalization is carried out within two $\log_2$ orders of magnitude. Greater fluctuations with respect to the

reference feature (printed IgG in our case) suggest that the obtained data are not reliable comparable to independent measurements.

11. Create suitable graphs to visualize your data.

4. Notes

1. All solutions are prepared in water with specific electrical resistance of 18.6 MΩ cm that has been passed through a 0.22-μm filter.
2. Spots of printed proteins are usually not detectable by the naked eye. We suggest to print bromophenol blue containing spots at the upper-left and very bottom-right corners this way one can easily check the position of printed grid on the slides. These spots will gradually fade out during the washing procedures. If fluorescently labeled proteins are also mixed to the bromophenol blue solution then these spots will appear during scanning and will be useful to align grids during analysis.
3. One can fabricate cheap, home-made slides by sticking an appropriately sized nitrocellulose blotting membrane (Sigma, St. Louis, MO) onto a conventional (yet good quality) microscopic glass slide with the help of a double-adherent tape. When making your own slides always use gloves and avoid touching the nitrocellulose membrane. Physical injuries disrupt the microstructure of the membrane and will lead to uneven spotting efficiency and background signals, and increased technical variation.
4. The list of solvent that are compatible with nitrocellulose membrane can be found at http://www.whatman.com/UserFiles/File/Selection%20Guides/Appendix%20B%20MemProdSelectChemicalComp.pdf.
5. Three types of MCP pins are available for use in the BioOdyssey Calligrapher miniarrayer: MCP100 and MCP360 are capillary pins for repetitive spotting of even high-viscosity protein sample and MCP310S is a solid pin for printing individual samples in single spots. Spot size depends on pins: around 100 and 360 μm for capillary pins and 400 μm for solid pins. Be careful because spot size depends on the type and viscosity of the printing buffer. For example, printing of samples containing butanol with capillary pins can result in very large diffused spots, requiring an increased distance between the spots.
6. We suggest straining of serum diluent buffer with a 0.45-μm filter to remove undissolved particles and aggregates that can cause increased and uneven background fluorescence. A 5% BSA solution will quickly clog a 0.22 μm filter, sterile

filtration is not necessary. Upon long-term storage repeated filtration can improve the quality of your detection.

7. Other immunoglobulin classes, like even IgM or IgA or IgG subclasses (IgG_1, IgG_2, IgG_3, IgG_4) are routinely detectable. Detection of antibodies of low serum concentration, e.g., IgE, may require the use of less diluted serum.

   Depending on the microarray scanner instrument, parallel detection of 2–4 types of Ig classes is also achievable by using different fluorescent dyes with nonoverlapping absorption or emission spectra for detection. In addition to antibody detection, the binding of other serum components, such as complement proteins can also be monitored using antigen arrays (18).

   The use of fragmented antibodies for detection is preferable to whole antibodies, since interactions via the Fc part can be excluded this way. This can be particularly important when the array contains materials with known Fc binding ability. Whatever format you use, determine background binding of the detection antibodies by using serum dilution buffer in place of serum in a pilot experiment.

8. Collect venous blood into a closed native blood collection system. Allow blood to clot for 1–2 h at room temperature, and then place it on ice for 1 h. Separate serum from the clot by spinning for 5 min at 2,000 × *g*. Aliquot serum and store at −70°C until use.

9. Nitrocellulose has a high binding capacity for most macromolecules we have studied and is our preferred surface material on the slides. Since it absorbs fluorescence these slides need to be scanned from the nitrocellulose side, not from the bottom, which some scanners do. Be sure to use a scanner that scans the surface.

10. As the source plate, we suggest to use a 384-well plate instead of 96-well plate even when your antigens would easily fit into a 96-well plate. There are two reasons for that: first, in a small diameter well the solution evaporates slower and your antigen will not dry out during printing; second, the volume required for the proper immersion of the printing pins and the lost volume at the end of the printing session is moderate.

11. When using home-made slides, place two 1 mm-wide polymer spacers at the ends of the slide, lay a clean glass slide over them. Fix the two slides, separated by the spacers, together with two silicon cooking bands, creating a small incubation chamber. This chamber can be filled up by slow but continuous pipetting.

12. Washing prior to serum treatment serves two purposes: removes excess material and rehydrates nitrocellulose-absorbed, dried proteins. Blocking is used in some protocols before serum treatment, in our experience this can reduce overall background

fluorescence but at the same time decreases background homogeneity and can even interfere with specific signals, depending on the blocking reagent used.

13. Proper agitation of the fluids on the microarray is indispensable for obtaining homogenous distribution and binding of the antibodies. Instead of using a simple one-dimensional shaking motion use combinations of orbital shaking, vibrating, and tilting motions. Simple movements can result in the formation of eddies and inhomogenous background signals.

14. Spinning the slides speeds up its drying. Alternatively, slides can be dried in a safety cabinet with the airflow on. What is important, follow the exact same procedure for each slide you want to compare and allow roughly the same time from detection to scanning. Fluorescence decreases exponentially as the slide dries out, so it is preferable to approach a low plateau by drying the slide thoroughly, instead of hurrying to the scanner.

15. As a general rule of thumb, spot diameter should be at least 10× pixel size in order to sample sufficient data for a quantitative analysis. Thus, if your scanner is set to 10 μm scanning resolution you will need spots with at least 100 μm diameter.

16. When imaging FAST Slides, the default imager parameters for glass slides will not be suitable for detection. Due to the higher binding capacity of FAST Slides, as well as the unique light scattering properties of the polymeric surface, laser power and/or PMT settings will need to be set lower than for glass slides. In order to take full advantage of the dynamic range of the scanner, signal intensities should be as high as possible without reaching or exceeding the maximum of the system (i.e., pixel intensity = 65,535 on a 16-bit system). Signals exceeding this limit cannot be analyzed quantitatively.

17. The GenePix Array List (GAL) file contains specific information for the layout of each block, and the identity of each feature within a block of microarray. GAL file is usually generated by the arrayer software or can also be created manually using Microsoft Excel.

18. Modification of image brightness or contrast does not have any effect on the final intensity value, it is only used to enhance visibility of the spots on the screen.

19. The GenePix Result (GPR) file contains general information about image acquisition and analysis, as well as the data extracted from each individual feature.

20. In the GPR file, the "Name" column contains the name of the feature; "B635 SD" column contains the standard deviation of the local background pixel intensities; "F635 Median – B635" column contains the median of feature pixel intensities minus the median of local background pixel intensities.

Acknowledgments

This work was supported by the Hungarian Academy of Sciences, grant KMOP-1.1.1-08/1-2008-0028 from the National Development Agency and NKTH-OTKA grant K68617. K.P. is supported by Janos Bolyai Research Fellowship. We thank Zoltán Szittner and Mariann Kremlitzka for their critical comments.

References

1. Jung D, Giallourakis C, Mostoslavsky R et al. (2006) Mechanism and control of V(D)J recombination at the immunoglobulin heavy chain locus *Annu. Rev. Immunol.* 24:541–70
2. Wu X, Feng J, Komori A et al. (2003) Immunoglobulin somatic hypermutation: double-strand DNA breaks, AID and error-prone DNA repair *J. Clin. Immunol.* 23:235–46
3. Prechl J, Papp K, Erdei A. (2010) Antigen microarrays: descriptive chemistry or functional immunomics? *Trends Immunol.* 31:133–7
4. Cahill DJ, Nordhoff E. (2003) Protein arrays and their role in proteomics *Adv. Biochem. Eng Biotechnol.* 83:177–87
5. Robinson WH. (2006) Antigen arrays for antibody profiling *Curr. Opin. Chem. Biol.* 10:67–72
6. Feinberg JG. (1961) A 'microspot' test for antigens and antibodies *Nature* 192:985–6
7. Graham KL, Robinson WH, Steinman L et al. (2004) High-throughput methods for measuring autoantibodies in systemic lupus erythematosus and other autoimmune diseases *Autoimmunity* 37:269–72
8. Lueking A, Huber O, Wirths C et al. (2005) Profiling of alopecia areata autoantigens based on protein microarray technology *Mol. Cell Proteomics.* 4:1382–90
9. Hueber W, Kidd BA, Tomooka BH et al. (2005) Antigen microarray profiling of autoantibodies in rheumatoid arthritis *Arthritis Rheum.* 52:2645–55
10. Kanter JL, Narayana S, Ho PP et al. (2006) Lipid microarrays identify key mediators of autoimmune brain inflammation *Nat. Med.* 12:138–43
11. Hiller R, Laffer S, Harwanegg C et al. (2002) Microarrayed allergen molecules: diagnostic gatekeepers for allergy treatment *FASEB J.* 16:414–6
12. Bacarese-Hamilton T, Mezzasoma L, Ingham C et al. (2002) Detection of allergen-specific IgE on microarrays by use of signal amplification techniques *Clin. Chem.* 48:1367–70
13. Gaseitsiwe S, Valentini D, Mahdavifar S et al. (2008) Pattern recognition in pulmonary tuberculosis defined by high content peptide microar ray chip analysis representing 61 proteins from M. tuberculosis *PLoS. One.* 3:e3840
14. Mezzasoma L, Bacarese-Hamilton T, Di CM et al. (2002) Antigen microarrays for serodiagnosis of infectious diseases *Clin. Chem.* 48:121–30
15. Davies DH, Liang X, Hernandez JE et al. (2005) Profiling the humoral immune response to infection by using proteome microarrays: high-throughput vaccine and diagnostic antigen discovery *Proc. Natl. Acad. Sci. USA* 102:547–52
16. Casiano CA, Mediavilla-Varela M, Tan EM. (2006) Tumor-associated antigen arrays for the serological diagnosis of cancer *Mol. Cell Proteomics* 5:1745–59
17. Madoz-Gurpide J, Kuick R, Wang H et al. (2008) Integral protein microarrays for the identification of lung cancer antigens in sera that induce a humoral immune response *Mol. Cell Proteomics* 7:268–81
18. Papp K, Szekeres Z, Terenyi N et al. (2007) On-chip complement activation adds an extra dimension to antigen microarrays *Mol. Cell Proteomics* 6:133–40

Chapter 15

Limited Proteolysis in Proteomics Using Protease-Immobilized Microreactors

Hiroshi Yamaguchi, Masaya Miyazaki, and Hideaki Maeda

Abstract

Proteolysis is the key step for proteomic studies integrated with MS analysis. Compared with the conventional method of in-solution digestion, proteolysis by a protease-immobilized microreactor has a number of advantages for proteomic analysis; i.e., rapid and efficient digestion, elimination of a purification step of the digests prior to MS, and high stability against a chemical or thermal denaturant. This chapter describes the preparation of the protease-immobilized microreactors and proteolysis performance of these microreactors. Immobilization of proteases by the formation of a polymeric membrane consisting solely of protease-proteins on the inner wall of the microchannel is performed. This was realized either by a cross-linking reaction in a laminar flow between lysine residues sufficiently present on the protein surfaces themselves or in the case of acidic proteins by mixing them with poly-lysine prior to the crosslink-reaction. The present procedure is simple and widely useful not only for proteases but also for several other enzymes.

Key words: Enzyme immobilization, Microfluidics, Microreactor, Protease, Proteolysis, Proteomics

1. Introduction

Proteolysis by sequence-specific proteases is the key step for positive sequencing in proteomic analysis integrated with MS (1). The conventional method of in-solution digestion by proteases is a time-consuming procedure (overnight at 37°C). The substrate/protease ratio must be kept high (generally >50) in order to prevent excessive sample contamination by the protease and its auto-digested products. But this leads to a relatively slow digestion. In addition, obtaining reliable peptide maps and meaningful sequence data by MS analysis requires not only the separation of the digested peptides but also

Michael Kaufmann and Claudia Klinger (eds.), *Functional Genomics: Methods and Protocols*,
Methods in Molecular Biology, vol. 815, DOI 10.1007/978-1-61779-424-7_15, © Springer Science+Business Media, LLC 2012

strictly defined proteolysis conditions (2, 3). Furthermore, peptide recovery from in-solution digestion is highly dependent on the structural properties of the target proteins because proteins with rigid structures, e.g., by disulfide bonds tend to be resistant to complete digestion. In fact, the typical preparation of a sample for proteolysis includes denaturation, reduction of disulfide bonds, and alkylation procedures to decrease the conformational stability. It is obvious that insufficient sequence coverage could compromise the accuracy of proteome characterization.

A microreactor is a suitable reaction system for handling small-volume samples (nl to μl) in a microchannel to perform chemical or enzymatic reactions. Enzyme-immobilized microreactors have been widely used in chemical and biotechnological fields (4–7). The protease-immobilized microreactor provides several advantages for proteolysis (5); e.g., low degree of auto-digestion even at high protease concentrations and a large surface and interface area that leads to rapid proteolysis. Furthermore, the immobilized proteases on the microchannel walls can be easily isolated and removed from the digested fragments prior to MS, which means elimination of the requirement to stop the reaction by chemical or thermal denaturation after digestion. These features can contribute to higher sequence coverage compared to the approach based on in-solution digestion. High sequence coverage is important to enhance the probability of identification of the protein and increase the likelihood of detection of structural variants generated by processes such as post-translational modifications.

Several methods for protease immobilization have been reported, wherein the proteases have been covalently bounded, trapped, or physically adsorbed onto different supports based either on silica and polymer particles or monolithic materials (5–8). However, preparations of these protease-immobilized microreactors require multistep procedures consuming considerable amounts of time and effort. Therefore, a facile preparation method of the enzyme-immobilized microreactor is desirable for the routine proteolysis step in proteomic analysis. In addition, reusability is also an important feature required for laboratory use. We developed the procedure for immobilizing enzymes on the internal surface of the poly-tetrafluoroethylene (PTFE) microtube by forming an enzyme polymeric membrane through a cross-linking reaction in a laminar flow between lysine residues on the protein surfaces (9) or between the mixture of proteins with isoelectric point $pI < 7.0$ and poly-lysine (10). The proteolysis method using the presented protease-immobilized microreactors is a simple and rapid approach for high-throughput analysis in proteomics.

2. Materials

2.1. Preparation of Protease-Immobilized Microreactors

1. The lengths of the PTFE microtubes (500 μm inner diameter (i.d.) and 1.59 mm outer diameter, (o.d.)) were cut to 5, 6, and 13 cm, respectively. The microcapillary (100 μm i.d. and 375 μm o.d.) was cut to a length of 5 cm (Fig. 1a). All materials were rinsed by 18.2 MΩ-cm water (Milli-Q water Purification System, Millipore, Bedford, MA, USA).

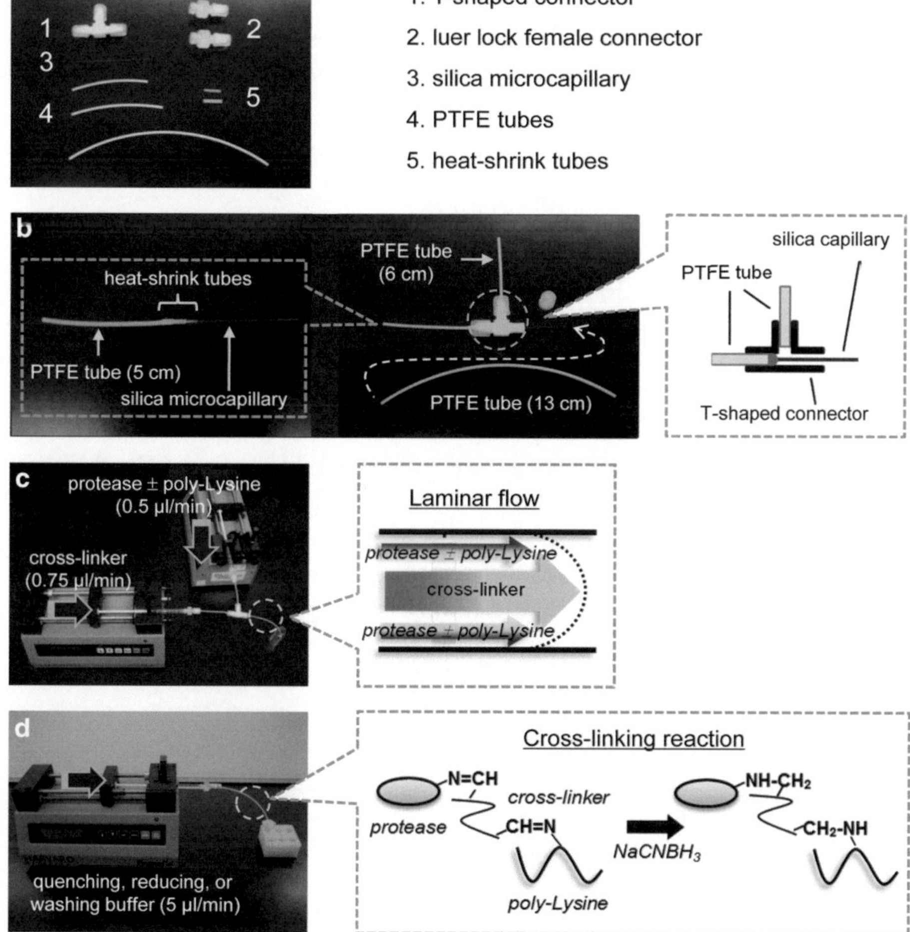

Fig. 1. Preparation of a protease-immobilized microreactor. (**a**) Materials. (**b**) A silica microcapillary attached to a PTFE microtube using a heat-shrink tubing. The silica microcapillary was set in a T-shape connector. (**c**) The assembled microflow system. The cross-linker solution was supplied to the substrate PTFE microtube through a silica capillary, corresponding to a central stream in the concentric laminar flow. A solution of proteases or a protease/poly-lysine mixture was supplied from another PTFE microtube connected to the T-shaped connector. Both solutions were introduced by syringe pumps. (**d**) Introducing the buffer solution to the protease-immobilized PTFE microtube.

2.2. Experimental Buffers (see Note 1)

2.2.1. Preparation of the Protease-Immobilized Microreactor

1. Reaction buffer: 50 mM phosphate buffer (PB), pH 8.0.
2. Cross-linking buffer: 4% (v/v) paraformaldehyde (PA) and 0.25% (v/v) glutaraldehyde (GA) in 50 mM PB, pH 8.0.
3. Quenching buffer: 1 M Tris–HCl, pH 8.0.
4. Reducing buffer: 50 mM sodium cyanoborohydride ($NaCNBH_3$) in 50 mM borate buffer, pH 9.0.
5. Washing and storage buffer: 50 mM PB, pH 7.5.

2.2.2. Enzymatic Reactions

1. Buffer for hydrolysis of small synthetic compound: 50 mM Tris–HCl, pH 8.0.
2. Buffer for proteolysis: 10 mM ammonium acetate, pH 8.5.

2.3. Biomaterials and Substrates

1. Protease: *N*-Tosyl-L-phenylalanyl chloromethyl ketone-treated trypsin (TY) and α-chymotrypsin (CT) were dissolved to concentrations of 10 and 5 mg/ml in 50 mM PB, pH 8.0, respectively (see Note 2).
2. Poly-lysine: Poly-L-lysine hydrobromides (poly-Lysine, MW 62,140) was dissolved to a concentration of 10 mg/ml in 50 mM PB, pH 8.0.
3. Substrates for kinetic parameters analysis: The stock solutions of 20 mM were prepared in dimethyl sulfoxide (DMSO) and were stored at −20°C until use. The final solutions of benzoyl-L-arginine *p*-nitroanilide (BAPA) of 0.1–0.5 mM and *N*-glutaryl-L-phenylalanine *p*-nitroanilide (GPNA) of 0.1–1 mM were prepared in 50 mM Tris–HCl, pH 8.0 (see Note 3).
4. Substrates for the chemical stability analysis: The BAPA (0.5 mM) and GPNA (1 mM) solutions were prepared in 50 mM Tris–HCl, pH 8.0, and 4 M urea.
5. Substrate for proteolysis: Substrate proteins (β-casein, cytochrome *c* (Cyt-C), and lysozyme) were dissolved to a concentration of 50–100 μg/ml in 10 mM ammonium acetate, pH 8.5. The proteins were filtered with a 0.45-μm polypropylene cellulose syringe filter.

3. Methods

A microfluidics-based enzyme-polymerization technique (9, 10) was used for the preparation of the protease-immobilized microreactor. Trypsin and chymotrypsin, both well-characterized and commonly used proteases in proteolysis experiments were employed. A common procedure for cross-linking proteases involves the activation of the primary amine groups of proteins (e.g., side-chain of Lys residue) with GA and PA to create aldehyde groups that can react readily with other primary amine groups of proteins (5, 6, 11). Because the cross-linking yields depend on the number of the Lys residue of proteins, the acidic or neutral proteins ($pI < 7.0$) cannot be efficiently cross-linked by the mere use of a cross-linker. To overcome this difficulty, poly-lysine was used as a booster for the improvement of the cross-linking yields of the acidic or neutral enzyme (10).

3.1. Preparation of the Protease-Immobilized Microreactors

1. The silica microcapillary (5 cm length) was connected with the PTFE microtube (5 cm length) and fixed by the heat-shrink tubing (1.1 mm o.d. and 1.9 mm o.d., before heating) (Fig. 1b).
2. The microcapillary as prepared above and the PTFE microtube (6 cm length) for introducing reagents as well as the PTFE microtube (13 cm length) used to introduce the substrate for the protease-immobilized microreactor were attached to the T-shaped connector (Fig. 1b) (see Note 4).
3. Preparation of the TY-microreactor: TY (700 μl, 10 mg/ml) and the cross-linker (900 μl, 4% PA and 0.25% GA, see Note 5) were supplied to the PTFE microtube by using the T-shape connector which was prepared as described above. All proteins were filtered through a 0.45-μm polypropylene cellulose syringe filter and charged into the 1 ml plastic syringes. Solution introduction was performed at different flow rates (0.75 μl/min for the cross-linker and 0.5 μl/min for trypsin) by a Pico Plus syringe pump (Harvard Apparatus, Inc., Holliston, MA, USA) (Fig. 1c). The cross-linking reaction was performed for 20 h at 4°C (see Note 6).
4. Preparation of the CT-microreactor: CT (5 mg/ml) and poly-lysine (10 mg/ml) were mixed in a volume ratio of 1:1 in 50 mM PB, pH 8.0 (see Notes 7 and 8). The cross-linker (200 μl) and the CT/poly-lysine mixture (150 μl) were introduced as described above. The reaction was performed for 3 h at 4°C.
5. After the cross-linking reaction, the protease-immobilized microtubes were removed from the T-shaped connector and rinsed with 200 μl of 1 M Tris–HCl (pH 8.0) by the syringe pump (5 μl/min for 40 min) at 4°C which simultaneously quenched any residual active aldehyde groups remaining on the cross-linked proteases (Fig. 1d).
6. To reduce the resulting Schiff base, the protease-immobilized microtubes were treated with 200 μl of 50 mM $NaCNBH_3$ buffer solution by the syringe pump (5 μl/min for 40 min) at 4°C.
7. The protease-immobilized microtubes were washed with 200 μl of 50 mM PB, pH 7.5 at 4°C (see Note 9). The obtained protease-immobilized microreactors were filled with 50 mM PB, pH 7.5, and stored at 4°C (see Note 10).

3.2. Hydrolytic Activity of the Protease-Immobilized Microreactors

1. The hydrolytic activities of the protease-immobilized microreactors were carried out in 50 mM Tris–HCl, pH 8.0 at 30°C. The syringe pump was used to deliver the substrates (Fig. 2a) (see Note 11). Synthetic compounds; BAPA for TY-microreactor, and GPNA for CT-microreactor were used as substrates. The reaction was evaluated as the amount of released *p*-nitroaniline calculated from the absorbance at 405 nm. The extinction coefficient of *p*-nitroaniline at 405 nm is 9,920 cm^{-1} M^{-1}. To determine the kinetic parameters, K_m and V_{max}, the initial velocities were measured at various substrate concentrations (12). The data were fitted to the Michaelis–Menten equa-

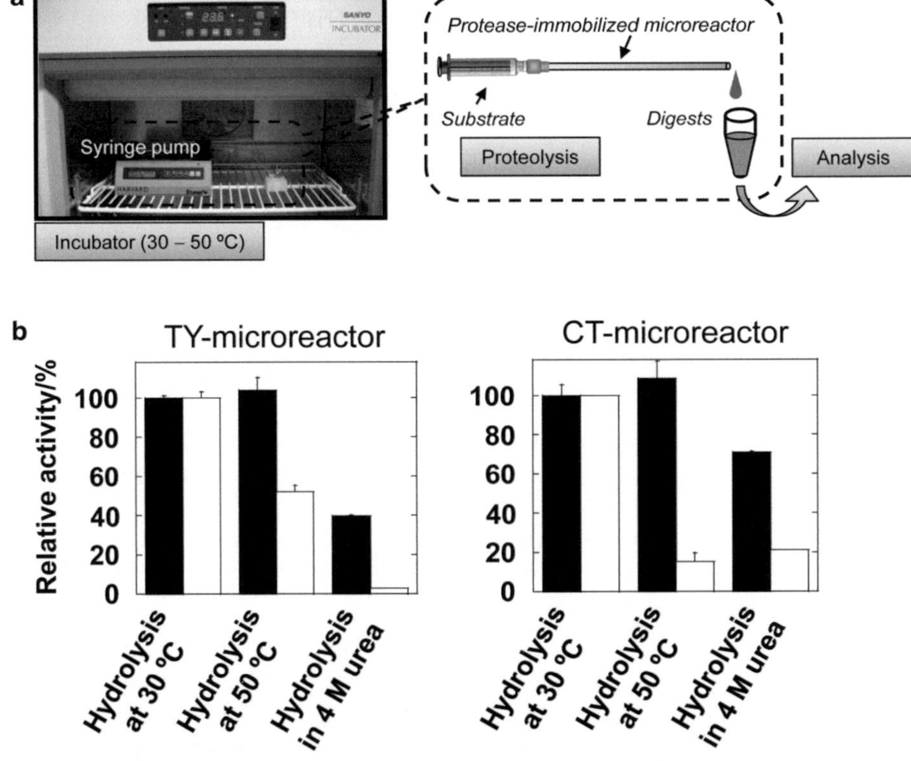

Fig. 2. (**a**) Schematic representation of the hydrolytic reaction by the protease-immobilized microreactor. The substrate was introduced through the microreactor from a syringe pump. Reaction temperature was kept in an incubator. The digests were collected in a sample tube and then analyzed by a spectrophotometer or ESI-TOF MS. (**b**) Comparison of stabilities of immobilized proteases (*closed bars*) and free proteases (*open bars*) against high temperature (50°C) and a chemical denaturant (4 M urea at 30°C). Concentrations of free protease, BAPA for TY, and GPNA for CT were 40 μg/ml, 0.5 mM, and 1 mM, respectively. The reaction time for the free-protease digestion was 5.2 min.

tion by KaleidaGraph 4.0 (Synergy Software, Reading, PA, USA). The estimated kinetic parameters were for K_m of 2.2 ± 0.3 mM and V_{max} of 338 ± 26 μM/min for TY-microreactor and K_m of 2.2 ± 0.2 mM and V_{max} of 468 ± 24 μM/min for CT-microreactor.

2. The thermal stability of the protease-immobilized microreactors was tested using the synthetic compounds in 50 mM PB, pH 8.0 at 30°C or 50°C (see Note 12). The flow rate of the substrate was 5.0 μl/min. The concentrations of BAPA and GPNA were 0.5 and 1 mM, respectively.
3. The chemical stability of the protease-immobilized microreactors was tested using the synthetic compounds in 50 mM Tris–HCl (pH 8.0) and 4 M urea at 30°C (see Note 13). The concentrations of BAPA and GPNA were 0.5 and 1 mM, respectively.
4. After reactions were performed, the protease-immobilized microreactors were washed with 200 μl of 50 mM PB, pH 7.5, and stored at 4°C.

3.3. Proteolysis by the Protease-Immobilized Microreactors

1. Proteolysis was carried out in 10 mM ammonium acetate buffer, pH 8.5 (see Note 14) at 30°C or 50°C. 50 μl of protein solution was pumped through the microreactor at a flow rate of 1.2–2.5 μl/min using the syringe pump. The digested peptides were collected in 1.5 ml test tubes for 20 min and directly analyzed by MS.
2. After the reactions were completed, the protease-immobilized microtubes were washed with 200 μl of 50 mM PB, pH 7.5, and stored at 4°C.
3. Mass spectra were measured using an ESI-TOF MS (Mariner; Applied Biosystems Inc., Foster City, CA, USA). MS measurements were performed with a scan range of 200–1,500 *m/z*. The digested samples were dissolved in 50% aqueous acetonitrile and 1% formic acid at a concentration of 50 μg/ml. The acceleration voltage was 4 kV. The electrospray signal was stabilized by a flow of nitrogen curtain gas set (1 l/min) and nitrogen nebulizer gas set (0.3 l/min).
4. Peptide fragments were assigned based on the Swiss-Prot database using PeptideMass of the ExPASy Proteomics Server (http://www.expasy.ch/tools/peptide-mass.html) with the following constraints: tryptic or chymotryptic cleavage and up to two missed cleavage sites (see Note 15 and Table 1).

Table 1
Summary of ESI-TOF MS results of the digests of Cyt-C, β-casein, and lysozyme

Digestion methods	Substrate protein	Reaction temperature (°C)	Reaction time	Identified amino acids	Sequence coverage (%)
TY-microreactor	Cyt-C	30	10.4 min	97/104	93[a]
CT-microreactor	Cyt-C	30	10.4 min	40/104	38[a]
TY (in-solution)[b]	Cyt-C	37	18 h	99/104	95[a]
CT (in-solution)[b]	Cyt-C	37	18 h	68/104	65[a]
TY-microreactor	β-Casein	30	10.4 min	30/209	14[a]
CT-microreactor	β-Casein	30	10.4 min	120/209	57[a]
TY (in-solution)[b]	β-casein	37	18 h	44/209	21[a]
CT (in-solution)[b]	β-Casein	37	18 h	95/209	45[a]
TY-microreactor	Lysozyme	30	21.7 min	7/129	5[c]
TY-microreactor	Lysozyme	50	21.7 min	126/129	98[c]
CT-microreactor	Lysozyme	30	21.7 min	11/129	9[c]
CT-microreactor	Lysozyme	50	21.7 min	54/129	42[c]
TY (in-solution)[b]	Lysozyme	37	18 h	99/129	77[c]
CT (in-solution)[b]	Lysozyme	37	18 h	66/129	51[c]

[a] Data from ref. 13

[b] In-solution digestion was carried out in 10 mM ammonium acetate buffer, pH 8.5 for 18 h. Concentrations of substrate and free proteases were 100 and 2 μg/ml, respectively

[c] Data from ref. 17

4. Notes

1. All solutions should be prepared in water that has a resistivity of at least 18.2 MΩ cm.
2. If the enzyme solution contains free primary amines (e.g., Tris or glycine), these compounds must be removed by dialysis or a gel filtration step.
3. The final DMSO concentrations of 0.5 mM BAPA solution and 1 mM GPNA solution were 2.5 and 5.0%, respectively. The hydrolysis activities of both proteases were not affected at these DMSO concentrations. However, at high DMSO concentrations (>50%), the hydrolytic activities were reduced (9).
4. In order to form a concentric laminar flow in the cross-linked proteases, the silica capillary was set in a T-shaped connector and positioned at the concentric position of the PTFE microtube that contains the immobilized protease on its internal surface.
5. Although high concentrations of GA (typically 5~10%) rapidly enables attachment of enzyme to the support, it is difficult to control the reaction, and the resulting reactor often shows low enzymatic activity (9). Due to this fact, we used the combination of GA (0.25%) and PA (4%) that provided better cross-linking yields between the enzymes maintaining their activities (9). If a novel enzyme is used for the preparation of the microreactors, the concentration of the cross-linker should be examined in relation to the cross-linking yields with its respective enzymatic activity.
6. When the cross-linking reaction was allowed to proceed at room temperature for 3 h, the resulting TY-microreactor had 74-fold lower hydrolytic activity for BAPA than that of the TY-microreactor prepared at 4°C for 20 h, indicating that auto-digestion of TY in bulk solution occurred at room temperature during the period of the cross-linking reaction. For an enzyme that is sensitive to long-term procedures (e.g., precipitation), the reaction should be performed at 4°C. Longer reaction times are acceptable when the concentration of the protein to be immobilized is increased. Protease immobilizations on PTFE tubes were analyzed by the Bradford method or by the absorbance at 280 nm using uncross-linking protease fractions. For example, 50 μg CT was formed by polymerizing on a 1-cm long PTFE tube by the presented procedure.
7. Because the pI value of CT (8.6) is close to the pH value of the reaction buffer (8.0), this probably leads to low cross-linking yields. Thus, poly-lysine supporting the cross-linking procedure was used for the preparation of the CT-microreactor. On the other hand, for the TY-microreactor, poly-Lysine was

omitted because the pI value of trypsin was 10.5 and poly-lysine was a substrate for TY.

8. The molecular weight of poly-lysine affects the interaction between some enzymes and poly-lysine. For example, when high molecular weight of poly-lysine (MW >60,000) which was intended for the CT-microreactor was used for an alkaline phosphatase (AP)-immobilized microreactor, an aggregation of protein was readily observed, suggesting that highly positively charged poly-lysine (MW >60,000) quickly reacted with the acidic AP protein by electrostatic interactions. Because our enzyme-polymeric membrane is formed on the inner wall of the microchannel (500 μm i.d.) in a laminar flow, the quickly aggregated enzyme and poly-lysine can get stucked on the microchannel during the cross-linking reaction. Therefore, the large poly-lysine molecule is not considered appropriate for the preparation of some proteins such as AP. To overcome this problem, a low molecular weight poly-lysine (e.g., MW 4,200) was used for the preparation of an AP-microreactor. As expected, with the use of the low molecular weight poly-lysine, quick aggregation was suppressed and the AP-microreactor was successfully prepared (13).
9. To confirm the elution of protease from PTFE microtubes, measurements of absorbance at 280 nm or MS should be performed. To our knowledge, TY and CT were not eluted from the microtubes during the presented procedures.
10. Both protease-immobilized microreactors retained over 90% of their hydrolytic activities against synthetic substrates over 60 days. In contrast, free proteases almost completely lost their activities at 25°C within a couple of days. The hydrolytic reactions by protease-immobilized microreactors were done repeatedly over 20 times intermitted by storage periods.
11. The substrate solution was pumped through the microreactor at a flow rate of 5.0 μl/min and yielded a reaction time of 5.2 min (13 cm of PTFE microtube volume: 26 μl). A reaction time is correlated with the flow rate of the substrate (12, 13). With an increase in flow rate from 2.5 to 50 μl/min, any free proteases or any cross-linked aggregations did not come off the protease-immobilized PTFE microtubes, demonstrating good mechanical and chemical stability.
12. Both immobilized proteases were more stable at high temperature than free proteases (Fig. 2b). At 50°C, free TY and free CT showed 15 and 52% of hydrolytic activities respectively, while the immobilized proteases kept at 30°C retained their activities.

13. Both immobilized proteases were more stable at 4 M of urea than free proteases (Fig. 2b). In addition, the proteolysis of Cyt-C by the CT-microreactor was also efficiently carried out even at high concentration of denaturant (3 M of guanidinium chloride) (12).

14. The ammonium acetate buffer is easily evaporated during an ESI-TOF MS measurement without the need for any desalting procedure. If an additional purification step using reversed-phase micropipette tips prior to MS analysis was necessary to remove excessive amounts of buffer salts, other buffer systems (e.g., PB and Tris) are acceptable for proteolysis but they could lead to sample losses especially of hydrophobic peptides due to their inherent affinity to reversed-phase surfaces, leading to lower sequence coverage.

15. Table 1 summarizes the ESI-TOF MS results of the protein digests using the protease-immobilized microreactors. The pI value of Cyt-C (horse, residues 2–104) is 9.6, suggesting that Arginine or Lysine residues locate on the protein surface with high probability. Therefore, Cyt-C was more efficiently digested by the TY-microreactor than by the CT-microreactor. The sequence coverage of Cyt-C by the TY-microreactor with 10 min of digestion time was similar to 18 h in-solution digestion time, indicating rapid and efficient proteolysis of the TY-microreactor. Similar results were obtained from the digested β-casein (bovine residue 16–224, pI = 5.1) by the CT-microreactor.

Thermally denatured proteins were efficiently digested by in-solution digestion (14) or by protease-microreactors using free proteases (15, 16). The protease-immobilized microreactors showed thermal stability (Fig. 2b). Lysozyme (chicken residue 19–147, pI = 9.3) which stabilizes its conformation by four disulfide bonds was thermally denatured during proteolysis and was efficiently digested by the immobilized but not by the free proteases. In addition, all four disulfide bonds on lysozyme were assigned from the digests by the TY-microreactor at 50°C (17).

Acknowledgments

The authors thank Dr. T. Honda for carrying out the initial experiments. Part of this work was supported by Grant-in-Aid for Basic Scientific Research (B: 20310074 and 23310092) from JSPS.

References

1. Aebersold, R, Mann, M. (2003) Mass spectrometry-based proteomics, *Nature* **422**, 198–207
2. Domon, B., Aebersold, R. (2006) Mass spectrometry and protein analysis, *Science* **312**, 212–217
3. Witze, E. S., Old, W. N., Resing, K. A., Ahn, N. G. (2007) Mapping protein post-translational modifications with mass spectrometry, *Nat. Methods* **10**, 798–806
4. Liu, Y., Liu, B., Yang, P., Girault, H. H., (2008) Microfluidic enzymatic reactors for proteome research, *Anal. Bioanal. Chem.* **390**, 227–229
5. Ma, J., Zhang, L., Liang, Z., Zhang, W., Zhang, Y. (2009) Recent advance in immobilized enzymatic reactors and their applications in proteome analysis, *Anal. Chim. Acta* **632**, 1–8
6. Miyazaki, M., Maeda, H. (2006) Microchannel enzyme reactors and their applications for processing, *Trends Biotechnol.* **24**, 463–470
7. Miyazaki, M., Honda, T., Yamaguchi, H., Briones, M. P. P., Maeda, H. (2008) in: Harding S. E. (Eds.), *Biotechnology & Genetic Engineering Reviews*, Nottingham University Press, Nottingham, **25**, 405–428
8. Ma, J., Zhang, L., Liang, Z., Zhang, W., Zhang, Y. (2007) Monolith-based immobilized enzyme reactors: Recent developments and applications for proteome analysis, *J. Sep. Sci.* **30**, 3050–3059
9. Honda, T., Miyazaki, M., Nakamura, H., Maeda, H. (2005) Immobilization of enzymes on a microchannel surface through cross-linking polymerization, *Chem. Commun.* 5062–5064
10. Honda, T., Miyazaki, M., Nakamura, H., Maeda, H. (2006) Facile preparation of an enzyme-immobilized microreactor using a cross-linking enzyme membrane on a microchannel surface, *Adv. Synth. Catal.* **348**, 2163–2171
11. Ma, J., Ziang, Z., Qiao, X., Deng, Q., Tao, D., Zhang, L., Zhang, Y. (2008) Organic-inorganic hybrid silica monolith based immobilized trypsin reactor with high enzymatic activity, *Anal. Chem.* **80**, 2949–2956
12. Yamaguchi, H., Miyazaki, M., Honda, T., Briones-Nagata, M. P., Arima, K., Maeda, H. (2009) Rapid and efficient proteolysis for proteomic analysis by protease-immobilized microreactor, *Electrophoresis* **30**, 3257–3264
13. Yamaguchi, H., Miyazaki, M., Kawazumi, H., Maeda, H. (2010) Multidigestion in continuous flow tandem protease-immobilized microreactors for proteomic analysis. *Anal. Biochem.* **407**, 12–18
14. Park, Z. -Y., Russell, D. H. (2000) Thermal denaturation: A useful technique in peptide mass mapping. *Anal. Chem.* **72**, 2667–2670
15. Sim, T. S., Kim, E. -M., Joo, H. S., Kim, B. G., Kim, Y. -K. (2006) Application of a temperature-controllable microreactor to simple and rapid protein identification using MALDI-TOF MS. *Lab. Chip* **6**, 1056–1061
16. Liu, T., Bao, H., Zhang, L., Chen, G. (2009) Integration of electrodes in a suction cup-driven microchip for alternating current-accelerated proteolysis. *Electrophoresis* **30**, 3265–3268
17. Yamaguchi, H., Miyazaki, M., Maeda, H. (2010) Proteolysis approach without chemical modification for a simple and rapid analysis of disulfide bonds using thermostable protease-immobilized microreactors. *Proteomics* **10**, 2942–2949

Chapter 16

Mass Spectrometry for Protein Quantification in Biomarker Discovery

Mu Wang and Jinsam You

Abstract

Major technological advances have made proteomics an extremely active field for biomarker discovery in recent years due primarily to the development of newer mass spectrometric technologies and the explosion in genomic and protein bioinformatics. This leads to an increased emphasis on larger scale, faster, and more efficient methods for detecting protein biomarkers in human tissues, cells, and biofluids. Most current proteomic methodologies for biomarker discovery, however, are not highly automated and are generally labor-intensive and expensive. More automation and improved software programs capable of handling a large amount of data are essential to reduce the cost of discovery and to increase throughput. In this chapter, we discuss and describe mass spectrometry-based proteomic methods for quantitative protein analysis.

Key words: Biomarkers, Proteomics, Mass spectrometry, Stable isotope labeling, Label-free protein quantification

1. Introduction

Quantitative proteomics has become a widely applied analytical tool for protein biomarker discovery (1–6). With improved software and computing tools for data processing, this technology has become a major force in pharmaceutical drug development and biomedical research in recent years. While two-dimensional gel electrophoresis (2DE) has gradually lost its popularity in proteomics for biomarker discovery due to its lack of the ability to widen the protein dynamic range and to analyze hydrophobic proteins or those with very high or low molecular weight, mass spectrometry (MS)-based shotgun proteomic technology has become the platform of choice for both biomarker discovery and validation (7–10). The performance of stable isotopic labeling and label-free protein

Michael Kaufmann and Claudia Klinger (eds.), *Functional Genomics: Methods and Protocols*,
Methods in Molecular Biology, vol. 815, DOI 10.1007/978-1-61779-424-7_16, © Springer Science+Business Media, LLC 2012

quantification technologies has improved significantly due to the improvement in instrumentation and algorithm development (11–15). These improvements provide the powerful tools needed to resolve and identify thousands of proteins from a complex biological sample in a high-throughput fashion. The MS-based approaches have been proven to be rapid and more sensitive than 2DE, and most importantly, they can also be automated and have the ability for high-throughput and large-scale proteomic analysis.

The stable isotopic labeling approach to catalog proteins in a complex biological sample was developed about a decade ago (11, 16–20), and it is now routinely used in proteome-wide studies as a method for the relative quantification of proteins to provide valuable information on the alterations in protein expression, interaction, and modification (21–23). More recently, absolute quantification using this approach has become possible (24–27). While a number of stable isotopic labeling approaches have been developed, three major methods are predominately utilized for biomarker discovery: (1) isotope-code affinity tags (ICAT), (2) isobaric peptide tags for relative and absolute quantification (iTRAQ), and (3) stable isotope labeling with amino acids in cell culture (SILAC). All of these methods employ differential stable isotopic labeling to create a specific mass tag that can be recognized by a mass spectrometer to provide relative quantification information. However, there are limitations associated with these methods including high reagent costs, an increased time for sample preparation, the complexity of samples, poor labeling efficiency, and a limit on the comparison of multiple samples (only up to eight) at the same time (see Note 1). Owing to these disadvantages, faster, cleaner, and simpler label-free shotgun technologies have generated increased interest for quantitative proteomic analysis.

The label-free relative protein quantification shotgun proteomics approach is a promising alternative to the stable isotope labeling approach and has been increasingly applied to the quantification of differentially expressed proteins from complex biological samples (28–31). This approach is simple and cost effective and has demonstrated high reproducibility and linearity when comparing peptide levels or protein abundance (32). One such strategy uses integrated ion abundances obtained from extracted ion chromatograms (XICs) to track peptide and protein expression levels across experimental samples. Several studies have demonstrated that XICs of selected peptide ions correlate well with protein abundances in complex biological samples (14, 15, 32). However, the application of this approach, especially to mammalian systems, requires more robust computing power and algorithms capable of handling both chromatographic peak alignments and peptide ion intensity measurements to analyze changes in protein abundances in complex biological samples. Another promising method for label-free protein quantification is spectral counting, where the number of mass spectra used for the identification of the protein is employed as an indicator to measure the protein abundance. Both of these label-free methods are compatible, with the peak area integration method showing a higher accuracy, and the spectral counting method having better sensitivity (33) (see Notes 2 and 3).

2. Materials

2.1. Stable Isotopic Labeling

2.1.1. Isotope-Coded Affinity Tags

1. Isotope-coded affinity tags (ICAT) labeling buffer: 0.05% SDS, 5–200 mM Tris–HCl, pH 8.3, 5 mM EDTA, 6 M urea (see Note 4).
2. ICAT reagents: both heavy and light ICAT labeling reagents are shown in Fig. 1 (Applied Biosystems/MDS SCIEX, Foster City, CA, USA).
3. Avidin Buffer pack and Avidin Affinity Cartridge (Applied Biosystems/MDS SCIEX, Foster City, CA, USA).

2.1.2. Isobaric Peptide Tags for Relative and Absolute Quantification

1. Sample homogenization buffer: 50 mM ammonium bicarbonate, pH 8, 0.1% SDS, and protease inhibitors.
2. Strong cation exchange (SCX) HPLC buffer:

   Buffer A: 10 mM KH_2PO_4, 25% acetonitrile (ACN), pH 3.0 with H_3PO_4, Milli-Q water to 1 L.

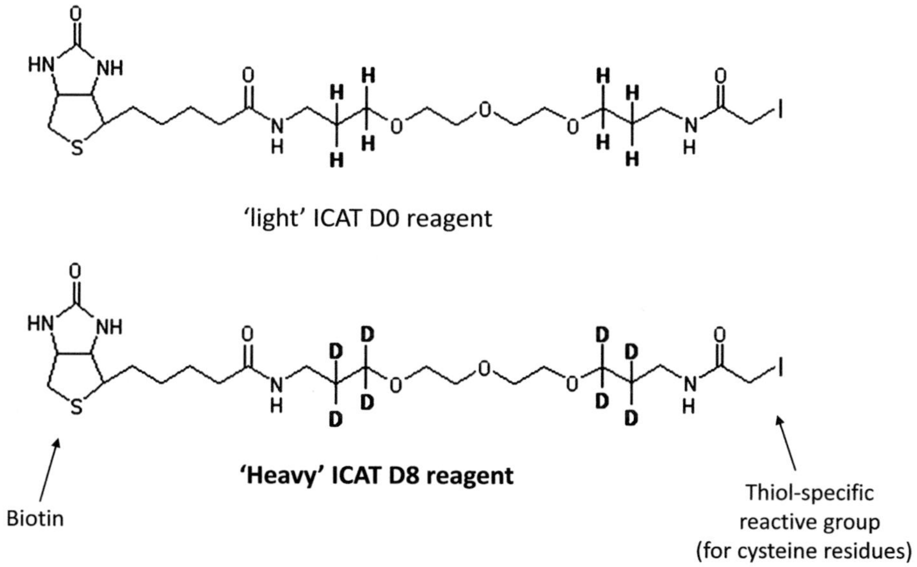

'light' ICAT D0 reagent

'Heavy' ICAT D8 reagent

Biotin

Thiol-specific reactive group (for cysteine residues)

Fig. 1. Structure of the ICAT reagent. The reagent consists of three functional groups: (1) an affinity tag (biotin), which is used to purify ICAT-labeled peptides; (2) a linker in which stable isotopes are incorporated ("heavy" or D8 – eight deuteriums and "light" or D0 – eight hydrogens), making the mass difference between the two reagents 8 Da; (3) a thiol-reactive group that can specifically react with thiol groups of cysteines for ICAT labeling.

Buffer B: 10 mM KH_2PO_4, 25% ACN, pH 3.0 with H_3PO_4, 500 mM KCl, Milli-Q water to 1 L.

Conditioning buffer: 0.2 M NaH_2PO_4, 0.3 M sodium acetate, Milli-Q water to 500 ml.

3. Reversed-phase (RP) HPLC buffer:

   Buffer A: 5% ACN, 0.1% trifluoroacetic acid (TFA), Milli-Q water to 1 L.

   Buffer B: 80% ACN, 0.1% TFA, Milli-Q water to 1 L.

   MALDI matrix solution: 3 mg/ml alpha-cyano-4-hydroxycinnamic acid, 70% ACN, 0.1% TFA, neurotensin for internal calibration of MALDI-MS/MS.

4. iTRAQ™ Reagents: Dissolution Buffer, Denaturant, Reducing Reagent, Cysteine Blocking Reagent, Cation Exchange Buffer-Load, Cation Exchange Buffer-Clean, Cation Exchange Buffer-Elute, and Cation Exchange Buffer-Storage are all from Applied Biosystems/MDS SCIEX (Foster City, CA, USA).

2.1.3. Stable Isotope Labeling with Amino Acids in Cell Culture

1. Colloidal Blue Staining Kit (Catalog # LC6025 Invitrogen, Carlsbad, CA, USA).
2. Amino-acid stock solutions: prepare concentrated stock solutions by dissolving amino acids in PBS or nonrestituted culture medium. Arginine (84 mg/ml), lysine (146 mg/ml) and methionine (30 mg/ml) are prepared as 1,000× concentration stocks for use in DMEM. Filter amino-acid solutions through a 0.22-μm syringe filter and store at 4°C for up to 6 months. Stocks can be frozen for long-term storage. Prepare stock solutions at a high concentration to minimize the dilution of the culture medium. Stable isotope-labeled amino-acid stock solutions are prepared in the same manner but the increased molecular weight of the amino acids bearing ^{13}C or ^{15}N should be taken into account to give equal molar amounts in both light and heavy media. For instance, L-arginine-$^{13}C_6$ is prepared at a concentration of 87.4 mg/ml.
3. Modified RIPA buffer for affinity purifications: 50 mM Tris–HCl, pH 7.8, 150 mM NaCl, 1% (*v/v*) NP-40, 0.25% sodium deoxycholate and 1 mM EDTA. Prepare as a stock solution and store at room temperature (25°C). Add protease inhibitors (e.g., Roche Complete tablet) right before use and chill on ice. Add the appropriate phosphatase and kinase inhibitors if preparing a sample for phosphorylation studies.
4. 4× SDS gel-loading buffer: 200 mM Tris–HCl (pH 6.8), 8% (*w/v*) SDS (electrophoresis grade), 0.4% (*w/v*) Colloidal Blue and 40% (*v/v*) glycerol. Use this 4× SDS buffer stock when preparing protein samples for SDS–PAGE analysis. Just before boiling the samples, add dithiothreitol (DTT) (from a 1 M stock solution) to give a final concentration of 100 mM.

2.2. Label-Free

1. Lysis buffer: 8 M urea and 10 mM DTT, freshly made every time 30 min before the start of the experiment.
2. Reduction/alkylation cocktail: 97.5% acetonitrile, 2% iodoethanol, and 0.5% triethylphosphine prepared in 200 μl aliquots and stored at −80°C.
3. Reagents for mass spectrometry: both acetonitrile and deionized water should be mass spec grade. Formic acid (FA) is dissolved in acetonitrile/water at 0.1%.

3. Methods

3.1. Stable Isotopic Labeling Experiments

3.1.1. ICAT (see Notes 5 and 6).

The ICAT workflow is shown in Fig. 2.

1. Dissolve protein in 5 mM Tris–HCl, pH 8.3, 10 mM EDTA and boil protein sample for 5 min; then, chill on ice for 20 min (see Note 7).
2. Add reducing agent [DTT, tributylphosphine (TBP), or Tris(2-carboxyethyl)phosphine (TCEP)] to a final concentration of

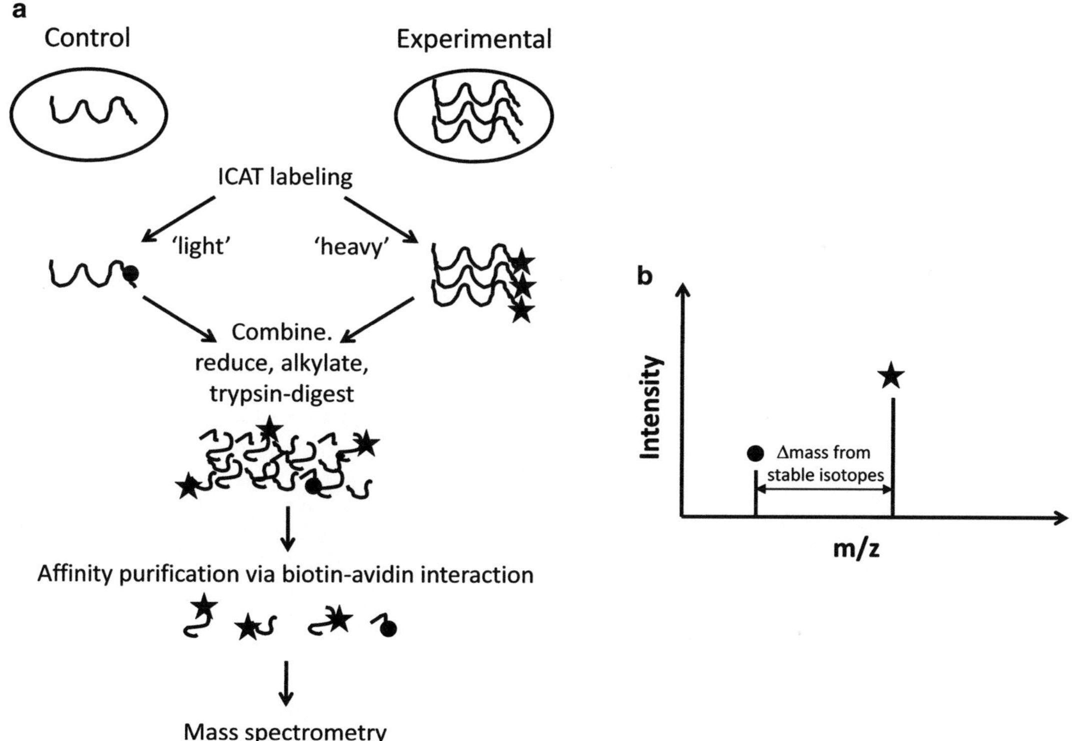

Fig. 2. ICAT strategy for relative quantification. (**a**) Two different conditions (control and experimental) are treated with "light" and "heavy" ICAT reagents, respectively. The protein mixtures are combined and proteolyzed to produce tryptic peptides. ICAT labeled peptides are purified by an avidin column and analyzed by MS. (**b**) Relative protein quantification is determined by the ratio of the peptides from "light" (control) and "heavy" (experimental) samples. Proteins are identified by a protein database search engine such as SEQUEST® or Mascot™.

5 mM (see Note 8). Incubate at 37°C for 20 min. If using DTT or TBP as a reducing agent, precipitate the proteins (to remove DTT) using cold acetone. Add 6–7 volumes of acetone stored at –20°C. Allow the protein to precipitate out of solution. Spin the sample, collect, dry the pellet, and resuspend in a minimal amount of 5 mM Tris–HCl, pH 8.3, 10 mM EDTA. If using TCEP, cold acetone precipitation is optional.

3. Dissolve the ICAT reagent in 20 μl of acetonitrile per tube. Add the sample to the ICAT tube at a minimum concentration of sixfold molar excess of ICAT reagent. This typically corresponds to one tube for every 200 μg of protein labeled. It is important that the ICAT reaction is conducted in as small of a volume as possible, ideally less than 150 μl per tube, or the ICAT reagent concentration will be too low to allow for adequate labeling (see Note 9).
4. Incubate the reaction at room temperature in the dark (put in a box covered with foil) for a minimum of 2 h. Agitation/gentle shaking is typically done.
5. Check ICAT labeling as a shift in the protein bands on SDS–PAGE for labeling efficiency.
6. The reaction can be stopped by adding 5 mM DTT and incubating at 55°C for 20 min.
7. Combine the ICAT labeled samples at the desired ratio and make sure that the final urea concentration is ~1 M (see Note 10).
8. Digest proteins with sequencing grade modified trypsin at 1:50 *w/w* (trypsin/sample) overnight at 37°C.
9. Prior to avidin purification, samples may be cleaned up by cation exchange separation.
10. Purify the ICAT labeled peptides by the avidin column according to the manufacturer's protocol.
11. Add 95 μl of cleaving reagent A and 5 μl of cleaving reagent B to each avidin purified fraction and incubate at 37°C for 2 h (see Note 11).
12. Add 8 μl of 0.2–0.4% acetic acid to each fraction, vortex and spin down. Transfer supernatant to a new tube for MS analysis.

3.1.2. iTRAQ

The iTRAQ workflow is shown in Fig. 3. For a duplex-type experiment, up to four samples can be prepared and analyzed in a single experiment.

1. Proteins are extracted from tissues, cultured cells, or biofluids in freshly made homogenization buffer (200 μl) by ten up and down strokes of a 27 G syringe.
2. Sonicate for 2 min.

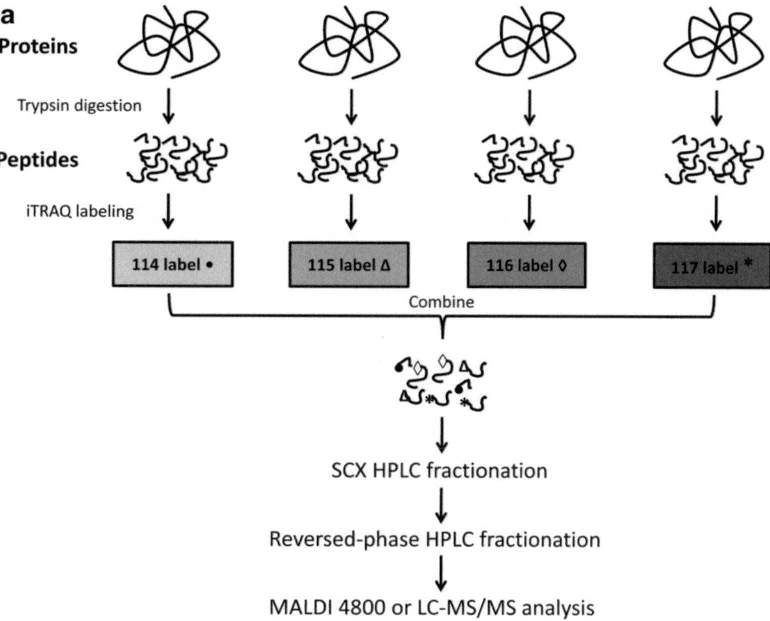

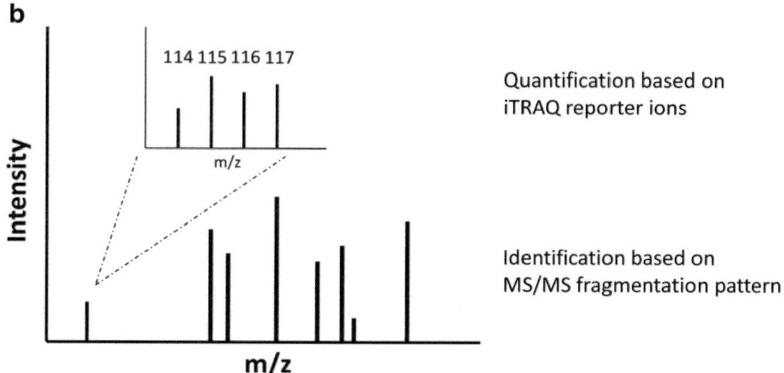

Fig. 3. iTRAQ workflow and data analysis process. (**a**) iTRAQ labeling and MS data acquisition. (**b**) Protein identification is carried out based on the MS/MS fragmentation patterns, while relative quantification is determined by comparing the peak intensities of the iTRAQ reporter ions.

3. Centrifuge at 12,000 × *g* to pellet insoluble materials.
4. Determine the total protein concentration using the Bradford assay (34).
5. It is generally recommended to clean up samples by acetone precipitation (to remove any undesired interfering agents, e.g., DTT and detergents). Add 6 volumes of cold acetone to the cold sample tube, invert the tube three times, incubate at −20°C until a precipitate forms (~1 h). Centrifuge and decant the acetone. Air-dry.
6. To each tube containing 5–100 μg of sample, add 20 μl Dissolution Buffer.

7. Add 1 μl of the Denaturant and vortex to mix.
8. To each sample tube, add 2 μl Reducing Reagent; vortex to mix and incubate at 60°C for 1 h; spin and add 1 μl Cysteine Blocking Reagent; mix and incubate at room temperature for 10 min.
9. Add 10 μl of the trypsin solution (final concentration of 5–10 ng/ml), mix, and incubate at 37°C overnight.
10. Add the iTRAQ™ reagents, which are individually dissolved in 70 μl of ethanol, to different sample tubes (e.g., the iTRAQ™ Reagent 114 to the sample #1 protein digest, and the iTRAQ Reagent 117 to the sample #2 protein digest), vortex and incubate at room temperature for 1 h (see Notes 12 and 13).
11. Combine the contents of each iTRAQ™ Reagent-labeled sample tube into one tube.
12. Mix and dilute the concentrations of the buffer salts and organics by tenfold with the Cation Exchange Buffer-Load.
13. Check the pH. If the pH is not between 2.5 and 3.3, adjust by adding more Cation Exchange Buffer-Load.
14. Use 1 ml of the Cation Exchange Buffer-Clean to condition the cartridge, followed by an injection of 2 ml of the Cation Exchange Buffer-Load.
15. Slowly inject (~1 drop/s) the diluted sample mixture onto the cation-exchange cartridge and collect the flow-through in a sample tube.
16. Inject 1 ml of the Cation Exchange Buffer-Load to wash the TCEP, SDS, $CaCl_2$, and excess iTRAQ™ Reagents. Save the flow-through until it is verified by MS/MS.
17. The peptides are eluted by the Cation Exchange Buffer-Elute. Collect the eluted peptides in a fresh 1.5 ml tube as a single fraction. Speedvac dry.
18. Clean the cartridge by injecting 1 ml of the Cation Exchange Buffer-Clean, and condition it by injecting 2 ml of the Cation Exchange Buffer-Load.
19. Repeat steps 14–18 if additional sample mixtures need to be prepared. Otherwise, the clean cartridge can be stored at 2–8°C after washing with 2 ml of the Cation Exchange Buffer-Storage.
20. Dissolve samples in Buffer A and transfer to injection vials for MS/MS analysis.

3.1.3. SILAC

SILAC is a simple and straightforward approach for in vivo incorporation of a label into proteins for MS-based quantitative proteomic analysis (the workflow is shown is Fig. 4). SILAC relies on the metabolic incorporation of a given "light" or "heavy" form

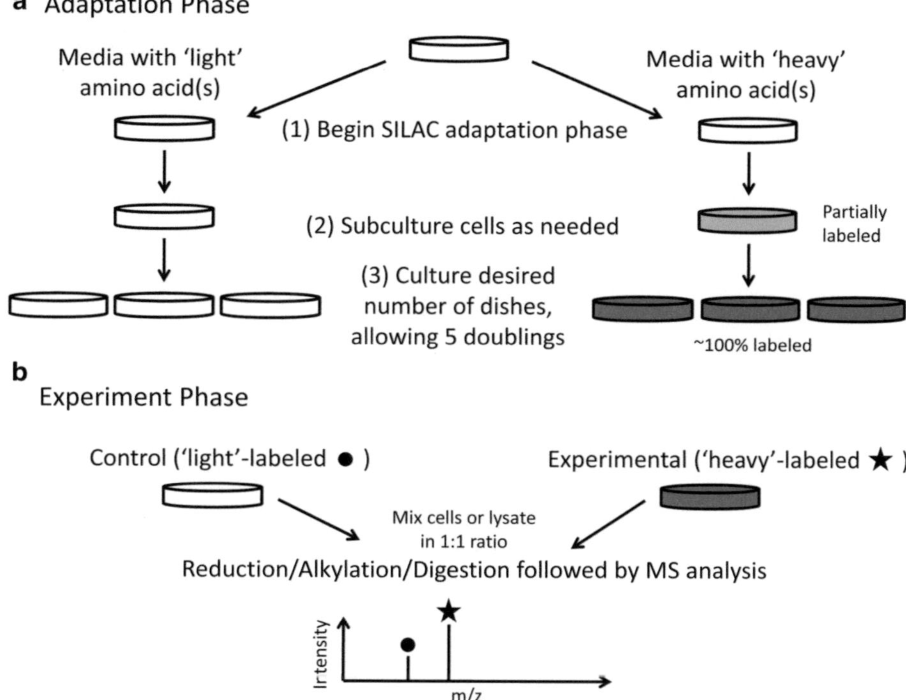

Fig. 4. SILAC experimental workflow. Two distinct phases are involved in SILAC: (**a**) Adaptation phase: cells are grown in DMEM media with either "light" or "heavy" amino acid(s) until the "heavy" cells have fully incorporated the "heavy" amino acid(s), which can be monitored by MS. This phase also includes the expansion of cells to reach the required number of dishes; (**b**) Experiment phase: two cell populations are mixed in this phase in a 1:1 ratio. If necessary, the subproteome can be either purified (e.g., membrane prep) or enriched (e.g., immune-affinity pull-down) after mixing. The resulting cell lysates are then reduced, alkylated and digested, and tryptic peptides are analyzed by MS for protein identification and quantification. The intensity of MS signals between "light" (control) and "heavy" (experimental) peptides allow for relative protein abundance determination.

of an amino acid into the proteins in a complex biological sample. Thus in a typical experiment, two cell populations are grown in culture media that are identical except that one of them contains a "light" and the other a "heavy" form of a particular amino acid (e.g. ^{12}C and ^{13}C labeled L-lysine, respectively). When the labeled analog of an amino acid is supplied to cells in culture instead of the natural amino acid, the labeled amino acid is incorporated into all newly synthesized proteins. After a number of cell divisions, each instance of this particular amino acid will be replaced by its isotope labeled analog. Since there is hardly any chemical difference between the labeled amino acid and the natural amino acid isotopes, the cells behave exactly like the control cell population grown in the presence of normal amino acids. It is efficient and reproducible as the incorporation of the isotope label is 100%.

1. Preparation of SILAC DMEM media: where necessary, restore common amino acids in the "light" and "heavy" SILAC medium to a 1-L bottle of custom-synthesized SILAC dropout medium.

For example, add normal stable isotope abundance, $^{12}C^{1}H_3$-methionine, to both "light" and "heavy" medium when SILAC labeling with arginine and lysine only (see Note 14).

2. Divide methionine-restituted, SILAC dropout medium in two equal volumes into the pre-filter containers of two 0.22-μm vacuum filter flasks.
3. Add appropriate amounts of "light" and "heavy" arginine and lysine to the "light" and "heavy" media, respectively, and filter media by applying a vacuum to the filter flasks (see Note 15).
4. Add supplementary antibiotics, glutamine and 10% dialyzed fetal bovine serum to both medium bottles.
5. Adaptation of cells from normal DMEM to SILAC medium: split the cells growing in normal DMEM medium formulation (80–90% confluency) into two culture dishes, one containing "light" and one containing "heavy" SILAC medium. Seed each dish with 10–15% of the cells from the original dish (or an appropriate cell density for your specific cell line) to allow at least two doublings in fresh SILAC medium.
6. Change medium (using either "light" or "heavy" SILAC medium) every 2–3 days if cells are not ready for subculture.
7. Subculture SILAC "light" or "heavy" cells in their respective media before cells reach confluent culture.
8. After the first passage, begin expansion of the "light" and "heavy" SILAC cell populations by culturing cells in larger dishes or a larger number of plates, as needed (see Note 16).
9. Subculture the cells at least twice in their respective SILAC medium and allow at least five cell doublings. This takes about 5 days assuming a doubling rate of 24 h (see Notes 17 and 18).
10. Checking for full incorporation of the SILAC amino acid: lyse an aliquot of cells at the end of the adaptation phase by adding 300 μl of a mix of 6 M urea and 2 M thiourea to a 60-mm culture dish (see Notes 19–21).
11. Scrape the dish with a cell scraper and pipette the lysate into a microcentrifuge tube; vortex the lysate intermittently for 5 min.
12. Pellet the debris by centrifuging for 10 min at 16,000–20,000 × *g* in a benchtop centrifuge at 18°C.
13. Collect the supernatant in a new microcentrifuge tube, taking care to avoid DNA and the cell pellet. DNA may not pellet completely and appears as a colorless gel-like clump, which can be easily removed when aspirating with the pipette.
14. Estimate the protein concentration using a standard protein assay (e.g., Bradford (34) or BCA (35)) and use about 25–50 μg of protein for the in-solution digest.

15. Reduce disulfide bonds by adding DTT to a final concentration of 1 mM and incubate the microcentrifuge tube at 37°C for 30 min (see Note 22).
16. Cool to room temperature, add iodoacetamide to a final concentration of 5 mM to alkylate cysteines, vortex, and incubate in the dark for 20 min.
17. Add 5 mM DTT to remove excess acetamide.
18. Add an equal volume of 50 mM ammonium bicarbonate to reduce the final concentration of urea to 3 M.
19. Add endoproteinase Lys-C at an enzyme to substrate ratio of 1:100 and incubate at 37°C for 2 h.
20. Dilute the lysate with an equal volume of 50 mM ammonium bicarbonate to make a urea concentration of less than 1.5 M.
21. Add trypsin at an enzyme to substrate ratio of 1:50 and incubate at 37°C for 4 h or overnight.
22. Stop digestion by acidifying until there is a final concentration of 1% TFA (see Note 23).
23. Analyze the sample by nanoflow LC–MS/MS to identify proteins and peptides.
24. Determine the degree of incorporation by looking for the presence of "light" peptides by MS.

3.2. Label-Free Protein Quantification Experiments

A general label-free protein quantification technology workflow is shown in Fig. 5. Special software or algorithms are required for data processing and quantification by either peak intensity or spectral counting.

1. Protein extraction, reduction, alkylation, and digestion: in general, proteins are extracted from tissues, cultured cells, or biofluids in freshly made lysis buffer containing 8 M urea and 10 mM DTT. For cells grown in 6-well plates, culture media are first removed by aspiration. Then, 100 μl of lysis buffer is added to each well and the cells are lysed by pipetting. Lysed cells are transferred to microcentrifuge tubes and incubated at room temperature with gentle agitation. Unlysed cells and insoluble particles are removed by centrifugation (12,000×g for 5 min) prior to the protein assay. In order to take the same amount of proteins from each sample, protein concentrations are measured by the Bradford assay (34). The same lysis buffer should be used as the background reference for the protein assay to obtain a relatively accurate measurement among all samples (due to the presence of urea in lysis buffer).
2. Resulting protein extracts are subsequently reduced and alkylated with DTT and iodoacetamide to block sulfhydryl groups in proteins. Alternatively, the volatile reagents triethylphosphine and iodoethanol can be used instead. This volatile reduction

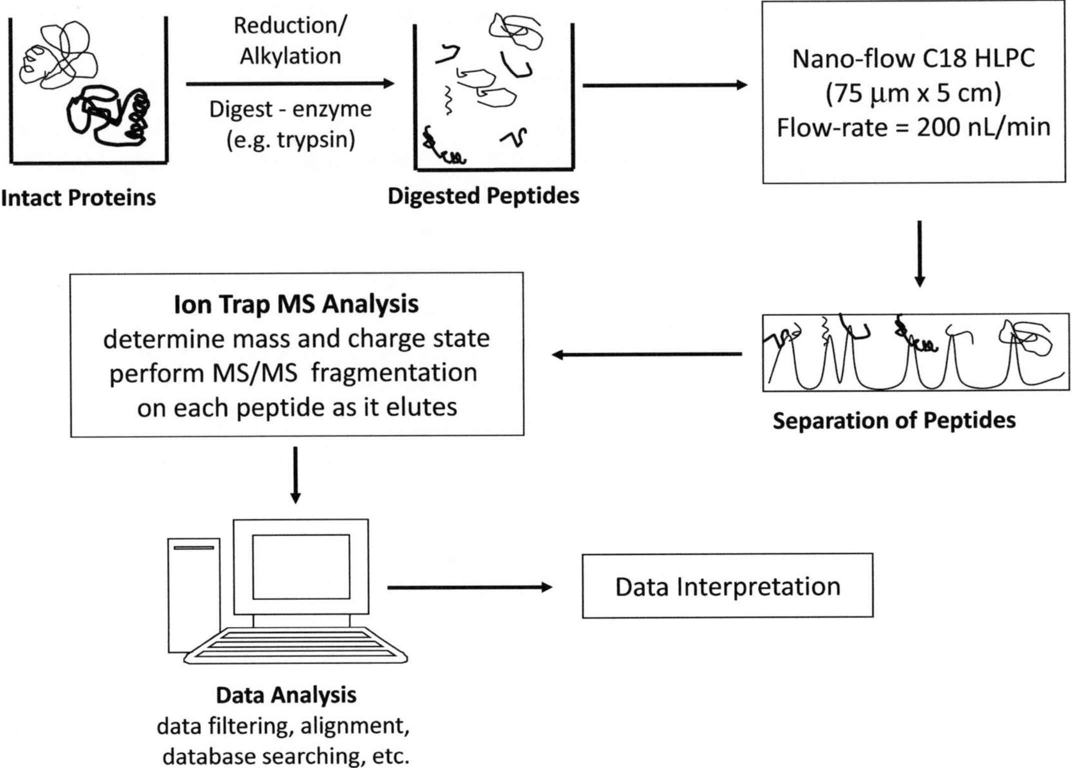

Fig. 5. The LC/MS-based label-free protein quantification technology workflow. Depending on the software or algorithms applied, relative protein quantification can be obtained by either spectral counting or chromatographic-peak-intensity (AUC) measurements.

and alkylation protocol has been described previously (36). The advantage of this protocol is that it allows all sample preparation steps to be carried out in one tube without washing, filtering, or sample transferring, which minimizes sample preparation variations. Briefly, 100 µg of protein are taken from each sample and pH values are raised by adding 100 µl of 100 mM ammonium carbonate (pH 11). Then, 200 µl of the reduction/alkylation cocktail (97.5% acetonitrile, 2% iodoethanol, and 0.5% triethylphosphine) are added to the protein samples and incubated for 1 h at 37°C. After incubation, protein samples are dried in a speedvac overnight to remove residual chemicals.

3. Protein mixtures are then digested by sequencing grade modified trypsin. Dried pellets are resuspended in 500 µl of 100 mM ammonium bicarbonate buffer containing 2 µg of trypsin and incubated at 37°C overnight. The protein to trypsin ratio is 50:1 (*w*/*w*), and urea concentration should be below 1.6 M. Tryptic digests are filtered with 0.45-µm spin filters.

3.3. Mass Spectrometric Analysis

1. All digested samples should be randomized for injection order to remove systematic bias from the data acquisition. Typically, up to 2 μg of the tryptic peptides are required for the injection onto a C_{18} nanoflow column (i.d. = 75 μm, length = 5 cm). Peptides are eluted with a linear gradient from 5 to 40% acetonitrile developed over 120 min at a flow rate of 200 nl/min, and effluent is electro-sprayed into a LTQ or LTQ-Orbitrap mass spectrometer (Thermo-Fisher Scientific).
2. The electro-spray ionization (ESI) source is operated with a 2 kV potential and a capillary temperature of 200°C. The instrument is tuned using an angiotensin I peptide. The max ion time is set to 200 ms for the parent ion scan and to 500 ms for the zoom scan and MS/MS scan. This method requires all the MS data be collected in the data-dependent "Triple-Play" mode (MS scan, Zoom scan, and MS/MS scan). Parent ion scans and MS/MS scans are collected in "Centroid" mode, and zoom scans are collected in "Profile" mode. Dynamic exclusion is set to a repeat count of one, an exclusion duration of 60 s, and rejection widths of −0.75 and +2.0 *m/z*.
3. Database searches against the International Protein Index (IPI) and/or the Nonredundant (NCBI) databases are carried out using SEQUEST®, X!Tandem, or Mascot algorithms, or a combination of two or three of these search engines. Protein identification confidence can be assessed using the algorithm described by Higgs et al. (37) or other publicly available algorithms (e.g., ProteinProphet™, which is an open source software available at http://proteinprophet.sourceforge.net/).

3.4. Protein Identification

1. Proteins identified by search engines such as SEQUEST® and X!Tandem are generally categorized into priority groups based on the confidence of the protein identification as shown in Table 1. Each algorithm compares the observed peptide MS/MS spectrum and a theoretically derived spectra from the database to assign quality scores (*XCorr* in SEQUEST® and *E-Score* in X!Tandem). These quality scores and other important predictors are combined in the algorithm that assigns an overall score, %ID confidence, to each peptide. The assignment is based on a random forest recursive partition supervised learning algorithm (38). The %ID confidence score is calibrated so that approximately *X*% of the peptides with %ID confidence > *X*% are correctly identified.
2. The confidence in protein identification is increased with the number of distinct amino acid sequences identified. Therefore, proteins are also categorized depending on whether they have only one or multiple unique sequences at the required confidence. A protein will be identified with a higher confidence if it has at least two distinct amino acid sequences with a required

Table 1
Prioritization of protein identification

Priority	Protein ID confidence	Multiple sequences
1	HIGH (90–100%)	Yes (≥2 unique sequences)
2	HIGH (90–100%)	No (single sequence)
3	MODERATE (75–89%)	Yes (≥2 unique sequences)
4	MODERATE (75–89%)	No (single sequence)

peptide ID confidence. Many researchers would view any protein identification with only a single amino acid sequence as questionable (39).

3.5. Protein Quantification

3.5.1. Stable Isotopic Labeling Approach

ICAT

ICAT protein quantification is carried out by integrating the peaks for both the "light" and "heavy" labeled peptide pairs identified on reconstituted chromatograms. Reconstituted chromatograms are obtained after the extraction of a specific mass (e.g., ±0.1 Da) from the LC–MS data using a commercial software package such as ProQuant (Applied Biosystems/MDS SCIEX) or Mascot (Matrix Science).

iTRAQ

1. The iTRAQ tags consist of a reporter group, a balance group and a peptide reactive group that covalently binds to the peptides. The balance group gives all of the tags the same mass during peptide mass fingerprinting. In the collision-induced dissociation (CID) stage of a tandem mass spectrometer, there is a neutral loss of the balance group, and the reporter groups are detected in the second MS. The tandem mass spectra include contributions from each sample, and the individual contributions of each sample can therefore be measured by the intensity of the reporter ion peaks (see Note 24).
2. Protein quantification for iTRAQ reporter ions requires special software packages. Different results may be obtained from different software packages depending on how the peak intensity is calculated for the peptide quantification and on how the peptide abundances are averaged for the protein quantification. Currently, iTRAQ data analysis can be carried out by Mascot (Matrix Science), I-Tracker (40), Libra (http://tools.proteomecenter.org/Libra.php), ProQuant (Applied Biosystems/MDS SCIEX), Peaks (http://www.bioinformaticssolutions.com), and SpectrumMill (Agilent), just to name a few.
3. All datasets should be exported in the mzData format and imported into the selected software package. In general,

quantification is carried out at the peptide level. Ratios for peptide matches (by comparing reporter ion intensities) are reported based on the peptides' PTMs, the minimum precursor charge, the strength of the peptide match, and the minimum number of fragment ion pairs, among others. A protein abundance ratio is an average or weighted average from a set of peptide ratios. Statistical considerations are usually incorporated into these software packages. Testing for outliers and reporting a standard deviation for the protein ratio can only be performed if the peptide ratios are consistent with a sample from a normal distribution. For each protein, all supporting peptides have their weights normalized to one, and then weighted averages are calculated.

SILAC

Relative protein quantification from the SILAC strategy is accomplished by comparing the intensities of the isotope clusters of the intact peptide in the survey spectrum to the full metabolic protein labeling by ^{13}C or ^{15}N (see Note 25).

3.5.2. Label-Free Approaches

Peak Intensity-Based Quantification

1. One of the key features of the algorithm described here for protein quantification based on ion intensity is the chromatographic peak alignment because large biomarker studies can produce chromatographic shifts due to multiple injections of the samples onto the same HPLC column. Un-aligned peak comparison will result in larger variability and inaccuracy in peptide quantification (Fig. 6) (38). All peak intensities are transformed to a $\log_2$ scale before quantile normalization (41) (see Note 26).
2. Quantile normalization is a method of normalization that essentially ensures that every sample has a peptide intensity histogram of the same scale, location and shape. This normalization procedure removes trends introduced by sample handling, sample preparation, total protein differences and changes in instrument sensitivity while running multiple samples. If multiple peptides have the same protein identification, then their quantile normalized $\log_2$ intensities are averaged to obtain $\log_2$ protein intensities. The average of the normalized $\log_2$ peptide intensities is a weighted average. A peptide is weighted proportionally to the peptide ID confidence for its protein category and receives a weight of zero if it is outside that category. For example, peptides with <90% confidence contribute zero quantitative weight to a "HIGH" category protein. The log transformation serves two purposes. First, relative changes in protein expression are best described by ratios. However, ratios are difficult to statistically model and the log transformation converts a ratio to a difference which is easier to model. Second, as is frequently the case in biology, the data better approximate

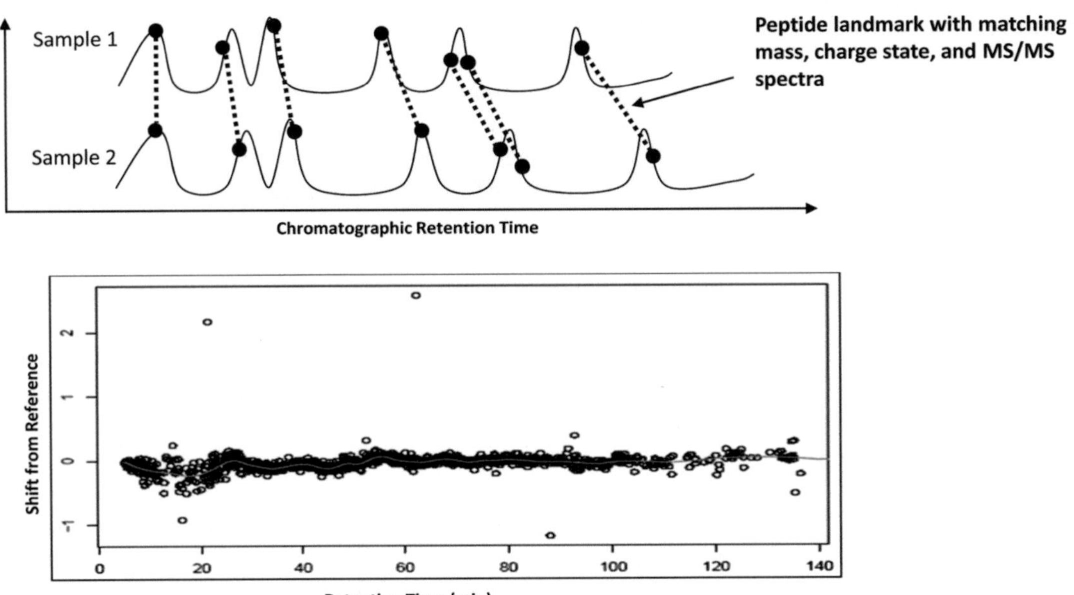

A time shifting function puts the 2 samples on the same time scale

Fig. 6. Chromatographic peak alignment requires that all landmark peaks match the peptide mass, charge state, MS/MS spectra, and retention time within a 1-min window.

the normal distribution on a log scale (42). This is important because normality is an assumption of the analysis of variance (ANOVA) statistical model used to analyze the data. The base of the log transform is arbitrary chosen as 2, the most common base with genomic data. Base 2 is popular because a twofold change (or doubling, or 100% increase) yielding an expression ratio of 2 is transformed to 1 on a log base 2 scale (e.g., a two-fold change is a unit change on the log base 2 scale). The $\log_2$ protein intensity is the final quantity that is fit by a separate ANOVA model for each protein shown below:

$$\begin{aligned}\log_2(\text{intensity}) \quad &= \text{overall mean} + \text{group effect (fixed)} \\ &+ \text{sample effect (random)} \\ &+ \text{replicate effect (random)}\end{aligned}$$

In this model, group effect refers to the effect caused by the experimental conditions or treatments being evaluated. Sample effect represents the random effects from individual biological samples. It also includes random effects from sample preparation. The replicate effect refers to the random effects from replicate injections of the same sample. All of the injections should be therefore randomized and the instrument be operated by the same operator for a particular study. The inverse $\log_2$ of each sample mean is calculated to determine the fold change between samples.

3. To qualify for protein quantification using this peak intensity label-free protein quantification method, each aligned peak must match precursor ion (MS data), charge state (zoom scan data), fragment ions (MS/MS data) and retention time (e.g., within a 1-min window). After alignment, the area-under-the-curves (AUCs) for individually aligned peaks from identified peptides from each sample are computed; the AUCs are then compared for relative protein abundances, followed by statistical analysis (e.g., ANOVA) to determine the significance of the changes (see Notes 27–29).

Spectral Counting-Based Quantification

1. Relative quantification by spectral count has been widely applied in different biological complexes (33, 43–49). One of the reasons that this method has become so popular is because it utilizes the same data processing steps that the general protein identification workflow in proteomics uses. In this approach, relative protein quantification is achieved by comparing the number of identified MS/MS spectra from the same protein in each of the multiple LC–MS/MS or LC/LC–MS/MS datasets. It has been shown that an increase in protein abundance typically results in an increase in the number of its proteolytic peptides, and vice versa. This increased number of tryptic peptides usually results in an increase in protein sequence coverage, the number of identified unique peptides, and the number of identified total MS/MS spectra (spectral count) for each protein (50) (see Note 30).
2. "RAW" files from LC–MS/MS analysis are processed using Xcalibur (Thermo-Fisher Scientific, Waltham, MA, USA) or Mascot Distiller (Matrix Science, Boston, MA, USA) to generate peak lists for database searching. Protein quantification based on spectral counting can be accomplished using commercially available software such as ProteoIQ (BioInquire, Athens, GA, USA) by importing the data generated from the database search algorithm (e.g., SEQUEST® or Mascot) and comparing by spectral count for large-scale studies (see Note 31).
3. Unlike XIC-based label-free method, which requires specialized software for peak alignment and comparison, no special software tools are required for spectral counting. However, normalization and statistical analysis are necessary to produce accurate and reliable data. A simple normalization method based on total spectral counts has been used extensively for this purpose (51).
4. Many different statistical tools have been applied to evaluate the significance of comparative quantification by spectral counting (52). The Fisher's exact test, Goodness-of-fit (*G*-test), AC test, Student's *t*-test, local-pooled-error (LPE), and most recently QSpec are a few examples.

3.6. Quality Assurance and Quality Control

To assess the stability of the HPLC system and MS instrument, a known purified standard protein is commonly spiked into every sample before tryptic digestion at a constant amount as an internal reference for assessment of technical variations. Several considerations should be given for the selection of the standard: (1) the protein must not come from the same species as the sample of interest; (2) a series of signature peptides should be easily detected and identified by the instrument; and (3) the amount of the standard protein spiked into each sample should be comparable to the amount of median abundant proteins in the sample. After global protein identification and quantification, these peptides and their relative quantities should be inspected for quality assurance (QA)/quality control (QC) purposes.

3.7. Statistical Analysis

The number of significant changes between groups, the fold changes and the variability (CV) for each Priority level can be determined from the ANOVA. The threshold for significance is set to control the False-Discovery-Rate (FDR) for each comparison at an investigator-desired percentile, normally 5% (53). The FDR is estimated by the *q*-value which is an adjusted *p*-value. The FDR is the proportion of significant changes that are false positives. If proteins with a *q*-value ≤ 0.05 are declared significant, it is expected that 5% of the declared changes will be false positives. For example, in the Peak Intensity-based label-free method, the *p*-value to *q*-value adjustment is done separately for Priority 1, Priority 2 and the MODERATE confidence categories (Table 1).

Fold change (FC) is computed from the means on the AUC scale (anti-log) as follows:

FC = Mean Treated Group/Mean Control Group

When Mean Treated Group ≥ Mean Control Group (up-regulation)

FC = – Mean Control Group/Mean Treated Group

When Mean Control Group > Mean Treated Group (down-regulation)

Absolute FC = |FC| = absolute or positive value of the FC

A fold change of 1 means there is no change. Also the median % Coefficient of Variation (%CV) for each Priority level is determined. The %CV is the standard deviation/mean on a % scale. The %CV is given both for the replicate variation as well as the combined replicate plus sample variation (see Note 32).

3.8. Pathway Analysis and Protein Classification

1. To understand the biological significance of the protein expression changes, which will help to identify biomarker candidates, the results including protein IDs and FC from the proteomic study can then be analyzed using protein–protein interaction and/or pathway analysis software packages (e.g., Pathway Studio® from Ariadne or Ingenuity Pathway Analysis® from Ingenuity Systems). These software packages allow for the

creation of protein–protein interaction networks, biological pathways, and gene regulation networks from a dataset, which can be helpful in better understanding specific biological processes that are involved in a particular study.

2. Proteins can also be classified into different categories based on their biological function, cellular location, and/or molecular pathway. It is desired that, when selecting the panel of biomarker candidates, the function(s) or biological pathway(s) they are involved in are known.

4. Notes

1. Relative quantification based on stable isotope labeling can be achieved by signal comparison in survey MS spectra (e.g., SILAC and ICAT) or tandem MS spectra (MS/MS, e.g., iTRAQ). However, much attention needs to be paid in the design and analysis of the experimental data. It is a generally valid assumption that stable isotopes do not alter the chemical or physical properties of the peptide, but it has been observed that deuterated peptides show small but significant retention time differences in reversed-phase HPLC as compared to their nonlabeled counterparts (54). This will make the ICAT data analysis part more complicated because the relative quantities of the two peptide species cannot be determined from the same spectrum, rather it requires integration across a window of the chromatographic time scale, which introduces additional variables for accurate quantification. Retention time shifts are less of a problem for ^{13}C, ^{15}N, or ^{18}O labeling.

2. Prior to any experiment, the study design must be developed with help from a statistician to ensure that the study answers the questions of interest and has sufficient technical and biological replicates to detect small but significant changes using appropriate statistical methods. A technical replicate is a replicate sample from the same biological sample. For example, split a single biological sample into two parts and run both replicates in the experiment. This will allow for assessment of instrument errors. Biological replicates are samples from independent experimental units (e.g., each of ten human plasma samples from different individuals). While a technical replicate estimates the precision for the assay itself, biological replicates provide an estimation of biological variation (55, 56). In general, biological replicates are more informative than technical replicates.

3. Group size determination depends on the size of effect to be detected (Fold-Change, FC), the sample-to-sample biological

Table 2
Group size determination

Proteins differentially expressed (%)	Group size (power = 95%, FDR = 5%, FC = 2, and CV = 20%)
5	4.25
10	3.81
15	3.54

variation expected (CV), and which error rates to be controlled. It is best to control the False-Discovery-Rate (FDR) instead of the False-Positive-Rate (FPR) when hundreds of proteins are analyzed. The FDR can be large (e.g., >0.05) even if the FPR is small (e.g., <0.05). If control of the FDR is chosen, then the proportion of proteins that will change (the prevalence) has to be estimated. With this information, the group size required for a given power (probability of determining a true change, e.g., the sensitivity) can be computed. Table 2 shows a suggested group size with given FC and %CV. As the percent of proteins expected to change varies, the group size required should be adjusted accordingly.

4. ICAT Sample solubilization: the Tris–HCl pH 8.3 concentrations can be from 5 to 200 mM depending on the sample type. Ultimately, the sample solution's pH should remain ~8.3 during the reduction and labeling steps. Sample type and history (e.g., sample solution ingredients) should dictate the ideal buffer to use.

5. The second generation ICAT kit (acid-cleavable ICAT®) is commercially available from Applied Biosystems/MDS SCIEX. Incorporation of ^{13}C rather than deuterium into the "heavy" reagent overcomes the deficiency associated with the deuterated "heavy" peptide, which often elutes at a different retention time than the "light" peptide, causing inaccurate peptide quantification measurements.

6. ICAT is not suitable for quantifying the significant number of proteins that do not contain any cysteines (10–20% of a given proteome depending on species) and is of limited use for the analysis of protein PTMs, splice isoforms, and mutations.

7. Protein concentration in ICAT: the highest total concentration of a protein from a complex protein mixture that has so far been successfully measured in an ICAT experiment is 4 mg/ml. To obtain the best results possible, all samples should be kept under identical conditions such as similar protein concentration, sample volume, etc.

8. Protein reduction in ICAT: alternative reducing reagents can be used. Two common ones are TCEP and DTT. TCEP is water soluble and less toxic than TBP. Thus, it is more user-friendly than TBP. However, due to its acidic nature, careful buffering to retain an ideal pH of ~8.3 is required. TCEP may also be more prone to react with the iodoacetamide group of the ICAT reagents than TBP. To counter these two potential problems, TCEP should be used at a relatively low concentration (1–5 mM recommended) and the pH of the solution should be checked and adjusted to ~8.3 if necessary. DTT is also water soluble and less toxic than TBP. It contains free sulfhydryl groups which will readily react with the ICAT reagent. Despite this impediment, it can be used as a reducing reagent in ICAT experiments by following this procedure:
   (a) Reduce with DTT at 5–100 mM for 10–30 min at 37°C.
   (b) Precipitate proteins by cold acetone.
   (c) Resolubilize pellets in labeling buffer + 5 mM TBP.
   (d) Continue on to labeling step.

   The advantage of reducing with DTT is that it will enable less cross-reactions of TBP with the ICAT reagent, although it will cause some sample loss due to the precipitation step.
9. The concentration of 1.2 mM ICAT labeling reagent is the recommended minimal concentration at which complex mixtures should be labeled.
10. ~1 M urea concentration should be maintained during digestion. Higher than 1 M urea concentration could inhibit tryptic digestion efficiency.
11. Cleaving reagent A is 95% TFA. Thus it should be used in a hood with a glass syringe or vial.
12. The iTRAQ method is based on the differential covalent labeling of peptides from proteolytic digests with one of the four iTRAQ reagents resulting in the incorporation of 144.1 Da to the peptides' N-termini and lysine residues. Peptides with different tags are indistinguishable by mass but can be differentiated by CID through release of a reporter ion, each of which has a different mass (114, 115, 116, or 117 Da). The analysis of the intensity of reporter ions allows for the simultaneous identification and quantification of the labeled peptides.
13. A mass shift of 4 Da or more is required for iTRAQ labeling reagents to minimize interfering peaks in quantification, especially in low-intensity spectra.
14. All the SILAC labeling steps should be carried out under a laminar flow hood. Store the media at 4°C for up to 2 months, if necessary.

15. In order to ensure that all tryptic peptides of a protein carry at least one labeled amino acid to produce a constant increase in mass over the unlabeled counterpart, $^{13}C_6$-Arginine and $^{13}C_6$-Lysine are often used in the media in an SILAC experiment.
16. The cells are always adapted to the custom SILAC medium for five passages to achieve complete incorporation of the isotope labeled amino acid before they are propagated to a scale needed for the experiment.
17. As long as the cells are monitored, one can use lower amounts of the heavy isotope labeled amino acids in the preparation of custom SILAC media than are present in the original formulation if the cost of heavy amino acids is a consideration. However, in such cases, it is advisable to present the cells with similar amounts of light amino acid.
18. Save a small number of cells for quality control tests in the labeling step of SILAC. This can be carried out by subculturing a small dish (60-mm) of cells specifically for this purpose.
19. Allow sufficient time for cells to fully incorporate the SILAC amino acid before proceeding with the next step in the experiment. This evaluation must be performed if this is the first time SILAC is used with this cell stock to avoid incomplete incorporation and potential errors in quantification.
20. It is known after testing many cell lines that SILAC labeling has no deleterious effect on cells in terms of growth and division, morphology, or biological responses.
21. The analysis of the SILAC amino acid incorporation efficiency does not have to be performed each time a SILAC experiment is carried out, but it should be performed when working with a new cell type or cell stock. The parameters for successful labeling can be recorded and applied to subsequent SILAC experiments.
22. During the disulfide bond reduction step of an SILAC experiment, incubation of solutions containing urea at temperatures higher than 37°C can lead to protein carbamylation and should be avoided.
23. After tryptic digestion, the resulting peptide mixture can be frozen at −80°C for up to a year. However, evaluation for incorporation should be performed before continuing with the SILAC experiment. If needed, undigested lysates can be saved at −80°C for a quality control analysis to be performed separately.
24. In iTRAQ, the contribution of peptidic or chemical background noise to the quantification does not depend on the mass resolution of the mass spectrometer but on the size of the *m/z* window chosen for the isolation of peptides for sequencing (usually 2–6 *m/z*). Thus, all ions present in this window will

contribute to the signal of the reporter ions. As a result, this can sometimes lead to a large underestimation of the true protein abundance changes, especially for very weak peptide signals (lower abundant ones).

25. In an SILAC experiment, a mass shift of 4 Da or more between the "light" and "heavy" reagents is required to distinguish the isotopomer clusters of the labeled and unlabeled forms of the peptide. Reporter ions used for quantification in MS/MS spectra should be designed such that interference by ordinary peptide fragments is minimal.

26. In a label-free quantification experiment, peak intensities of the peptides can vary from run to run, resulting in larger experimental variations or technical CVs. This is primarily caused by variations in sample preparation and sample injection. Normalization is required to minimize these kinds of variations.

27. Drifts in retention time will significantly affect the accuracy of quantification in a label-free quantification experiment. These drifts may occur as a result of multiple sample injections onto the same reversed-phase LC column. Unaligned peak comparison will result in inaccurate quantification and large overall CV. Thus, highly reproducible LC–MS (e.g., use a larger column such as a 2-mm inner diameter microflow column instead of a 75 μm nanoflow column for sample size >40) and careful chromatographic peak alignment (e.g., software package that allows for aligning thousands of peaks automatically) are required in this approach.

28. Automation is required for data analysis (e.g., peak alignment and peak intensity integration) to ensure an "apples-to-apples" comparison and an accurate quantitative measurement in a label-free quantification experiment. Thus, capable computer software is a critical part of using this platform for a large-scale proteomic study for biomarker discovery.

29. "Triple-Play" mode is applied for data acquisition in a label-free quantification experiment because the data processing algorithm requires matches in precursor ion (MS scans), charge state (zoom scans), fragment ion (MS/MS scans), and chromatographic peak retention time within a certain time window (typically 1-min window) to pass the filter for quantification and ensure the peptide charge status (38).

30. In a spectral counting-based protein quantification experiment, the number of spectral counts can be normalized to protein length allowing for the relative quantification of two different proteins. Since large proteins tend to contribute more peptides and thus spectra than small ones, a normalized spectral abundance factor (NSAF) is often defined to account for the effect

of protein length on spectral count (44, 57). NSAF is calculated as the number of spectral counts (SpC) identifying a protein, divided by the length of the protein (L), divided by the sum of SpC/L for all proteins in the experiment. NSAF thus allows the comparison of the abundance of individual proteins in multiple samples (44, 58).

31. The linearity of protein abundance is not the same for every protein in a spectral counting experiment. In fact, the spectral counts are different for every peptide due to different retention times and peak widths even if their abundance may be the same. Therefore, a reasonable number of spectra are required for accurate quantification of a given protein in complex biological mixtures. It requires as low as four spectra to detect a threefold protein abundance change (33), but this number increases exponentially for smaller changes (e.g., 15 spectra for a twofold change). A potential problem of higher spectral counts is saturation effects.
32. In theory, there is no limit to the number of samples that can be compared by label-free approaches. This is certainly an advantage over stable isotopic labeling technologies that are typically limited to eight samples. Practically, however, 100-sample comparisons may be a limit because of the LC column reliability and durability. In addition, the linearity and accuracy of this approach remains unclear over a large sample set.

Acknowledgment

The authors would like to thank Ms. Heather Sahm for critical reading of this book chapter.

References

1. Blackstock, W. P. and Weir, M. P. (1999) Proteomics: quantitative and physical mapping of cellular proteins. *Trends Biotechnol.* **17**, 121–127.
2. Gygi, S. P., Rist, B., and Aebersold, R. (2000) Measuring gene expression by quantitative proteome analysis. *Curr. Opin. Biotechnol.* **11**, 396–401.
3. Rabilloud, T. (2002) Two-dimensional gel electrophoresis in proteomics: old, old fashioned, but it still climbs up the mountains. *Proteomics* **2**, 3–10.
4. Conrads, T. P., Issaq, H. J., and Veenstra, T. D. (2002) New tools for quantitative phosphoproteome analysis. *Biochem. Biophys. Res. Commun.* **290**, 885–890.
5. Ong, S. E., Foster, L. J., and Mann, M. (2003) Mass spectrometric-based approaches in quantitative proteomics. *Methods* **29**, 124–130.
6. Tao, W. A. and Aebersold, R. (2003) Advances in quantitative proteomics via stable isotope tagging and mass spectrometry. *Curr. Opin. Biotechnol.* **14**, 110–118.
7. McDonald, W. H. and Yates, J. R., 3 rd. (2002) Shotgun proteomics and biomarker discovery. *Dis. Markers* **18**, 99–105.
8. Wu, C. C., MacCoss, M. J., Howell, K. E., and Yates, J. R., 3 rd. (2003) A method for the comprehensive proteomic analysis of membrane proteins. *Nat. Biotechnol.* **21**, 532–538.
9. Washburn, M. P., Ulaszek, R., Deciu, C., Schieltz, D. M., and Yates, J. R., 3 rd. (2002)

Analysis of quantitative proteomic data generated via multidimensional protein identification technology. *Anal. Chem.* **74**, 1650–1657.

10. Washburn, M. P., Wolters, D., and Yates, J. R., 3 rd. (2001) Large-scale analysis of the yeast proteome by multidimensional protein identification technology. *Nat. Biotechnol.* **19**, 242–247.
11. Gygi, S. P., Rist, B., Gerber, S. A., Turecek, F., Gelb, M. H., and Aebersold, R. (1999) Quantitative analysis of complex protein mixtures using isotope-coded affinity tags. *Nat. Biotechnol.* **17**, 994–999.
12. Yan, W. and Chen, S. S. (2005) Mass spectrometry-based quantitative proteomic profiling. *Brief. Funct. Genomic. Proteomics* **4**, 27–38.
13. Zhang, B., VerBerkmoes, N. C., Langston, M. A., Uberbacher, E., Hettich, R. L., and Samatova, N. F. (2006) Detecting differential and correlated protein expression in label-free shotgun proteomics. *J. Proteome Res.* **5**, 2909–2918.
14. Wang, G., Wu, W. W., Zeng, W., Chou, C-L., and Shen, R-F. (2006) Label-free protein quantification using LC-coupled ion trap or FT mass spectrometry: reproducibility, linearity, and application with complex proteomes. *J. Proteome Res.* **5**, 1214–1223.
15. Ono, M., Shitashige, M., Honda, K., Isobe, T., Kuwabara, H., Matsuzuki, H., et al. (2006) Label-free quantitative proteomics using large peptide data sets generated by nanoflow liquid chromatography and mass spectrometry. *Mol. Cell. Proteomics* **5**, 1338–1347.
16. Li, J., Steen, H., and Gygi, S. P. (2003) Protein Profiling with Cleavable Isotope-coded Affinity Tag (cICAT) Reagents: The Yeast Salinity Stress Response. *Mol. Cell. Proteomics* **2**, 1198–1204.
17. Ross, P. L., Huang, Y. N., Marchese, J. N., Williamson, B., Parker, K., Hattan, S., et al. (2004) Multiplexed Protein Quantitation in Saccharomyces cerevisiae Using Amine-reactive Isobaric Tagging Reagents. *Mol. Cell. Proteomics* **3**, 1154–1169.
18. Oda, Y., Huang, K., Cross, F. R., Cowburn, D., and Chait, B. T. (1999) Accurate Quantitation of Protein Expression and Site-Specific Phosphorylation. *Proc. Natl. Acad. Sci. USA.* **96**, 6591–6596.
19. Ong, S., Blagoev, B., Kratchmarova, I., Kristensen, D. B., Steen, H., Pandey, A., et al. (2002) Stable Isotope Labeling by Amino Acids in Cell Culture, SILAC, as a Simple and Accurate Approach to Expression Proteomics. *Mol. Cell. Proteomics* **1**, 376–386.
20. Ong, S., Kratchmarova, I., and Mann, M. (2003) Properties of ^{13}C-substituted Arginine in Stable Isotope Labeling by Amino Acids in Cell Culture (SILAC). *J. Proteome Res.* **2**, 173–181.
21. Moulder, R., Lonnberg, T., Elo, L. L., Filen, J. J., Rainio, E., Corthals, G., et al. (2010) Quantitative proteomics analysis of the nuclear fraction of human CD4+ cells in the early phases of IL-4-induced Th2 differentiation. *Mol. Cell. Proteomics* **9**, 1937–1953.
22. Collier, T. S., Sarkar, P., Rao, B., and Muddiman, D. C. (2010) Quantitative Top-down Proteomics of SILAC Labeled Human Embryonic Stem Cells. *J. Am. Soc. Mass. Spectrom.* 21, 879–889.
23. Imami, K., Sugiyama, N., Tomita, M., and Ishihama, Y. (2010) Quantitative proteome and phosphoproteome analyses of cultured cells based on SILAC labeling without requirement of serum dialysis. *Mol. Biosyst.* 6, 594–602.
24. Gerber, S. A., Rush, J., Stemman, O., Kirschner, M. W., and Gygi, S. P. (2003) Absolute quantification of proteins and phosphoproteins from cell lysates by tandem MS. *Proc. Natl. Acad. Sci. USA.* **100**, 6940–6945.
25. Rappsilber, J., Ryder, U., Lamond, A. I., and Mann, M. (2002) Large-scale proteomic analysis of the human spliceosome. *Genome Res.* **12**, 1231–1245.
26. Ishihama, Y., Oda, Y., Tabata, T., Sato, T., Nagasu, T., Rappsilber, J., et al. (2005) Exponentially modified protein abundance index (emPAI) for estimation of absolute protein amount in proteomics by the number of sequenced peptides per protein. *Mol. Cell. Proteomics* **4**, 1265–1272.
27. Lu, P., Vogel, C., Wang, R., Yao, X., and Marcotte, E. M. (2007) Absolute protein expression profiling estimates the relative contributions of transcriptional and translational regulation. *Nat. Biotechnol.* **25**, 117–124.
28. Griffin, T. J., Lock, C. M., Li, X. J., Patel, A., Chervetsova, I., Lee, H., et al. (2003) Abundance ratio-dependent proteomic analysis by mass spectrometry. *Anal. Chem.* **75**, 867–874.
29. Bondarenko, P. V., Chelius, D., and Shaler, T. A. (2002) Identification and relative quantitation of protein mixtures by enzymatic digestion followed by capillary reversed-phase liquid chromatography-tandem mass spectrometry. *Anal. Chem.* **74**, 4741–4749.
30. Chelius, D., and Bondarenko, P. V. (2002) Quantitative profiling of proteins in complex mixtures using liquid chromatography and mass spectrometry. *J. Proteome Res.* **1**, 317–323.
31. Wang, W., Zhou, H., Lin, H., Roy, S., Shaler, T., Hill, L., et al. (2003) Quantification of

proteins and metabolites by mass spectrometry without isotopic labeling or spiked standards. *Anal. Chem.* **75**, 4818–4826.

32. Bantscheff, M. and Schirle, M. (2007) Quantitative mass spectrometry in proteomics: a critical review. *Anal. Bioanal. Chem.* **389**, 1017–1031.
33. Old, W. M., Meyer-Arendt, K., Aveline-Wolf, L., Pierce, K. G., Mendoza, A., Sevinsky, J. R., et al. (2005) Comparison of label-free methods for quantifying human proteins by shotgun proteomics. *Mol. Cell. Proteomics* **4**, 1487–1502.
34. Bradford, M. M. (1976) A rapid and sensitive method for the quantitation of microgram quantities of protein utilizing the principle of protein–dye binding. *Anal. Biochem.* **72**, 248–254.
35. Hill, H. D. and Straka, J. G. (1988) Protein determination using bicinchoninic acid in the presence of surfhydryl reagents. *Anal. Biochem.* **170**, 203–208.
36. Hale, J. E., Butler, J. P., Gelfanova, V., You, J. S., and Knierman, M. D. (2004) A simplified procedure for the reduction and alkylation of cysteine residues in proteins prior to proteolytic digestion and mass spectral analysis. *Anal. Biochem.* **333**, 174–181.
37. Higgs, R. E., Knierman, M. D., Freeman, A. B., Gelbert, L. M., Patil, S. T., and Hale, J. E. (2007) Estimating the statistical significance of peptide identifications from shotgun proteomics experiments. *J. Proteome Res.* **6**, 1758–1767.
38. Higgs, R.E., Knierman, M.D., Gelfanova, V., Butler, J.P., and Hale, J.E. (2005) Comprehensive label-free method for the relative quantification of proteins from biological samples. *J. Proteome Res.* **4**, 1442–1450.
39. Carr, S., Aebersold, R., Baldwin, M., Burlingame, A., Clauser, K., and Nesvizhskii, A. (2004) The need for guidelines in publication of peptide and protein identification data: Working Group on Publication Guidelines for Peptide and Protein Identification Data. *Mol. Cell. Proteomics* **3**, 531–533.
40. Shadforth, I. P., Dunkley, T. P., Lilley, K. S., and Bessant, C. (2005) i-Tracker: for quantitative proteomics using iTRAQ. *BMC Genomics* **6**, 145–150.
41. Bolstad, B. M., Irizarry, R. A., Astrand, M., and Speed, T. P. (2003) A comparison of normalization methods for high density oligonucleotide array data based on variance and bias. *Bioinformatics* **19**, 185–193.
42. Limpert, E., Stahel, W. A., Abbt, M. (2001) Log-normal Distributions across the Sciences: Keys and Clues. *BioScience* **51**, 341–352.
43. Zybailov, B., Coleman, M. K., Florens, L., and Washburn, M. P. (2005) Correlation of relative abundance ratios derived from peptide ion chromatograms and spectrum counting for quantitative proteomic analysis using stable isotope labeling. *Anal. Chem.* **77**, 6218–6224.
44. Florens, L., Carozza, M. J., Swanson, S. K., Fournier, M., Coleman. M. K., Workman, J. L., et al. (2006) Analyzing chromatin remodeling complexes using shotgun proteomics and normalized spectral abundance factors. *Methods* **40**, 303–311.
45. Pang, J. X., Ginanni, N., Dongre, A. R., Hefta, S. A., and Opiteck, G. J. (2002) Biomarker discovery in urine by proteomics. J. Proteome Res. 1, 161–169.
46. Rao, P. V., Reddy, A. P., Lu, X., Dasari, S., Krishnaprasad, A., Biggs, E., et al. (2009) Proteomic identification of salivary biomarkers of type-2 diabetes. *J. Proteome Res.* **8**, 239–245.
47. Pan, J., Chen, H. Q., Sun, Y. H., Zhang, J. H., and Luo, X. Y. (2008) Comparative proteomic analysis of non-small-cell lung cancer and normal controls using serum label-free quantitative shotgun technology. *Lung* **186**, 255–261.
48. Asara, J. M., Christofk, H. R., Freimark, L. M., and Cantley, L. C. (2008) A label-free quantification method by MS/MS TIC compared to SILAC and spectral counting in a proteomics screen. *Proteomics* **8**, 994–999.
49. Seyfried, N. T., Huysentruyt, L. C., Atwood III, J. A., Xia, Q., Seyfried, T. N., and Orlando, R. (2008) Up-regulation of NG2 proteoglycan and interferon-induced transmembrane proteins 1 and 3 in mouse astrocytoma: a membrane proteomics approach. *Cancer Letters* **263**, 243–252.
50. Liu, H., Sadygov, R. G., and Yates, J. R., III (2004) A model for random sampling and estimation of relative protein abundance in shotgun proteomics. *Anal. Chem.* **76**, 4193–4201.
51. Dong, M. Q., Venable, J. D., Au, N., Xu, T., Park, S. K., Cociorva, D., et al. (2007) Quantitative mass spectrometry identifies insulin signaling targets in *C. elegans. Science* **317 (5838)**, 660–663.
52. Zhang, B., VerBerkmoes, N. C., Langston, M. A., Uberbacher, E., Hettich, R. L., and Samatova, N. F. (2006) Detecting differential and correlated protein expression in label-free shotgun proteomics. *J. Proteome Res.* **5**, 2909–2918.
53. Reiner, A., Yekutieli, D., and Benjamini, Y. (2003) Identifying differentially expressed genes using false discovery rate controlling procedures. *Bioinformatics* **19**, 368–375.
54. Zhang, R., Sioma, C. S., Wang, S., and Regnier, F. E. (2001) Fractionation of isotopically

labeled peptides in quantitative proteomics. *Anal. Chem.* 73, 5142–5149.

55. Yang, Y. H. and Speed, T. (2003) Design issues for cDNA microarray experiments. *Nat. Rev. Genet.* **19**, 649–659.
56. Simon, R., Radmacher, M. D., and Dobbin, K. (200) Design of studies using DNA microarrays. *Genet Epidemilo.* **23**, 21–36.
57. Zybailov, B., Mosley, A. L., Sardiu, M. E., Coleman, M. K., Florens, L., and Washburn, M. P. (2006) Statistical analysis of membrane proteome expression changes in *Saccharomyces cerevisiae*. *J. Proteome Res.* 5, 2339–2347.
58. Paoletti, A. C., Parmely, T. J., Tomomori-Sato, C., Sato, S., Zhu, D., Conaway, R. C., et al. (2006) Quantitative proteomic analysis of distinct mammalian Mediator complexes using normalized spectral abundance factors. *Proc. Natl. Acad. Sci. USA.* **103**, 18928–18933.

Part V

Protein Analysis II: Functional Characterization

Chapter 17

High-Throughput Microtitre Plate-Based Assay for DNA Topoisomerases

James A. Taylor, Nicolas P. Burton, and Anthony Maxwell

Abstract

We have developed a rapid, high-throughput assay for measuring the catalytic activity (DNA supercoiling or relaxation) of DNA topoisomerases. The assay utilizes intermolecular triplex formation between an immobilized triplex-forming oligo (TFO) and a triplex-forming region inserted into the plasmid substrate (pNO1), and capitalizes on the observation that supercoiled DNA forms triplexes more readily than relaxed DNA. Thus, supercoiled DNA is preferentially retained by the TFO under triplex-forming conditions while relaxed DNA can be washed away. Due to its high speed of sample analysis and reduced sample handling over conventional gel-based techniques, this assay can be used to screen chemical libraries for novel inhibitors of topoisomerases.

Key words: Topoisomerase, DNA gyrase, Triplex formation, Supercoiling, Relaxation, High-throughput screening

1. Introduction

DNA topoisomerases are essential enzymes that control the topological state of DNA in all living cells (1). The crucial nature of their role, plus the fact that they must cleave DNA as part of their mechanism, has made them effective targets for antimicrobial and anti-cancer drugs as well as potential targets for herbicides, anti-virals, and anti-protozoal agents. All topoisomerases can relax supercoiled DNA, while only DNA gyrase is capable of introducing negative supercoils (2). Topoisomerases can also catenate/decatenate DNA and introduce/remove knots from DNA, to greater or lesser degrees (3).

The traditional assay for topoisomerase activity is based on the principle that different DNA topoisomers have different mobilities on an agarose gel. This assay is not only information-rich, but is also slow and requires intensive handling. As such, it is not suitable for the high-throughput format necessary for large-scale screening for

Michael Kaufmann and Claudia Klinger (eds.), *Functional Genomics: Methods and Protocols*,
Methods in Molecular Biology, vol. 815, DOI 10.1007/978-1-61779-424-7_17, © Springer Science+Business Media, LLC 2012

novel inhibitors. To address the limitations of this assay, a microtitre plate-based assay was developed (4) (see Note 1), which capitalizes on the fact that supercoiled plasmids form DNA triplexes more readily than relaxed plasmids (5, 6). Reactions are carried out in streptavidin-coated microtitre wells, which have had a single-stranded biotinylated oligo immobilized on their surfaces. This oligo can form triplexes with a target plasmid by the addition of a low pH triplex formation (TF) buffer, which also stops the topoisomerase reaction. Supercoiled DNA will be retained in the wells after washing, whereas relaxed, open circle and linear DNA will be removed. The amount of DNA retained can be quantified with a fluorescent dye, and directly correlates with the level of enzyme activity (Fig. 1a).

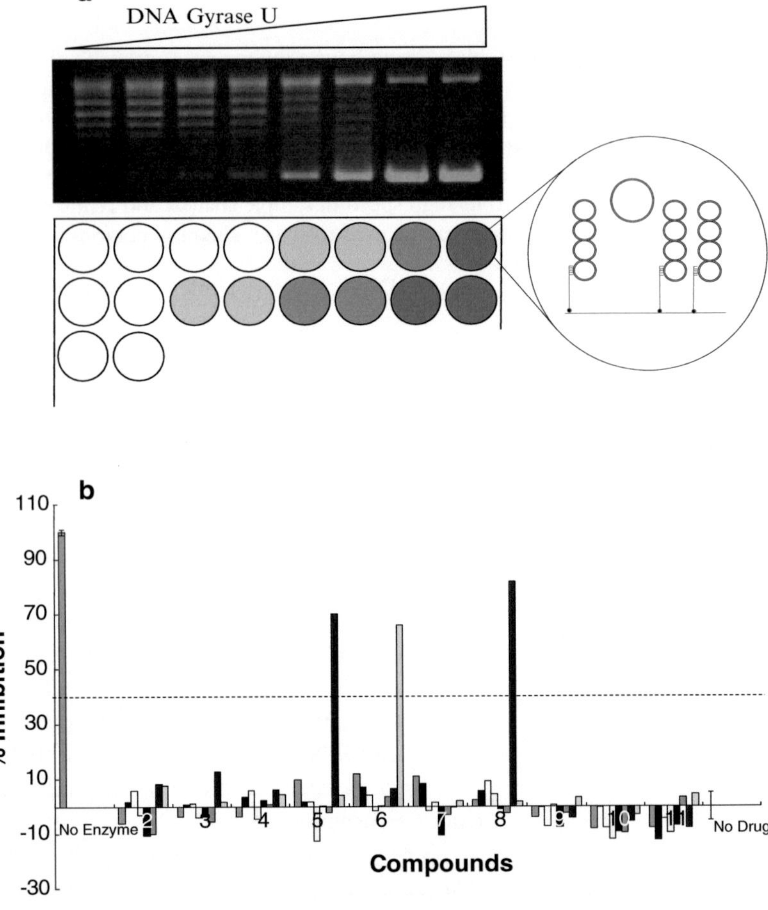

Fig. 1. (**a**) Sample agarose gel assay for increasing units of DNA gyrase with a *cartoon* representation of its equivalence in the high-throughput assay. Conversion of relaxed to supercoiled DNA increases the amount of DNA retained in the well due to triplex formation and subsequently the intensity of the fluorescent signal after SYBR Gold staining. (**b**) Sample data for a DNA gyrase inhibitor screen, performed in duplicate in a 96-well plate. Each *bar* represents the percentage inhibition of a single compound, calculated from the negative and positive controls for enzyme activity (of which there were 16 repeats each). The *dotted line* denotes the hit threshold, as calculated from the standard deviation of the controls. Most of the compounds fall well below this threshold while three are clearly above it.

2. Materials

The highest purity of materials and ultrapure water with a resistivity of >18 MΩ.cm should be used throughout. Unless otherwise stated, materials were purchased from Sigma. The wash, triplex formation, and T10E1 buffers should be filtered with a 0.2-μm cellulose nitrate membrane before use (see Note 2).

2.1. Buffers and Solutions

1. DNA gyrase dilution buffer: 50 mM Tris–HCl (pH 7.5), 100 mM KCl, 2 mM DTT, 1 mM EDTA, 10% (*w/v*) glycerol. Store at –20°C.
2. DNA gyrase supercoiling buffer: 35 mM Tris–HCl (pH 7.5), 24 mM KCl, 4 mM $MgCl_2$, 2 mM DTT, 1.8 mM spermidine, 1 mM ATP, 6.5% (*w/v*) glycerol, 0.1 mg/ml bovine serum albumin. The buffer is stored as a 5× concentrate at –20°C.
3. PDM medium: 7.9 g tryptone, 4.4 g yeast extract, 0.5 g NH_4Cl and 0.24 g $MgSO_4$ in 880 ml of water. After autoclaving add 20 ml 50% (*w/v*) glucose and 100 ml 10× phosphate buffer (see Note 3).
4. Phosphate buffer (10×): 128 g $Na_2HPO_4{\cdot}7H_2O$, 30 g KH_2PO_4 in 1 L of water. Autoclave and store at room temperature.
5. STEB (2×): 40% sucrose, 100 mM Tris–HCl (pH 8.0), 100 mM EDTA, 0.5 mg/ml bromophenol blue.
6. T10E1 buffer: 10 mM Tris–HCl (pH 8.0), 1 mM EDTA. Store at room temperature.
7. TAE: 40 mM Tris–Acetate (pH 8.0), 1 mM EDTA.
8. Topoisomerase I dilution buffer: 10 mM Tris–HCl (pH 7.5), 1 mM DTT, 1 mM EDTA, 50% (*v/v*) glycerol, 100 μg/ml bovine serum albumin.
9. Topoisomerase I relaxation buffer: 20 mM Tris–HCl (pH 8), 200 mM NaCl, 0.25 mM EDTA, 5% glycerol. The buffer is stored as a 10× concentrate at –20°C.
10. Topoisomerase VI dilution buffer: 20 mM HEPES pH 7.5, 10% (*v/v*) glycerol. Store at –20°C.
11. Topoisomerase VI relaxation buffer: 20 mM bis-Tris propane pH 6.5, 100 mM potassium glutamate, 10 mM $MgCl_2$, 1 mM DTT, and 1 mM ATP. The buffer is stored as a 5× concentrate at –20°C.
12. Triplex column buffer: 75 mM calcium acetate (pH 4.7). Store at room temperature.
13. Triplex column elution buffer: 10 mM Tris–HCl (pH 8.0), 100 mM EGTA. Store at room temperature.
14. Triplex formation buffer: 75 mM magnesium acetate (pH 4.7). Store at room temperature (see Note 4).

15. Wash buffer: 20 mM Tris–HCl (pH 7.6), 137 mM NaCl, 0.01% (*w/v*) bovine serum albumin (acetylated), 0.05% (*v/v*) Tween-20. Store at 4°C.

2.2. DNA

1. Plasmid pNO1 is a modified version of the high-copy number plasmid pBR322* (7) containing a triplex-forming insert (4). It was prepared by ligation of oligos TFO1W and TFO1C (Table 1) into the AvaI site of the plasmid. Supercoiled pNO1 is prepared by transforming it into competent *Escherichia coli* cells (e.g. Top10, Invitrogen), growing cells overnight in PDM media (8) (see Note 3) containing 100 μg/ml ampicillin (see Note 5) at 37°C, 200 rpm. The DNA can then be purified using a Qiagen giga-prep kit (or similar) or by a cesium chloride density gradient.
2. Relaxed pNO1 is prepared by incubating the supercoiled form with chicken erythrocyte topoisomerase I (~40–50 μg plasmid with 200 U topoisomerase I in topoisomerase I relaxation buffer) for 1 h at 37°C. The DNA is extracted with two phenol/chloroform extractions and purified by ethanol precipitation.
3. TFO1 oligo (Table 1) with a 5′ biotin tag (e.g. Sigma-Genosys or Bioneer) is resuspended to 10 μM in T10E1 buffer and stored at –20°C.

2.3. Enzymes

1. *E. coli* DNA gyrase subunits GyrA and GyrB are expressed in *E. coli* and purified according to established protocols (9). The subunits are stored separately in DNA gyrase dilution buffer at –80°C (see Note 6). The complete enzyme is reconstituted by mixing equimolar amounts of GyrA and GyrB and stored on ice prior to use.
2. The chicken erythrocyte topoisomerase I used was a gift from Alison Howells of Inspiralis Ltd., and was produced using the published protocol (10).

Table 1
Oligonucleotides used in the high-throughput assay

Name	Sequence (5′–3′)	5′ Modification
TFO1	TCTCTCTCTCTCTCTC	Biotin
TFO1W	TCGGAGAGAGAGAGAGAGAG	
TFO1C	CCGATCTCTCTCTCTCTCTC	

2.4. Equipment and DNA Staining

1. Black, streptavidin-coated, high binding capacity, 96-well plates (Pierce) are used for the screen. Plates should be stored covered at 4°C.
2. DNA is stained with SYBR Gold (Invitrogen), which is stored as a 10,000× concentrate at 20°C. The dye is diluted 10,000-fold with T10E1 buffer to reach the working concentration. This should be prepared fresh for each use.
3. Fluorescence measurements are made using a SPECTRAmax Gemini fluorimeter and Softmax Pro Software. Alternative fluorimeters and software can be used.

3. Methods

3.1. DNA Gyrase Microtitre Plate-Based Supercoiling Assay

1. Wash microtitre plate wells with 3× 200 μl of wash buffer (see Note 7).
2. Load 100 μl 500 nM biotinylated TFO1 oligo (diluted from stock in wash buffer) into wells and allow immobilization to proceed for 2 min.
3. Remove oligo solution and wash carefully with 3× 200 μl volumes of wash buffer (see Note 7).
4. The DNA gyrase supercoiling reaction is performed in the wells in a 30 μl reaction volume containing: 1–6 μl reconstituted DNA gyrase [1–2 U (see Note 8); the total volume of DNA gyrase is made up to 6 μl with DNA gyrase dilution buffer], 1 μl of 1 μg/μl relaxed pNO1, 6 μl 5× DNA gyrase reaction buffer and H_2O to 30 μl (see Note 9). Incubate the reaction at 37°C for 30 min (this can be carried out in the plate reader if temperature control is available).
5. The reaction is stopped with the addition of 100 μl TF buffer, which lowers the pH, and the plate is incubated at room temperature for 30 min to allow triplex formation to occur. Supercoiled DNA becomes immobilized on the plate while relaxed DNA remains in solution.
6. Remove unbound relaxed and linear plasmid by washing the wells thoroughly with 3× 200 μl volumes of TF buffer.
7. Remove all the liquid from the wells and stain DNA with 200 μl SYBR Gold (diluted in T10E1 buffer). The plate is incubated for further 20 min at room temperature. After incubation, mix the contents of the wells and then read the fluorescence in a fluorimeter.

3.2. Screening a Compound Library for DNA Gyrase Inhibitors

1. Wash the plates and immobilize the TFO1 as described in Subheading 3.1, steps 1–3 (see Note 10).
2. Add 23.5 μl of a DNA Mix (see Note 11) containing 4.7 μl of 5× DNA gyrase reaction buffer, 2 μl of 1 μg/μl relaxed pNO1 and H_2O to each well by a multichannel pipette.
3. Add 3 μl of compound to each well (the assay is tolerant of up to 10% DMSO in the reaction). Include positive and negative control wells to which only DMSO is added.
4. Add 33.5 μl of Control Mix containing 12 μl DNA gyrase dilution buffer in 1x DNA gyrase supercoiling buffer to the negative control wells.
5. Add 33.5 μl of Enzyme Mix containing 1.8 U of DNA gyrase (where 1 U is defined as the amount of enzyme required to supercoil 1 μg of relaxed pBR322; the enzyme is diluted in 12 μl DNA gyrase dilution buffer) in 1× DNA gyrase supercoiling buffer to the sample and the positive control wells.
6. Quickly mix (e.g. using the plate reader if it has such a facility), and incubate at 37°C for 30 min.
7. Follow steps 5–7 of Subheading 3.1 to form DNA triplexes and visualize retained plasmid.

3.3. Data Processing and Hit Validation of Hits Using Agarose Gel Assay

Once data has been collected from the screen (Fig. 1b), the fluorescence signals for the duplicates should be averaged and converted into percentage inhibition using the positive and negative controls (see Note 10). The standard deviation of the controls should also be calculated and converted into percentage inhibition. There are various ways to decide what constitutes a hit such as (i) taking compounds with a percentage inhibition over three standard deviations of the negative control, (ii) take an arbitrary threshold such as anything over 25 or 50% inhibition, or (iii) or take a certain proportion of the compounds (e.g. the top twenty). The suitability of the methods has to be decided on an individual basis, depending on the logistics of validating the number of hits which would be selected. It is also possible to calculate the *Z* factor for the assay (11) to determine the quality of the data (in our hands $Z = \sim 0.75$ for *E. coli* DNA gyrase). Hits obtained from the screen should be verified using an independent secondary assay, for example agarose gel electrophoresis. It is advisable to use fresh stocks of compounds for verification, rather than solutions taken from the library to ensure that the inhibition seen is not due to any cross-contamination or degradation which may have occurred during library storage.

1. The supercoiling reaction is conducted in 1.5-ml microcentrifuge tubes containing the following: 500 ng of relaxed pNO1, 6 μl 5× DNA gyrase supercoiling buffer, 0.5 U of reconstituted DNA gyrase in 6 μl dilution buffer, up to 1.5 μl of compound in 100% DMSO at the desired concentration and H_2O to 30 μl.

2. Incubate at 37°C for 30 min.
3. Stop the reaction with an equal volume chloroform:isoamyl alcohol (24:1), and an equal volume of 2× STEB, vortex vigorously to form an emulsion. Centrifuge at 16,000 × *g* for 10 min.
4. Load 15 μl of the top phase onto a 1% TAE agarose gel. Run the gel at 80 V for 2–3 h or until the gel has run for at least 7 cm, or 15–30 V overnight. The further that the gel is run, the better the topoisomers will be resolved.
5. Stain gel with 2 mg/ml ethidium bromide for 10 min and visualize under UV.

3.4. Measuring Relaxation Activity with the Microtitre Plate-Based Assay

The plate assay can also be used to measure the activity of enzymes which relax DNA, and consequently used to screen to novel inhibitors for them. Below is a sample protocol for using the assay to measure *Methanosarcina mazei* topoisomerase VI, but it is easily adapted for other enzymes (such as human topoisomerase I and II, and bacterial topoisomerase IV). Likewise, the screen protocol described for gyrase above can be modified to screen against such enzymes.

1. Wash the plates and immobilize the TFO1 as described in Subheading 3.1, steps 1–3.
2. The topoisomerase VI relaxation reaction is performed in the wells in a 30 μl reaction volume containing: 1–6 μl reconstituted topoisomerase VI [1–2 U (see Note 8); the total volume of topoisomerase VI is made up to 6 μl with topoisomerase VI dilution buffer], 1 μl of 1 μg/μl supercoiled pNO1, 6 μl 5× topoisomerase VI reaction buffer and H_2O to 30 μl (see Note 9). Incubate the reaction at 37°C for 30 min (this can be carried out in the plate reader if temperature control is available).
3. Follow steps 5–7 of Subheading 3.1 to form DNA triplexes and visualise retained plasmid. The enzyme activity is denoted by a drop in signal as supercoiled substrate is converted to relaxed product and lost from the wells during subsequent wash steps.

3.5. Purification of Supercoiled Plasmid via Triplex Affinity Columns and Other Applications for the High-Throughput Assay

As well as its utility in high-throughput screening, the assay and the principles behind it have potential to be exploited in a range of other settings. Previously, it has been used for the rapid determination of inhibitor IC_{50}s (12), which is useful for the characterization of hits determined during screening. It is also possible to use the assay for measuring the activity (and inhibition) of any enzyme which alters the linking number of the plasmid substrate (such as restriction enzymes). In addition, the protocol above could be altered to screen for compounds which either promote or inhibit the formation of DNA triplexes, simply by omitting the enzyme. The compounds should be included in the TF buffer wash steps

(Subheading 3.1, step 6), as well as in the TF buffer for the triplex formation step. Finally, the capture of supercoiled DNA plasmids by triplex-forming oligos has been utilized to create an affinity column for the specific purification of supercoiled plasmid (13).

1. Add 1 ml of 4% agarose streptavidin coated beads to an empty gravity flow column (see Note 12).
2. Wash three times with 1 ml T10E1 Buffer.
3. Stop the flow and add 1 ml of 100 μM TFO1 in T10E1 buffer. Incubate at room temperature for 1 min.
4. Wash three times with 1 ml T10E1 Buffer.
5. Equilibrate the column by washing three times with 1 ml triplex column buffer.
6. Add 1 mg of DNA in 1 ml triplex column buffer. Retain the flow through for subsequent steps.
7. Elute bound DNA with 2 ml triplex column elution buffer.
8. Re-equilibrate the column by washing three times with 1 ml triplex column buffer.
9. Recirculate the flow through from step 6.
10. Repeat steps 6–9 until all plasmid is recovered.
11. Combine eluents and concentrate through ethanol precipitation.
12. After elution of the DNA, the column should be washed with three column volumes of T10E1 buffer and then stored in fresh T10E1 buffer at 4°C.

4. Notes

1. The triplex assay described is protected by patent application WO06/051303. Commercial performance of the assay requires a license, available from Plant Bioscience Ltd. (Norwich, UK; http://www.pbltechnology.com/). The assay is available as a kit from Inspiralis Ltd. (Norwich, UK; http://www.inspiralis.com/) who also supply pNO1 and a range of topoisomerase enzymes, including several bacterial DNA gyrases and topoisomerase IVs, and human topoisomerases I and II.
2. The presence of suspended particles will result in light scattering, which will reduce the quality of data and may introduce variation. It is also recommended that the microtitre plates are protected with a cover where possible to prevent airborne particles entering the wells.

3. Alternatively Luria-Bertani (LB) or Terrific Broth (TB) can be used, although lower yields and increased amounts of nicked plasmid may be observed.
4. The assay was originally developed using a TF buffer of 50 mM sodium acetate pH 5.0, 50 mM sodium chloride, and 50 mM magnesium chloride. However, 75 mM magnesium acetate (pH 4.7) was found to give slightly improved results. Certain DNA-binding compounds can promote or inhibit the formation of triplexes, resulting in them appearing either as false negatives or positives in the assay. Although the inclusion of proper controls and hit validation should minimize the impact of these compounds on the screen results, we have found that substituting magnesium with another divalent metal (e.g. calcium) can reduce the effect of intercalators upon triplex formation. However, no one metal ion has emerged as a panacea for the problem.
5. Alternatively, carbenicillin can be used, which is broken down more slowly than ampicillin, resulting in more sustained selective pressure which may increase plasmid yields.
6. Repeated freeze thaw cycles will reduce enzyme activity, so it is recommended that the subunits are aliquoted before freezing. The GyrB subunit can precipitate if stored at concentrations higher than 1 mg/ml.
7. It is essential to wash the wells thoroughly before and after TFO1 addition as unbound TFO1 can interfere with triplex formation. Buffer should be thoroughly removed after the final wash to prevent it interfering with subsequent steps. Residual buffer can be removed by pipetting or aspiration.
8. It is highly advisable to calculate the concentration of DNA gyrase equivalent to 1 U with the triplex assay prior to screening. One unit of supercoiling activity is defined as the amount of enzyme required to just fully supercoil 0.5 μg of relaxed pNO1 at 37°C in 30 min. The extent of supercoiling is determined by the inclusion of a control containing 1 μg supercoiled pNO1 without enzyme. Conversely, a unit of relaxation activity is defined as the amount of enzyme required to just relax 0.5 μg of supercoiled pNO1 at 37°C in 30 min. Units of enzyme activity are quantified by performing the reaction over a range of enzyme concentrations. For example, to quantify units of DNA gyrase supercoiling activity using the triplex assay, supercoiling would be assayed with DNA gyrase between 2 and 20 nM in triplicate, and the amount of enzyme corresponding to 1 U can then be extrapolated from a linear regression of these data.
9. The assay can be carried out with less DNA (e.g. 0.75 μg) and correspondingly less enzyme. The enzyme should be added to

the reaction mix last. It is possible to make a Master Mix of water, gyrase supercoiling buffer and relaxed pNO1, which is then aliquoted into the microtitre plate wells. The enzyme is then added to bring the volume to 30 μl and it starts the reaction.

10. A larger reaction volume (60 μl) is suggested here for performing the screen since although it uses more materials, it can give more consistent results. When designing the screen, the first and last column of each plate should be reserved for negative controls (substrate plasmid in identical buffer conditions to sample wells but lacking compounds and enzyme) and positive controls (substrate plasmid in identical buffer conditions to sample wells, including enzyme but without compounds). It is advisable to perform the screen in duplicate to ensure confidence in the results.

11. It is advisable to make both DNA and Enzyme Mixes as single stocks for the entire screen; these can be frozen at –20°C in appropriately sized aliquots if necessary. However, the reconstituted DNA gyrase should only be added to the Enzyme Mix immediately before use, since repeated freeze/thaw cycles and storage at low concentrations can result in loss of activity. The amount of each mix required should be calculated on the number of wells required, plus 10% to allow for reservoir dead volume.

12. Using the protocol provided, 1 ml of streptavidin coated agarose beads with the TFO1 oligo immobilized on them can capture ~0.2 mg of supercoiled plasmid. By recirculating the flow through after elution of the bound plasmid, more plasmid can be captured. We have found that five such rounds of elution/recirculation result in a yield of ~60% with high purity. Higher yields may be obtainable with a matrix with a larger pore size (since it is theorized that plasmids are too large to gain access to the binding sites within the pores of 4% agarose), larger column volumes or with automated pumping.

References

1. Bates, A. D., and Maxwell, A. (2005) *DNA Topology*, Oxford University Press, Oxford.
2. Nollmann, M., Crisona, N. J., and Arimondo, P. B. (2007) Thirty years of Escherichia coli DNA gyrase: from in vivo function to single-molecule mechanism, *Biochimie* **89**, 490–499.
3. Schoeffler, A. J., and Berger, J. M. (2008) DNA topoisomerases: harnessing and constraining energy to govern chromosome topology, *Q. Rev. Biophys.* **41**, 41–101.
4. Maxwell, A., Burton, N. P., and O'Hagan, N. (2006) High-throughput assays for DNA gyrase and other topoisomerases, *Nucleic Acids Res* **34**, e104.
5. Hanvey, J. C., Shimizu, M., and Wells, R. D. (1988) Intramolecular DNA triplexes in supercoiled plasmids, *Proc. Natl. Acad. Sci. USA* **85**, 6292–6296.
6. Sakamoto, N., Akasaka, K., Yamamoto, T., and Shimada, H. (1996) A triplex DNA structure of the polypyrimidine:polypurine stretch in the 5′

flanking region of the sea urchin arylsulfatase gene, *Zoolog. Sci.* **13**, 105–109.

7. Boros, I., Pósfai, G., and Venetianer, P. (1984) High-copy-number derivatives of the plasmid cloning vector pBR322, *Gene* **30**, 257–260.
8. Danquaha, M. K. (2007) Growth medium selection and its economic impact on plasmid DNA production, *J. Biosci. Bioeng.* **104**, 490–497.
9. Maxwell, A., and Howells, A. J. (1999) Overexpression and purification of bacterial DNA gyrase, In *DNA Topoisomerase Protocols I. DNA Topology and Enzymes* (Bjornsti, M.-A., and Osheroff, N., Eds.), pp 135–144, Humana Press, Totowa, New Jersey.
10. Tricoli, J. V., and Kowalski, D. (1983) Topoisomerase I from chicken erythrocytes: purification, characterization, and detection by a deoxyribonucleic acid binding assay, *Biochemistry* **22**, 2025–2031.
11. Zhang, J. H., Chung, T. D., and Oldenburg, K. R. (1999) A simple statistical parameter for use in evaluation and validation of high throughput screening assays, *J Biomol Screen* **4**, 67–73.
12. Anderle, C., Stieger, M., Burrell, M., Reinelt, S., Maxwell, A., Page, M., and Heide, L. (2008) Biological activities of novel gyrase inhibitors of the aminocoumarin class, *Antimicrob. Agents Chemother.* **52**, 1982–1990.
13. Schluep, T., and Cooney, C. L. (1999) Immobilization of oligonucleotides on a large pore support for plasmid purification by triplex affinity interaction, *Bioseparation* **7**, 317–326.

Chapter 18

Microscale Thermophoresis as a Sensitive Method to Quantify Protein: Nucleic Acid Interactions in Solution

Karina Zillner, Moran Jerabek-Willemsen, Stefan Duhr, Dieter Braun, Gernot Längst, and Philipp Baaske

Abstract

Microscale thermophoresis (MST) is a new method that enables the quantitative analysis of molecular interactions in solution at the microliter scale. The technique is based on the thermophoresis of molecules, which provides information about molecule size, charge, and hydration shell. Since at least one of these parameters is typically affected upon binding, the method can be used for the analysis of each kind of biomolecular interaction or modification of proteins or DNA. Quantitative binding parameters are obtained by using a serial dilution of the binding substrate. This section provides a detailed protocol describing the analysis of DNA–protein interactions, using the AT-hook peptides as a model system that bind to short double-stranded DNA.

Key words: Binding assay, Dissociation constant, DNA–protein interactions, Microscale thermophoresis, Interaction affinity

1. Introduction

AT-hooks are short peptide motifs that bind to the minor groove of AT-rich DNA sequences. The binding of the AT-hooks to DNA results in changing the regular B-form structure of DNA (1). The core motif of a canonical AT-hook is a GRP tripeptide flanked by basic amino acid patches. The motif is highly conserved from bacteria to mammals and crucial for the DNA binding properties of a wide variety of proteins, ranging from transcription factors to chromatin remodelers (2). The well-characterized HMGA class of proteins, belonging to the "High Mobility Group" (HMG) family, solely contains AT-hooks as DNA binding domains. HMG proteins

Michael Kaufmann and Claudia Klinger (eds.), *Functional Genomics: Methods and Protocols*,
Methods in Molecular Biology, vol. 815, DOI 10.1007/978-1-61779-424-7_18, © Springer Science+Business Media, LLC 2012

are involved in many DNA dependent biological processes, involving transcription, replication, and repair. Furthermore, they play a crucial role in the regulation of pathological processes, such as viral infection and tumor formation. For example, HMGA overexpression is correlated with neoplastic transformation and tumor progression in breast cancer, colorectal, and lung carcinoma (3, 4). It is suggested that molecules inhibiting the binding of AT-hooks to its target sites could serve as potential antiviral and anticancer drugs (5). The screening of potential molecular targets in pharmaceutical research strongly depends on quick and sensitive methods to measure binding affinities.

The chromatin remodeling complex NoRC, involved in the regulation of the ribosomal RNA genes, contains one subunit, Tip5, with multiple AT-hooks (6, 7). This subunit is required for the specific repression of rRNA gene expression and the AT-hooks are implicated in tethering the rRNA gene to the nucleolar matrix (Längst, unpublished results). Tip5, the large regulatory subunit of NoRC, has multiple AT-hooks and two of them are in close proximity in the linear protein sequence. For the microscale thermophoresis (MST) experiments presented here, GST fusion constructs were cloned encompassing only the first AT-hook (GST-AT1) and the two neighboring AT-hooks (GST-AT1 + 2).

MST is based on the directed movement of molecules along temperature gradients, an effect termed thermophoresis. A spatial temperature difference ΔT leads to a depletion of molecule concentration in the region of elevated temperature, quantified by the Soret coefficient S_T: $c_{hot}/c_{cold} = \exp(-S_T \Delta T)$ (8, 9). Figure 1 shows the setup of the apparatus to determine MST.

MST depends on the interface between molecule and solvent. Under constant buffer conditions, thermophoresis probes the size, charge, and solvation entropy of the molecules. Upon binding of a nucleic acid to a protein at least one of these parameters changes and thus the binding can be quantified by measuring the change in the MST signal (10). In order to analyze the binding of the fluorescently labeled dsDNA to the peptide, the measurement is performed at various concentrations of the unlabeled peptide. In a typical MST-experiment, the concentration of the fluorescently labeled molecule (in this case Cy3-labeled-dsDNA) is kept at a constant concentration and the unlabeled molecule (here the unlabeled peptide) is titrated until the saturation of the binding is achieved (11).

The depicted protocol compares the binding properties of the single and dimeric AT-hooks in a typical MST experiment and shows that the dimeric AT-hooks (GST-AT1 + 2) have a higher binding affinity than the single AT-hooks. The experimental scheme outlined below can be adapted for most protein–nucleic acid binding measurements.

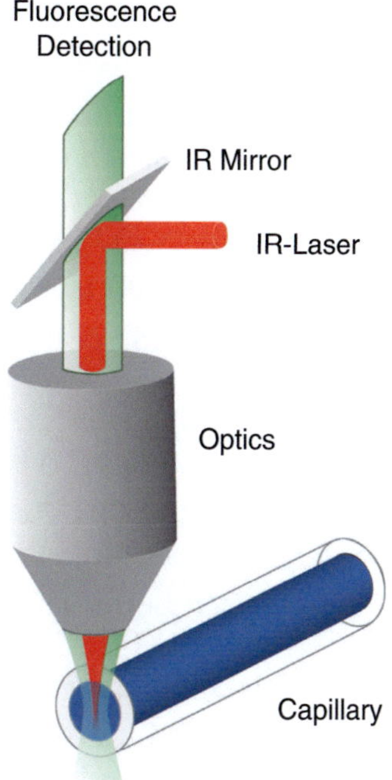

Fig. 1. Setup of the MST assay. The aqueous solution inside the capillary is locally heated with a focused IR-laser, which is coupled into an epifluorescence setup using an IR mirror.

2. Materials

2.1. Preparation of DNA Template

1. Fluorescently labeled oligonucleotides (Metabion) were ordered with the following sequences: 5′-Cy3-GGG AAA TTC CTC-3′ and the complementary strand 5′-GAG GAA TTT CCC-3′ (12). Once dissolved in water (see Note 1) oligonucleotides were stored in light protected vials at –20°C.
2. Annealing buffer (10×): 200 mM Tris–HCl pH 7.4, 20 mM $MgCl_2$, 500 mM NaCl. Stored at room temperature.
3. TBE buffer (0.4×): 35 M Tris, 35 mM Boric acid, 0.8 mM EDTA pH 8.0.
4. Thirty percent acrylamide/bis-acrylamide solution (37.5:1, Roth). Avoid unnecessary exposures, as the unpolymerized solution is neurotoxic.
5. 99% p.a. *N,N,N,N′*-tetramethylethylenediamine for electrophoresis (TEMED, Roth). Stored at 4°C.
6. Ammonium persulfate (APS): Prepare 20% solution in water and store aliquots at –20°C.
7. Gel chamber system, such as XCell Sure Lock™ System (Invitrogen).

8. GeneRuler™ Ultra Low Range DNA Ladder (Fermentas).
9. Glycerol >99.5% p.a. (Roth).
10. Ethidium bromide (Roth) stored at room temperature and a dark place: Prepare a fresh 1:10,000 solution in water before use. Beware that the chemical is toxic and mutagenic, so avoid contact and wear adequate protection.
11. FLA-5100 Fluorescence Imager (Fujifilm).
12. UV/VIS Spectrophotometer, such as Nanodrop (Peqlab).

2.2. Microscale Thermophoresis

1. NanoTemper's Microscale Thermophoresis instrument Monolith NT.115.
2. Monolith NT™ capillaries purchased from NanoTemper Technologies GmbH (Standard treated, Hydrophobic or Hydrophilic).
3. MST buffer (5×): 250 mM Hepes pH 7.4, 25 mM $MgCl_2$, 500 mM NaCl, and 0.25% (v/v) NP-40. When stored at 4°C, it is stable for 1 month.
4. GST-AT1, GST-AT1+2 peptides, and GST protein as interaction partners and negative control, respectively. Stored in 10% glycerol at −20°C.

3. Methods

3.1. Annealing of Oligonucleotides

Double-stranded DNA substrate molecules were annealed from single-stranded oligonucleotides. It is crucial for the experiment that the fluorescently labeled oligonucleotide is quantitatively incorporated into the DNA substrate. This is achieved by adding the unlabeled oligonucleotide at a 1.1-fold molar ratio with respect to the labeled oligonucleotide to the annealing reaction. The efficiency of the annealing reaction can be determined on a 15% native polyacrylamide (PAA) gel that is first analyzed on a fluorescence imager to reveal nonincorporated, fluorescently labeled oligonucleotides and second, poststained with ethidium bromide.

1. Dissolve oligonucleotides according to the manufacturer's instructions and measure the nucleic acid concentration using a UV/VIS Spectrophotometer.
2. Mix 550 pmol unlabeled oligonucleotides with 500 pmol Cy3-labeled oligonucleotide. Then, add 5 μl annealing buffer (10×) and adjust the volume to 50 μl with ddH_2O to finally obtain a 10 μM solution of double-stranded DNA (see Note 2 for general instructions how to work with fluorescence dyes).
3. Incubate the mixture for 15 min at 95°C on a thermoblock, then switch off the thermoblock, and allow the reaction to slowly cool down until it reaches room temperature. The reaction can now be stored at −20°C.

4. A 15% native PAA gel is prepared by the following scheme and quickly poured into an assembled gel chamber: 9 ml 30% bis-acrylamide, 9 ml 0.4× TBE, 25 μl APS, 5 μl TEMED. Position a 10-well comb in the top of the gel. After the gel polymerized (60 min), place the chamber into the running cell, remove the comb and fill it with 0.4× TBE running buffer. To remove unpolymerized acrylamide, prerun the gel for 60 min at 120 V.
5. 15 pmol of the annealing reaction, as well as 15 pmol of the single-stranded oligonucleotides are individually mixed with glycerol to reach a final concentration of 5% (v/v) glycerol. This will weigh down the sample and prevent the solution to mix with the buffer in the well. Load carefully all samples (see Note 3) together with the DNA ladder onto the prerun gel, connect it to a power supply and run it at 4°C for 90 min at 120 V. Bromophenol blue (usually present in the DNA marker) can be used as an indicator, as it migrates ahead of the single-stranded oligonucleotides with an apparent molecular weight corresponding to an oligonucleotide of about 10 nt in length.
6. The gel is visualized with a fluorescence imager. The fuzzy oligonucleotide band has to be quantitatively shifted up in the annealing reaction, migrating as a defined band representing the double-stranded oligonucleotide.
7. Optionally, the efficiency of the annealing reaction is monitored by ethidium bromide staining. The gel is placed in the aqueous ethidium bromide solution and shaked for 10 min at room temperature. The gel can subsequently be visualized by a UV screen.
8. If free, labeled oligonucleotides are visible, the annealing reaction has to be repeated with an increased ratio of the fluorescently labeled oligonucleotide.

3.2. Preparation of the Titration Series

A titration series consists of up to 16 capillaries which are measured in a single thermophoresis run. Notice that pipetting the samples and filling the capillaries will take about 30 min in total. In order to ease pipetting, the DNA substrate is diluted to a final concentration of 750 nM. The binding reaction contains 50 nM fluorescently labeled DNA with varying protein concentrations (see Note 4). The concentration of the fluorescently labeled molecule should be close to the expected K_D and in the range of 80–1,500 fluorescence counts. Dilutions of the unlabeled peptide should start at a concentration about 20-fold higher as the expected K_D, being diluted to concentrations of about 0.01-fold of the expected K_D. The individual binding reactions should be prepared with a volume of 15 μl, for the ease of pipetting and the minimization of experimental errors. However, a volume of only 5 μl would be sufficient to fill the capillary. All solutions are mixed carefully by pipetting the reaction up and down, rather than vortexing the solution.

1. To ensure high accuracy between the individual samples, a Master Mix (MM) is prepared, just lacking the protein. An individual reaction mix contains 3 μl MST buffer (5×), 1 μl annealed oligonucleotides (750 nM) and 1 μl water to adjust the volume to 5 μl. The MM has to be protected from light and can be stored on ice for several days. Prepare a serial dilution reducing the protein concentration 1.5-fold with every dilution step (see Note 5).
2. It is crucial to assure that the correct protein amount is pipetted because there is no possibility to normalize for protein concentrations in contrast to the fluorescent DNA. Therefore, it is beneficial to add firstly 5 μl of MM and then 10 μl of the protein solution for each dilution step. Mix well by pipetting up and down the reaction mix.
3. Before filling a capillary, the reaction must be in equilibrium which only takes 1 min for the presented kind of interaction. From now on, all steps including the work with capillaries should be performed with powder free gloves to prevent impurities and adverse effects on the glass surface (see Note 6). In addition, touch only the ends and not in the middle part of the capillaries where the observation field is located. To find out which type of capillary should be used (see Note 7).
4. To fill a capillary, dip it into the reaction sample. Care should be taken that the capillary does not touch the surface of the reaction tube, since adhering molecules may falsify the measurement. Loaded capillaries are sealed on both ends by shortly sticking them into wax (provided with the capillaries).
5. All capillaries are then inserted into the metal tray and analyzed using the NanoTemper Monolith and the MST-data acquisition software ("Titration").

3.3. Measurement and Analysis of the Binding Affinity

After starting the NanoTemper Software "Titration" the LED "green" is selected for Cy3-dyes and then, the capillaries are automatically identified by clicking on the respective button "find capillaries" (initial settings: LED power: 50%, IR-laser power 10%). During capillary scan, the fluorescence signal should be in the range between 100 and 2,000 fluorescence units. If the value is below 100 fluorescence units, please refer to Note 8. Since all samples should contain the same concentration of fluorescently labeled DNA, individual differences in intensity should be maximally 10% (see Note 9 for trouble shooting if the aberration is more than this value). If the experiments should be performed at a specific reaction temperature, then see Note 10 for instructions. After the identification of capillary positions, the IR-laser is set to be "off" for 10 s and to be "on" for 40 s, guarantying sufficient time for thermophoretic movement and optimal thermophoretic resolution.

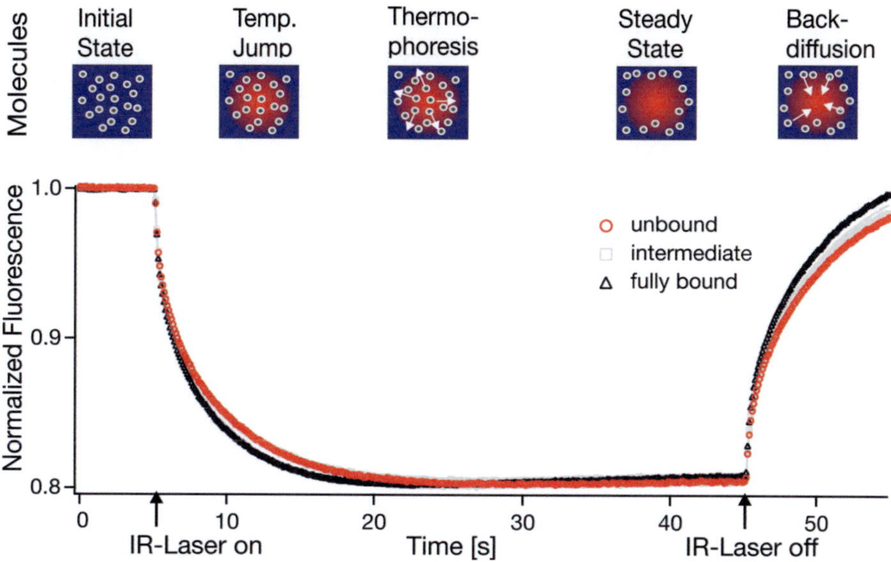

Fig. 2. DNA binding to GST-AT1. The concentration of the fluorescently labeled DNA is kept constant and GST-AT1 is titrated. The fluorescence inside the capillary is measured for each different GST-AT1 concentration and the normalized fluorescence in the heated spot is plotted against time. The IR-laser is switched on at $t=5$ s and the fluorescence changes as the temperature increases. There are two effects, separated by their time scales, contributing to the new fluorescence distribution: the fast temperature jump (time scale ≈ 1 s) and the thermophoretic movement (time scale ≈ 10 s). Both effects show the binding of the labeled DNA to its target: the temperature jump signal increases upon binding of GST-AT1, whereas the thermophoresis decreases upon binding. Once the IR-laser is switched off ($t=45$ s) the molecules diffuse back.

1. After updating the software with the unlabeled protein concentrations, select the destination folder to save the experiment and initiate the measurement. With the stated settings, a run will be completed in about 10–15 min. When opening the Start-window, it is still possible to adjust the number of measurements and the strength/level of the temperature field induced by the IR-laser: 10% (low IR-laser power), 40% (moderate), 100% (high). Immediately after starting the measurement, the data is ready for analysis (on the fly data analysis) with the provided MST-analysis software. A binding curve is being plotted using the normalized fluorescence of the labeled dsDNA at different concentrations of the unlabeled peptide. A normalized fluorescence data set is shown in Fig. 2. If the time traces of recorded fluorescence traces look different, please refer to Note 11.

2. For data analysis, at first the different effects of Microscale Temperature Jump (MST T-Jump) and MST have to be identified according to their different time scales, see Fig. 2: The fast MST T-Jump equilibrates within 1 s after switching on the IR-Laser. This time scale is determined by the heat conductivity of water and the thickness of the capillaries. Thermophoresis equilibration is accomplished only after the establishment of the temperature difference. The time scale of MST equilibration takes about 10 s and is determined by the diffusion constant of

the labeled molecules. The analysis software automatically detects the temperature jump signal and the thermophoresis signal and the respective data points are plotted against the concentration of the binding partner (Fig. 3a). see Note 12 for choosing the "right" signal intensities.

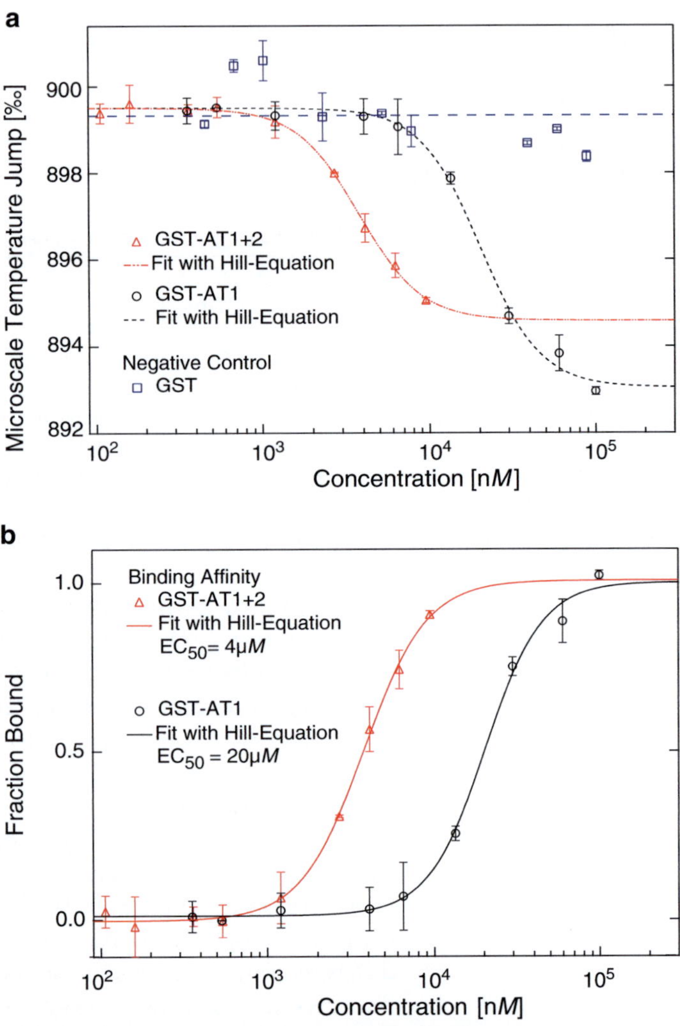

Fig. 3. Binding analysis. The concentration of the fluorescently labeled DNA is kept constant and the target is titrated. (**a**) The temperature jump data for different targets (GST-AT1 + 2, GST-AT1, and GST) is plotted against the titrated target concentrations. GST-AT1 and GST-AT1 + 2 show a sigmoidal binding curve with different amplitude and different affinity, whereas GST shows no binding to the DNA. The measured values are fitted with the Hill-equation to determine the "fully bound" state which is, in the case of GST-AT1 + 2, not reached with the highest titrated concentration. (**b**) For a better comparison, the binding curves of GST-AT1 and GST-AT1 + 2 are normalized to the fraction of bound molecules. The value 1 denotes that 100% of the fluorescently labeled DNA is bound to its target. The plot indicates that GST-AT1 + 2 has a five times higher affinity (EC_{50} = 4 μM) to the DNA than GST-AT1 (EC_{50} = 20 μM). Both curves were fitted with the Hill-equation with a Hill-coefficient of $n = 2.5$.

3. For quantifying the interaction affinity, the value of the dissociation constant K_D or the EC_{50} value are calculated (Fig. 3b) using the NanoTemper Analysis software:

   Push the "load folder" button in order to import the measured data.

   Push "select all" button.

   By pushing the thermophoresis, temperature jump, and fluorescence button, the appropriate parameter will be selected and analyzed.

   By using the Fit-window, the data using the parameter of choice can be fitted. To determine the dissociation constant K_D, the data is fitted using MST-standard fit algorithms (law of mass action) or Hill-algorithms. Often the data is best fitted using the Hill-equation (Fig. 3b) because the Hill-equation provides one more free parameter n (the Hill-coefficient, see Note 13). From the Hill equation, the EC_{50}-affinity value can be determined.

   The fitted data can be saved as an image or alternatively be exported as a text file.

   For comparison, the data can be normalized (Fig. 3b) to the fraction of complexed molecules (FB) by the following equation:

   $FB = (value(c) - free)/(complexed - free)$, where value($c$) is the MST-value measured for the concentration c, *free* is the MST-value for the unbound state (lowest concentration) and *complexed* is the MST-value for the fully bound state (see Note 14).

4. Notes

1. Unless otherwise stated, water refers to a standard having 18.2 $M\Omega \times cm$ and organic content less than five parts per billion.
2. Fluorescence dyes are sensitive to light. Thus, avoid direct exposure to light radiation. Fluorescently labeled molecules are best kept in a dark cup that shields the dye from bleaching effects.
3. Do not use loading dye instead of glycerol since we observed that bromophenol blue quenches the fluorescence. Carefully load the gel in a way that the sample is not diluted by the running buffer. Place the tip of the pipette at the bottom of the well and slowly fill it. Do not blow bubbles out of the tip which push the sample out of the well. Native gels do not have a "stacking gel," therefore compact loading is crucial to form sharp bands in the gel. In addition, the gel should be kept cool

for optimal resolution. However, running at room temperature is also possible.

4. Any kind of solution may be used if most suitable for the application. But take into consideration that the pH of some buffers, such as Tris-based ones are more sensitive to temperature than others and the IR-laser creates a temperature increase of up to 20 K in its centre, at high IR-laser power.
5. When an electrophoretic mobility shift assay (EMSA) was already performed with the binding partners, the binding affinity can be estimated on the gel and the used concentrations of the interaction partners can be adjusted accordingly. If there is no EMSA experiment existing, start with the protein stock solution and make a serial dilution of 1:1.5 in water to find at which protein concentration the binding takes place.
6. It is recommended to wear powder free gloves when working with fluorescence because the powder used in laboratory gloves can fluoresce and scatter light and hence, impair the measurement.
7. Depending on the binding partners and experimental set up, standard treated capillaries may not always be the first choice. To find out which capillary is the best for your experiment have a closer look to the capillary scan. If all maxima of the scanned capillaries are shaped like an inverse "U," the capillaries are well suited. If the capillaries show an "M" like (splitted pick pattern) fluorescence intensity, the fluorescent molecules are sticking to the capillary walls. In this case, test the samples with the hydrophobic and hydrophilic capillaries and also vary salt- and detergent concentration until you get close to the inversed "U" form.
8. When the fluorescence intensities are below 100, there are several possibilities to augment the value. First, the concentration of the fluorescently labeled oligo can be increased. Is this not feasible or desired, turn on the LED power up to 90%. It is always better to work at low LED power to reduce bleaching effects. Depending on the thermophoretic amplitude of binding, a fluorescence value of 30 can also be used.
9. In case the fluorescence signals of single capillaries are highly divergent, there are several points that can be considered. First, solutions were not adequately mixed and care should be taken next time to efficiently merge all solutions. Second, the capillary might have touched the surface of the tube and thus, fluorescence that stuck onto the cup was sucked in. Third, it is also possible that the pipette is imprecise and another one should be tried in next experiments. However, a gradual increase in the signal with higher protein amounts indicates already a binding event and is not due to technical errors.

10. Especially when measurements are performed at different temperatures than room temperature, place the metal block with the capillaries onto the tray, switch on the temperature control and wait for at least 10 min to achieve even temperature distribution in all capillaries.
11. The raw data of the titration series may look different than described in Fig. 2. In most cases, a positive MST and MST T-Jump signal is observed, i.e., the fluorescence of the time traces is decreasing. It is also possible that a negative MST-signal and/or MST T-Jump signal is observed. This is a kind of fingerprint for the molecule – buffer – interaction system.
12. Some interactions show a good binding signal in both temperature jump and thermophoresis, for these interactions choose the option "thermophoresis + temperature jump" for analysis. In some cases, the temperature jump increases upon binding and the thermophoresis decreases upon binding (Fig. 2) or vice versa. In these cases, both effects cancel out each other when choosing the option "thermophoresis + temperature jump" and no binding is detected. Thus, it is important to look at temperature jump and thermophoresis separately.
13. The Hill-equation $1/[1+(EC_{50}/c)^\wedge n]$ should be used if the slope of the binding curve is too steep ($n>1$) or too smooth ($n<1$) for using a law of mass action (K_D) fit. A Hill-coefficient $n<1$ or $n>1$ can indicate that the binding reaction is more complex than a simple one: one molecule A in one conformation binds to one molecule B in one conformation. The binding affinity EC_{50} is the value at which 50% of the labeled molecules are bound to their targets, thus reasonable EC_{50} values have to be equal or bigger than 50% of the provided constant concentration of labeled molecule.
14. For some interaction, it is not possible to reach the fully bound (100% of labeled molecule is bound to its titrated target) state (Fig. 3a, GST-AT1 + 2). In these cases, the fully bound value can be calculated with the fitting algorithm of the NanoTemper analysis software as this algorithm also uses the slope of the binding curve to determine the bound state.

    If the MST-value of the free state is higher than the MST-value of the complexed state the following equation for calculating the fraction of the complexed molecules has to be used: $FB = 1-[(\text{value}(c) - \text{free})/(\text{complexed} - \text{free})]$.

Acknowledgments

The authors would like to thank Christoph J. Wienken the fruitful comments and suggestions for data analysis.

References

1. Reeves, R. and Nissen, M.S. (1990) The A.T-DNA-binding domain of mammalian high mobility group I chromosomal proteins. A novel peptide motif for recognizing DNA structure. *J. Biol. Chem.* **265**, 8573–8582.
2. Aravind, L. and Landsman, D (1998) AT-hook motifs identified in a wide variety of DNA-binding proteins. *Nucleic Acids Res.* **26**, 4413–4421.
3. Reeves, R. (2001) Molecular biology of HMGA proteins: hubs of nuclear function. *Gene* **277**, 63–81.
4. Reeves, R. (2010) Nuclear functions of the HMG proteins. *Biochim. Biophys. Acta* **1799**, 3–14.
5. Susbielle, G., et al. (2005) Target practice: aiming at satellite repeats with DNA minor groove binders. *Curr. Med. Chem. Anticancer Agents* **5**, 409–420.
6. Strohner, R., et al. (2001) NoRC--a novel member of mammalian ISWI-containing chromatin remodeling machines. *EMBO J.* **20**, 4892–4900.
7. Németh, A., et al. (2004) The chromatin remodeling complex NoRC and TTF-I cooperate in the regulation of the mammalian rRNA genes in vivo. *Nucleic Acids Res.* **32**, 4091–4099.
8. Duhr, S. and Braun, D. (2006) Why molecules move along a temperature gradient. *Proc. Natl. Acad. Sci. USA* **103**, 19678–19682.
9. Baaske, P., et al. (2007) Extreme accumulation of nucleotides in simulated hydrothermal pore systems. *Proc. Natl. Acad. Sci. USA* **104**, 9346–9351.
10. Baaske, P., et al. (2010) Optical thermophoresis for quantifying the buffer dependence of aptamer binding. *Angew. Chem. Int. Ed.* **49**, 2238–2241.
11. Wienken, C. J., et al. (2010) Protein-binding assays in biological liquids using microscale thermophoresis. *Nat. Commun.* **1**:100 doi: 10.1083/ncomms1093(2010).
12. Huth, J.R., et al. (1997) The solution structure of an HMG-I(Y)-DNA complex defines a new architectural minor groove binding motif. *Nat. Struct. Biol.* **4**, 657–665.

Chapter 19

Bioluminescence Resonance Energy Transfer: An Emerging Tool for the Detection of Protein–Protein Interaction in Living Cells

Søren W. Gersting, Amelie S. Lotz-Havla, and Ania C. Muntau

Abstract

In the field of proteomics, numerous advanced technologies have evolved that aim to provide the molecular data necessary for an in-depth understanding of biological processes. Protein–protein interactions (PPI) are at the heart of cellular function and a milestone yet to be achieved is the mapping of a complete human interactome. Bioluminescence resonance energy transfer (BRET) has become a popular technique to investigate PPI. As BRET enables the detection of PPI in living cells, problems associated with in vitro biochemical assays can be circumvented, thus making BRET a powerful tool for monitoring interactions of virtually all kinds of protein species.

Key words: Bioluminescence resonance energy transfer, Protein–protein interaction, Living cells, *Renilla* luciferase, Yellow fluorescent protein, Nucleofection

1. Introduction

The analysis of PPI is of central importance for a comprehensive understanding of cellular processes. The concept of proteins exerting their function as a part of larger functional complexes is increasingly appreciated and the network of all PPI, the human interactome, is estimated to cover several hundreds of thousands of binary interactions (1–4). Among the numerous methods currently applied for the determination of PPI, some are more appropriate for large-scale high-throughput screens while others proved particularly advantageous for individual approaches. Precise knowledge on the weaknesses and strengths of a given

Michael Kaufmann and Claudia Klinger (eds.), *Functional Genomics: Methods and Protocols*, Methods in Molecular Biology, vol. 815, DOI 10.1007/978-1-61779-424-7_19, © Springer Science+Business Media, LLC 2012

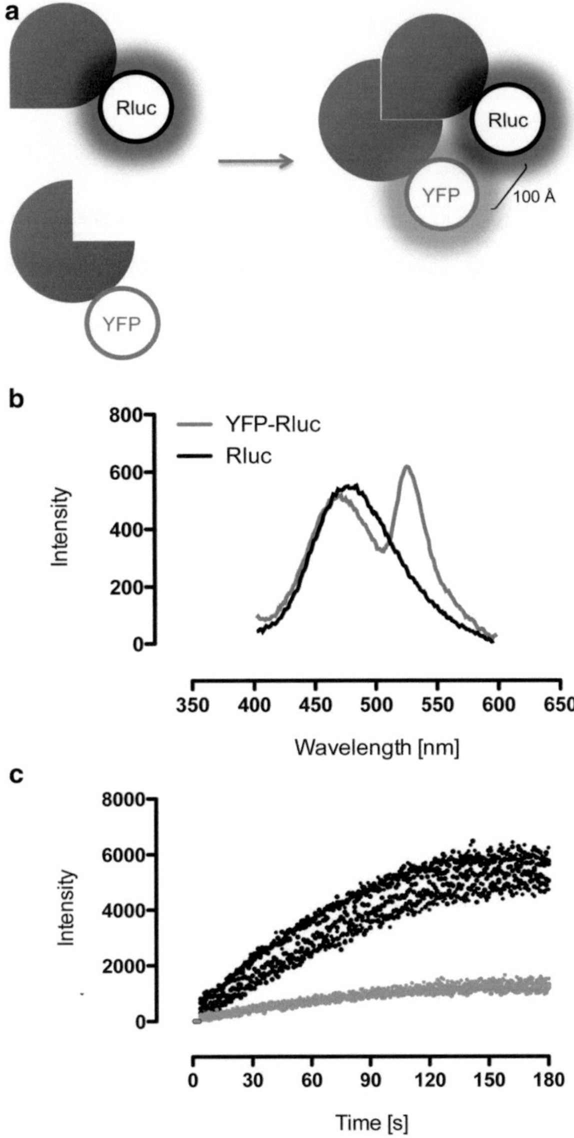

Fig. 1. The principle of protein–protein interaction (PPI) analyses based on bioluminescence resonance energy transfer (BRET). (**a**) The scheme depicts the interaction of two proteins of interest fused to *Renilla* luciferase (Rluc) as energy donor and yellow fluorescent protein (YFP) as energy acceptor, respectively (BRET[1]). Energy transfer results in acceptor luminescence (BRET signal) when the proteins come into close proximity (Förster radius <100 Å), indicative for a positive PPI. (**b**) Emission spectra of Rluc luminescence and of BRET of the YFP–Rluc fusion protein are shown, with peaks at 485 nm for Rluc and 535 nm for YFP. The spectral overlap of Rluc and YFP may result in a bleed-through of donor emission in the detection channel of the acceptor signal. (**c**) Luminescence signals at 485 nm (*black symbols*) and 535 nm (*gray symbols*) detected in COS7 cells expressing Rluc as well as YFP fusion constructs. Upon addition of coelenterazine at time point 0, the luminescence signals increase constantly over 120 s and then remain stable over a period of at least 60 s.

technique applied to answer a specific question determines the biological significance that can be derived from the experiment.

The detection of binary PPI in living eukaryotic cells allows for expression in the physiological cellular environment permitting posttranslational modifications and trafficking to the correct subcellular compartment. To carry out BRET analyses, two proteins of interest are genetically fused to an energy donor protein (a luciferase) or to an energy acceptor (a fluorescent protein), respectively, constituting a BRET pair (5). If the proteins of interest interact with each other the BRET-tags come into close proximity and energy transfer occurs after oxidation of a luciferase substrate. The resulting acceptor energy emission can be detected (see Fig. 1a) and is given in proportion to the donor luminescence (BRET ratio). BRET involves nonradiative transfer of energy from an excited-state energy donor to a suitable fluorescent acceptor situated within a distance of 100 Å (6). This enables the detection of binary PPI as well as the detection of positive PPI being part of higher order protein complexes. As the transfer of energy is inversely proportional to the sixth power of the distance between donor and acceptor dipoles (6), any change in the spatial orientation of the BRET-tags can be detected at high sensitivity. As in BRET the measure for a PPI is given as the ratio of acceptor to donor signals, this can be exploited to additionally determine the relative affinity of interaction by performing BRET saturation experiments (7, 8).

Several different BRET-techniques are known comprising different luciferase species and derivatives of green fluorescent protein as well as various substrates (5). The most common applications are based on *Renilla* luciferase (Rluc), where $BRET^1$ uses the common coelenterazine substrate and yellow fluorescent protein (YFP) whereas $BRET^2$ exploits a special DeepBlueC substrate and green fluorescent protein, for better spatial separation of donor and acceptor emission (9). Besides the improved separation of emission spectra of $BRET^2$, the $BRET^1$ method bears the advantage of higher quantum yields and thus provides better sensitivity of signal detection. Here, we describe the use of $BRET^1$ for the detection of PPI in living cells in a multi-well format allowing for reasonable throughput at high sensitivity and validity.

2. Material

2.1. Cell Culture

1. RPMI 1640 Medium with L-glutamine, with phenol red, supplemented with 10% fetal bovine serum (FCS GOLD) and 1% antibiotic–antimycotic solution (corresponding to 100 U/ml penicillin, 0.1 mg/ml streptomycin, and 0.25 μg/ml amphotericin B).

2. Solution of trypsin (0.125%) and EDTA (0.05%).
3. Dulbecco's PBS (1×) without calcium or magnesium.
4. HYPER*Flask* Cell Culture Vessel (Corning, New York, USA).
5. COS7 cells (DSMZ ACC 60, Braunschweig, Germany) (see Note 1).

2.2. Transfection (Electroporation)

1. Electroporation system, for example Amaxa 96-well Shuttle Device and Nucleofector II Device (Lonza, Cologne, Germany).
2. Electroporation buffer and cuvettes. Here, the SE Cell Line 96-well Nucleofector Kit (96 RCT) (Lonza, Cologne, Germany) for the use in 96-cuvette nucleofection plates is described. The ready-to-use Nucleofector Solution SE is prepared by adding the supplement to the solution, and is then stable for 3 months at 4°C.
3. White-wall 96-well plates, clear bottom, TC-treated.
4. RPMI 1640 Medium with L-glutamine, without phenol red, supplemented with 10% FCS and 1% antibiotic–antimycotic solution (corresponding to 100 U/ml penicillin, 0.1 mg/ml streptomycin, and 0.25 µg/ml amphotericin B).
5. BRET expression constructs based on plasmids for eukaryotic expression coding for proteins of interest that are N- and C-terminally fused to Rluc or YFP, respectively. A plasmid coding for a YFP–Rluc fusion protein (positive control), as well as a plasmid coding for YFP only (correction factor), have to be cloned in advance.

2.3. Detection of BRET Signals

1. Coelenterazine native dissolved in methanol (1 mg/ml) and stored in aliquots at −80°C for long-term stability. Coelenterazine is sensitive to light.
2. *Renilla* luciferase assay buffer: NaCl 64.28 g/L, Na_2EDTA 0.82 g/L, KH_2PO_4 29.92 g/L, BSA 0.44 g/L, sterile filtered, and stored at 4°C.
3. Dulbecco's PBS (1×) without calcium or magnesium.
4. Luminescence multi-well plate reader equipped with two filters for simultaneous detection of the emission of Rluc (475 ± 30 nm) and YFP (535 ± 30 nm), for example LUMIstar OPTIMA (BMG LABTECH, Offenburg, Germany).

3. Methods

Since the first description of BRET (10), this technology has evolved steadily. The following protocol focuses on the use of the first generation of BRET ($BRET^1$), using Rluc as energy donor, YFP as energy acceptor, and coelenterazine as the Rluc-specific substrate (see Note 2 and Fig. 1). Hence, to study the interaction of two proteins by BRET[1], one protein of interest is genetically fused to Rluc (energy donor) and the other to YFP (energy acceptor). Both proteins are then coexpressed in cells, and the BRET signal is detected after the addition of coelenterazine. To enable large-scale transfection of cells at high efficiency, an electroporation system in 96-well format is applied.

3.1. Cell Culture

1. To gain a sufficient number of cells, COS7 cells are cultured under monolayer conditions in an HYPER*Flask* cell culture vessel. For this, ten million COS7 cells are seeded into one vessel and cultured in RPMI 1640 medium at 37°C and 5% CO_2 for 1 week (see Note 1).
2. On the day of transfection, cells are detached by trypsinization and the cell number is determined in a conventional Neubauer counting chamber. Usually, a total of 240 million COS7 cells can be harvested at this juncture (see Note 3).

3.2. Transfection (Electroporation)

1. For electroporation, the use of a 96-well nucleofection device is described resulting in sufficient transfection efficiency and providing transfection at adequate throughput (see Note 4).
2. For nucleofection in 96-well format, 2×10^5 COS7 cells, 1 μg of plasmid DNA and 20 μl of 96-well Nucleofector Solution SE are needed per well.
3. To analyze binary PPI by BRET, cotransfection of BRET expression vectors coding for the two proteins of interest is performed at a ratio of 3:1 of acceptor (YFP) to donor (Rluc) constructs (see Note 5).
4. Each protein pair is tested in duplicate and two independent experiments are performed (see Note 6).
5. Several control experiments are performed (in triplicate) for every plate. A plasmid coding for a YFP–Rluc fusion protein serves as a positive control and always gives similar intra-assay results (~1.0). As a device-specific negative control, a construct expressing the Rluc-tagged protein of interest with a YFP construct in the absence of the protein of interest is cotransfected. The BRET ratio measured will be used as correction factor (cf) and subtracted from every BRET pair (see Subheading 3.4). Here, the light emission detected in the acceptor channel (535 nm) predominantly results from a bleed-through of donor emission that is specific for the filter set used (see Fig. 1b). In addition, a background control with non-transfected cells is included to ensure stable assay conditions.
6. DNA is prepared in a 96-well V-bottom plate (sterile) or conventional PCR-stripes (sterile) with 0.65 μg for the donor construct and 1.95 μg for the acceptor construct (0.25 and 0.75 μg,

respectively, multiplied by 2 for duplicates and an additional dead volume factor of 1.3).

7. The required total number of cells (2×10^5 COS7 cells multiplied by the respective number of wells and the dead volume factor of 1.3) is pelleted at 200 × *g* for 5 min at 37°C and resuspended in the corresponding volume of prewarmed (37°C) electroporation buffer (20 μl solution multiplied by the respective number of wells and the dead volume factor of 1.3) (see Note 7).
8. A volume of 52 μl of the cell suspension (20 μl multiplied by 2 for duplicates and the dead volume factor of 1.3) is then added to the prepared DNA and mixed by pipetting up and down.
9. The DNA-cell solution mix (20 μl) is then transferred to the wells of the 96-well nucleofection plate in duplicates for each sample.
10. For electroporation using the nucleofection system, the appropriate program optimized for COS7 cells (DSMZ ACC 60) is FP-100 (see Note 1).
11. After transfection, prewarmed RPMI 1640 without phenol red (80 μl) is added to each well and mixed thoroughly. A volume of 50 μl of each well is then transferred accordingly into a 96-well white plate with clear bottom, prepared in advance with prewarmed 150 μl of RPMI media without phenol red per well (see Note 8).
12. Cells are then incubated at 37°C and 5% CO_2 (see Note 9).

3.3. Detection of BRET Signals

1. BRET signals can be detected 24 h after transfection (see Note 9).
2. To prepare plates for BRET measurement, aspirate the culture medium (170 μl) and place the plate into the luminescence plate reader (see Note 10).
3. Coelenterazine solution has to be prepared at least 15 min before the measurement. To prepare coelenterazine solution for the measurement of one plate, 127 μl of coelenterazine native suspended in methanol are added to 1 ml of *Renilla* luciferase assay buffer to obtain a 300 μM solution. Immediately prior to the start of measurement, a volume of 1.1 ml of the 300 μM solution is diluted with 6.6 ml PBS (equivalent to the total volume for one plate: 70 μl/well × 96 wells + 1 ml dead volume for priming of the injection pump).
4. After washing the injection pump with pure water, the pump is primed with the coelenterazine solution.
5. For BRET measurement, a protocol has to be set up, starting with a sequential injection of 70 μl of the coelenterazine solution to each well (resulting in a concentration of 30 μM), followed by an incubation time of 2 min. Signals are then

detected using the dual emission option at 485 nm (Rluc-signal) and 535 nm (BRET-signal) over 60 s with a total of 60 intervals (see Note 11 and Fig. 1c).

3.4. Calculation of BRET Ratios

1. To allow data evaluation, Rluc signals at 485 nm for transfected cells should exceed the interval of the mean value and the ninefold standard error of the nontransfected background control.
2. The BRET-ratio is calculated based on the equation: $R=(I_A/I_D)-\mathrm{cf}$, where R is the BRET ratio, I_A is the intensity of acceptor luminescence emission at 535 nm, I_D is the intensity of donor luminescence emission at 485 nm, and cf is a correction factor ($\mathrm{BRET}_{\mathrm{control}}/\mathrm{Rluc}_{\mathrm{control}}$) with the control being the cotransfection of donor fusion-proteins with YFP in the absence of the second protein of interest.
3. As a positive control, the YFP–Rluc fusion protein should result in BRET ratios around 1.0.
4. A positive interaction of two investigated protein pairs is assumed, if at least one out of eight tested tag combinations results in a BRET ratio above a method-specific threshold of 0.1 (see Note 12). An example of a positive PPI tested for all eight combinations of BRET-tags as well as a network of multiple binary PPI is shown in Fig. 2.

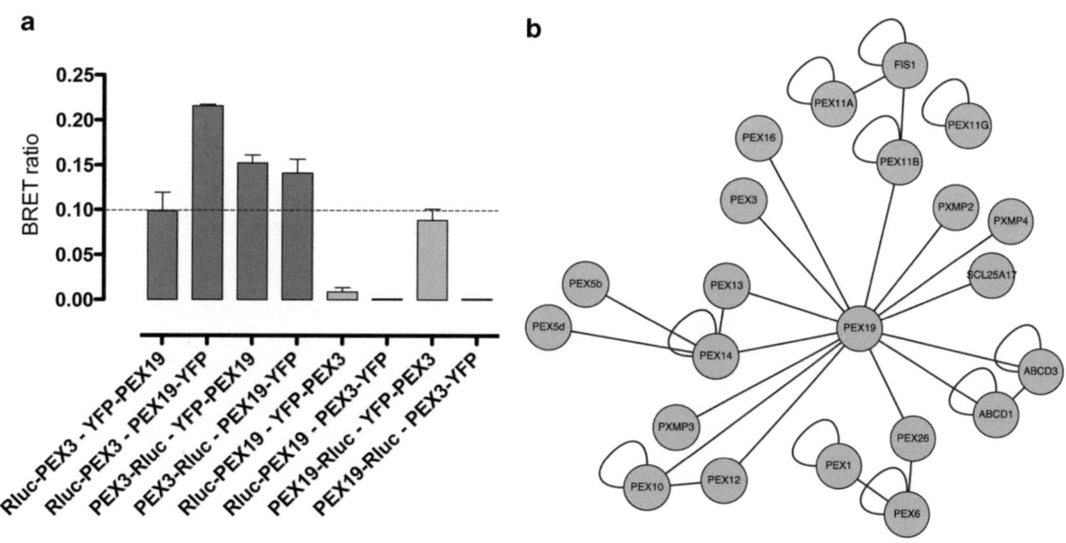

Fig. 2. Determination of binary PPI of peroxisomal proteins by BRET. (**a**) Peroxisomal proteins PEX3 and PEX19 coexpressed in COS7 cells with varying Rluc and YFP tag combinations resulted in BRET ratios above the threshold of 0.1 (*dashed line*) in four out of eight possible tag combinations, confirming that PEX3 and PEX19 interact in living cells. Note that for all four tag combinations of YFP fused to PEX3 no positive BRET ratio was detected. (**b**) Binary PPI of a set of 22 peroxisomal membrane proteins were assayed by BRET. The results are depicted as a network with each line connecting two nodes representing a positive interaction of the respective proteins.

4. Notes

1. BRET experiments can be performed in all primary and stable adherent and suspension cell lines allowing for efficient transfection and good expression of the BRET constructs. Transfection conditions have to be optimized for the respective cell lines. Optimization of nucleofection as described here may also be required for COS7 cells of different sources than the cell line used in this protocol.
2. In alternative to the BRET[1] system (5, 11) with Rluc protein as energy donor and YFP as energy acceptor using coelenterazine as substrate, several variations of BRET have been established: (1) the use of ViviRen (Promega) as substrate, where a cytosolic substrate activation ensures signal detection confined to living cells (12), (2) the use of EnduRen (Promega) as substrate allowing for time-dependent detection of BRET signals over a period of up to 24 h (13), (3) BRET[2] that is taking advantage of distinct spectral properties of Rluc upon oxidation of DeepBlueC substrate (emission wavelength 400 nm) using a modified blue-shifted GFP (emission wavelength 511 nm) (14), and (4) BRET[3] that is based on energy transfer between Rluc using coelenterazine as the substrate and a mutant red fluorescent protein (mOrange, peak wavelength 564 nm) (15).
3. COS7 cells should be used for transfection up to a maximum of 20 cell passages.
4. A high efficiency transfection system allowing for adequate throughput is recommended. Electroporation has proven to provide transfection efficiencies of 80–90% for many different cell lines (16, 17). However, alternative transfection methods may be equally useful. Please note that for 96-well live cell BRET measurement, a minimal transfection efficiency of about 50% is required.
5. For binary PPI, a sequential increase in the ratio of proteins carrying the YFP tag (energy acceptor) over proteins carrying the Rluc tag (energy donor) results in hyperbolic behavior of the BRET ratios reaching a plateau ($BRET_{max}$) when all donor proteins are saturated with the energy acceptor (BRET saturation experiments) (7, 18). In contrast, in the case of nonspecific interactions resulting from random collision, the "bystander" BRET signal increases almost linearly with increasing acceptor to donor ratios, making BRET saturation experiments a suitable tool to distinguish positive and false positive interactions (7, 8). A relative binding affinity index can be determined by the use of the YFP to Rluc ratio (acceptor/donor ratio) at half-maximal BRET ($BRET_{50}$). Examples for BRET saturation experiments are demonstrated in Fig. 3.

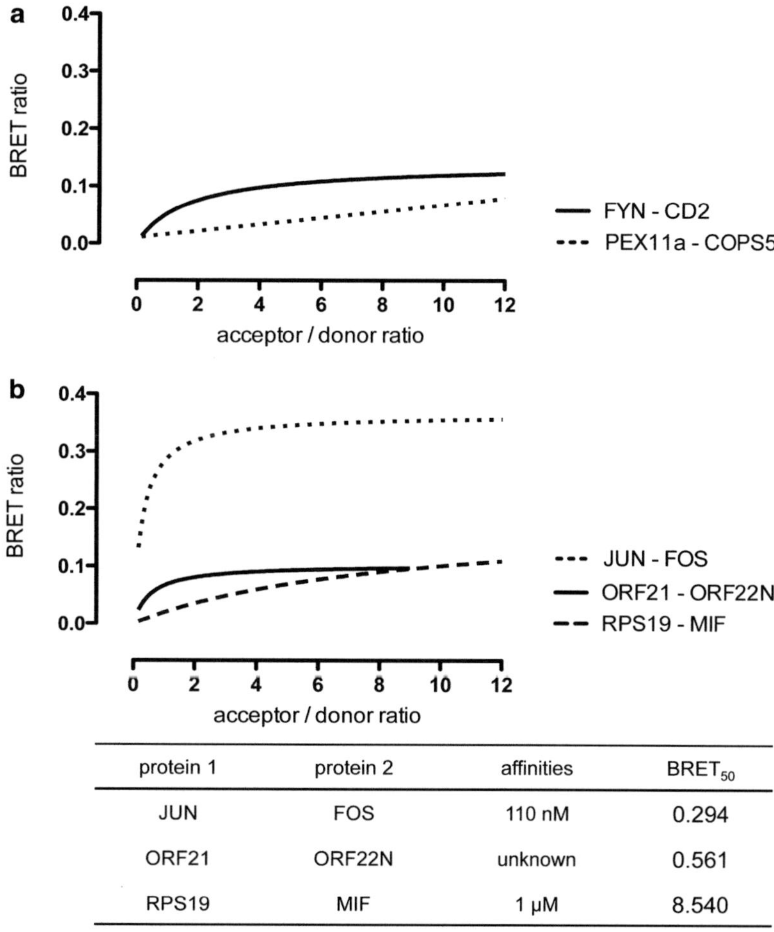

protein 1	protein 2	affinities	$BRET_{50}$
JUN	FOS	110 nM	0.294
ORF21	ORF22N	unknown	0.561
RPS19	MIF	1 μM	8.540

Fig. 3. Characterization of PPI using BRET saturation experiments. Protein pairs were coexpressed in COS7 cells with the respective donor (Rluc) and acceptor (YFP) fusion proteins at increasing acceptor to donor ratios. (**a**) Coexpression of FYN and CD2, which are known to interact (21), resulted in a hyperbolic BRET curve consistent with a positive PPI. BRET ratios for PEX11a-COPS5 increased linearly with increasing acceptor to donor ratios, demonstrating "bystander" BRET of a putative negative PPI. (**b**) The relative binding affinity ($BRET_{50}$) and the maximal BRET ratio ($BRET_{max}$) of known protein interactions (22–24) were determined by nonlinear regression analyses and set into context with data of protein binding affinity derived from the literature.

6. Tag orientation of proteins can influence the BRET signal or the PPI itself. The strategy used here to circumvent this specific problem is the variation of N- and C-terminal fusion constructs for both proteins of interest (19), resulting in eight possible tag combinations for every tested protein pair (see Fig. 2).
7. The number of wells for each plate can be calculated by the following equation: $(n \times 8 \times 2 + 9)$, where n is the number of protein pairs tested, multiplied by eight tag combinations for

each protein pair, multiplied by 2 for duplicates and nine wells for the control experiments are added.

8. To achieve high luminescence signals, particularly when luciferase signals are low due to low protein expression, it is recommended to use white 96-well plates. However, the application of white-wall clear bottom plates provides the advantage of using cell microscopy to determine density and viability of cultured cells as well as the possibility for fluorescence imaging to analyze subcellular distribution of YFP-tagged proteins.
9. When Rluc signals are low, longer incubation periods following transfection (up to 48 h) and different temperature conditions (e.g., 30°C) may be helpful to achieve sufficient protein expression.
10. For BRET experiments performed in suspension cells, particular attention should be given to accidental elimination of cells while preparing plates for the BRET measurement. A crucial step is the aspiration of culture medium prior to the addition of the luciferase substrate. One possible approach is to centrifuge the plate and to subsequently take off the medium carefully. Alternatively, cells can be cultured in a total of 150 μl medium per well and coelenterazine at higher concentration can be added directly to each well. To allow for sufficient dilution of the media and the added coelenterazine, plates should be shaken in the plate reader while injecting the coelenterazine.
11. Depending on the number of wells that are to be measured, an interlaced sequence of injection, incubation, and detection as well as a reduction of the detection period up to a minimum of 10 s may be required to reduce total measurement time and thereby cell stress.
12. The method-specific threshold of 0.1 was determined in our laboratory based on a supervised approach. Using the setting as reported in these instructions or comparable setups, the threshold is assumed to be transferable. Nevertheless, for evaluation of large datasets or in cases of major system modifications, we recommend to redetermine the threshold based on a reference data set consisting of well-documented pairs of interacting human proteins and randomly chosen protein pairs, respectively (20).

Acknowledgments

Financial support from the Bavarian Genome Research Network (BayGene) and the LMUexcellent grant 42595-6 to A.C.M. is gratefully acknowledged.

References

1. Rual JF, Venkatesan K, Hao T et al (2005) Towards a proteome-scale map of the human protein–protein interaction network. Nature 437:1173–1178
2. Venkatesan K, Rual JF, Vazquez A et al (2009) An empirical framework for binary interactome mapping. Nat Methods 6:83–90
3. Stumpf MP, Thorne T, de Silva E et al (2008) Estimating the size of the human interactome. Proc Natl Acad Sci USA 105:6959–6964
4. Hart GT, Ramani AK, and Marcotte EM (2006) How complete are current yeast and human protein-interaction networks? Genome Biol 7:120
5. Pfleger KD, and Eidne KA (2006) Illuminating insights into protein–protein interactions using bioluminescence resonance energy transfer (BRET). Nat Methods 3:165–174
6. Wu P, and Brand L (1994) Resonance energy transfer: methods and applications. Anal Biochem 218:1–13
7. Mercier JF, Salahpour A, Angers S et al (2002) Quantitative assessment of beta 1- and beta 2-adrenergic receptor homo- and heterodimerization by bioluminescence resonance energy transfer. J Biol Chem 277:44925–44931
8. Hamdan FF, Percherancier Y, Breton B et al (2006) Monitoring protein–protein interactions in living cells by bioluminescence resonance energy transfer (BRET). Curr Protoc Neurosci Chapter 5:Unit 5 23
9. Ayoub MA, and Pfleger KD (2010) Recent advances in bioluminescence resonance energy transfer technologies to study GPCR heteromerization. Curr Opin Pharmacol 10: 44–52
10. Xu Y, Piston DW, and Johnson CH (1999) A bioluminescence resonance energy transfer (BRET) system: application to interacting circadian clock proteins. Proc Natl Acad Sci USA 96:151–156
11. Hamdan FF, Audet M, Garneau P et al (2005) High-throughput screening of G protein-coupled receptor antagonists using a bioluminescence resonance energy transfer 1-based beta-arrestin2 recruitment assay. J Biomol Screen 10:463–475
12. Promega (2010) ViviRenTM Live Cell Substrate. Promega Technical Resources TM064
13. Promega (2009) EnduRenTM Live Cell Substrate Promega Technical Resources TM244
14. Bertrand L, Parent S, Caron M et al (2002) The BRET2/arrestin assay in stable recombinant cells: a platform to screen for compounds that interact with G protein-coupled receptors (GPCRS). J Recept Signal Transduct Res 22:533–541
15. De A, Ray P, Loening AM et al (2009) BRET3: a red-shifted bioluminescence resonance energy transfer (BRET)-based integrated platform for imaging protein–protein interactions from single live cells and living animals. FASEB J 23:2702–2709
16. Gresch O, Engel FB, Nesic D et al (2004) New non-viral method for gene transfer into primary cells. Methods 33:151–163
17. Hamm A, Krott N, Breibach I et al (2002) Efficient transfection method for primary cells. Tissue Eng 8:235–245
18. James JR, Oliveira MI, Carmo AM et al (2006) A rigorous experimental framework for detecting protein oligomerization using bioluminescence resonance energy transfer. Nat Methods 3:1001–1006
19. Lin H, Hutchcroft JE, Andoniou CE et al (1998) Association of p59(fyn) with the T lymphocyte costimulatory receptor CD2. Binding of the Fyn Src homology (SH) 3 domain is regulated by the Fyn SH2 domain. J Biol Chem 273:19914–19921
20. Claret FX, Hibi M, Dhut S et al (1996) A new group of conserved coactivators that increase the specificity of AP-1 transcription factors. Nature 383:453–457
21. Filip AM, Klug J, Cayli S et al (2009) Ribosomal protein S19 interacts with macrophage migration inhibitory factor and attenuates its pro-inflammatory function. J Biol Chem 284: 7977–7985
22. Vizoso Pinto MG, Villegas JM, Peter J et al (2009) LuMPIS – a modified luminescence-based mammalian interactome mapping pull-down assay for the investigation of protein–protein interactions encoded by GC-low ORFs. Proteomics 9:5303–5308
23. Bacart J, Corbel C, Jockers R et al (2008) The BRET technology and its application to screening assays. Biotechnology journal 3:311–324
24. Braun P, Tasan M, Dreze M et al (2009) An experimentally derived confidence score for binary protein–protein interactions. Nat Methods 6:91–97

Chapter 20

LuMPIS: Luciferase-Based MBP-Pull-Down Protein Interaction Screening System

María G. Vizoso Pinto and Armin Baiker

Abstract

Analyzing the putative interaction partners of an individual protein is one approach to elucidate its function. In the LuMPIS protocol, bait and prey proteins are expressed with N-terminal maltose binding protein (MBP)- and eGFP-luciferase (eGFP-luc) tags, respectively. Positive protein–protein interactions (PPIs) can be detected after pull-down of the MBP-tagged prey protein using amylose beads followed by the bioluminescence detection of the bound eGFP-luc-tagged bait protein. The LuMPIS technology offers the following advantages: the PPIs are detected in the mammalian cell context, the use of two long protein tags (i.e., MBP and eGFP-luc) increases the expression levels of genes whose gene expression levels are known to be frequently impaired, the use of amylose beads for pull-down is much more economic as compared to sepharose beads in combination with monoclonal antibodies and finally, the use of an eGFP-luc-tag enables the qualitative control of transfection efficiencies by fluorescence microscopy prior to starting the assay.

Key words: Protein–protein interactions, High-throughput, Luciferase detection, Maltose binding protein

1. Introduction

Analyzing the putative interaction partners of an individual protein is one approach to elucidate its function. Due to its simplicity, high-throughput compatibility and cost-effectiveness, the yeast-two-hybrid (Y2H-) technology has been established as the method of choice for the genome-wide analysis of protein–protein interactions (PPIs). The Y2H-technology, however, is limited by its high rate of false-positive and false-negative PPIs (1) making the validation of the respectively detected PPIs in independent assay systems extraordinarily important.

Michael Kaufmann and Claudia Klinger (eds.), *Functional Genomics: Methods and Protocols*, Methods in Molecular Biology, vol. 815, DOI 10.1007/978-1-61779-424-7_20,

The "luminescence-based MBP pull-down interaction screening system" (LuMPIS-) technology represents a modification of the "luminescence-based mammalian interactome mapping" (LUMIER-) technology originally described by Barrios-Rodiles et al. (2). In the original LUMIER protocol, bait and prey proteins are expressed with N-terminal Flag- and *Renilla* luciferase-(Rluc-) tags, respectively. Positive PPIs can be detected after pull-down of the Flag-tagged prey protein using sepharose beads loaded with Flag-tag-specific monoclonal antibodies followed by the bioluminescence detection of the bound Rluc-tagged bait protein (Fig. 1, left panel). In case of LuMPIS, the Flag-tag has been replaced by maltose binding protein (MBP) and the Rluc-tag by eGFP-luciferase (eGFP-luc). Positive PPIs can be detected after pull-down of the MBP-tagged prey protein using amylose beads and bioluminescence detection of the bound eGFP-luc-tagged bait protein after elution with maltose (Fig. 1, right panel). The LuMPIS-specific modifications offer three main advantages over the original LUMIER-technology. First, the use of two long protein tags (i.e., MBP and eGFP-luc) increases the expression levels of genes whose gene expression levels are known to be frequently impaired (e.g., genes with low GC-contents). Second, the use of amylose beads for pull-down is much more economic as compared to sepharose beads and monoclonal antibodies. And third, the use of an eGFP-luc-tag enables the qualitative control of transfection efficiencies by fluorescence microscopy prior to starting the assay (3).

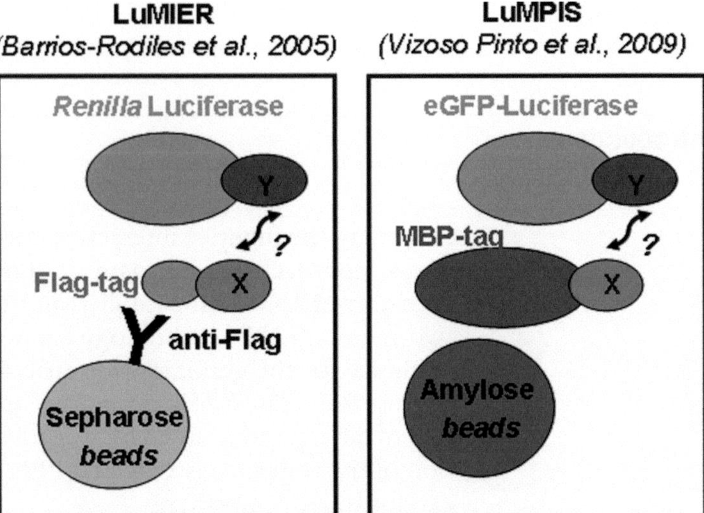

Fig. 1. Principle of LuMPIS compared with LuMIER. Prey proteins are tagged with *Flag* or *MBP* and bait proteins with *Renilla* or *eGFP*-luciferase, respectively. In the case of LuMIER, the PPI complex is pulled down with *anti-Flag* antibodies coupled to *sepharose beads*, whereas in LuMPIS, the complex is pulled down with *amylose beads*. In both cases, the detection of the preys is done by measuring luminescence after addition of the respective substrates.

In summary, the LuMPIS-technology represents an economic and easy-to-handle mammalian cell-based assay system for the qualitative, but not semiquantitative detection of PPIs. LuMPIS provides a highly sensitive technology, since PPIs with binding affinities of as low as $K_D = 20$ μM could be detected (3).

Jun and Fos form a heterodimeric transcription factor, which plays a fundamental role in the physiology and pathology of the mammalian cell (4). These two components have been shown to interact via a ZIP motif region, which makes them ideal control proteins for the development of new PPI assays. In our LuMPIS assays, jun and fos, as well as Δfos, whose ZIP motif has been deleted, are used as positive and negative controls (5).

2. Materials

2.1. Recombinatorial Cloning

1. *Pfu* polymerase (Promega, Germany), stored at −20°C.
2. QIAquick Gel Extraction Kit (Qiagen, Germany).
3. BP-clonase II enzyme mix (Invitrogen, Germany), stored in 5–10 μl aliquots at −80°C.
4. LR-clonase II enzyme mix (Invitrogen), stored in 5–10 μl aliquots at −80°C.
5. LB broth and agar.
6. Chemically competent *Escherichia coli* DH5α, stored in 300 μl aliquots at −80°C.
7. pDONR207 (Invitrogen). This plasmid can be propagated in *E. coli* DB 3.1 and selected with gentamycin (Roth, Germany) and/or chloramphenicol (Roth, Germany).
8. pCR3-eGFPLuc-N-[rfB] and pCR3-MBP-N-[rfB]. Customized destination vectors based on pCR3.1 (Invitrogen). These vectors contain the reading frame B [rfB] Gateway cassette. They can be propagated in *E. coli* (DB 3.1) and selected with ampicillin, kanamycin, and/or chloramphenicol.
9. QIAprep Spin Miniprep Kit (Qiagen).
10. *Ban*II and *Eco*RV (New England Biolabs, NEB, Germany).

2.2. Cell Culture and Transfection

1. Dulbecco's Modified Eagle's Medium (DMEM) (Gibco, Germany) supplemented with 10% heat-inactivated fetal bovine serum (FBS) from Invitrogen.
2. Solution of 0.25% trypsin and 1 mM EDTA from Invitrogen.
3. 12-Well tissue culture plates (Falcon, Germany).

2.3. LuMPIS

1. PBS (Gibco).
2. LuMPIS buffer: 20 mM Tris–HCl, 200 mM NaCl, 1 mM EDTA. Stored at 4°C.

3. Lysis buffer: 0.05% Tween 20, 5 μg/ml leupeptin (Sigma-Aldrich, Germany), 5 μg/ml DNase (Sigma-Aldrich), and 2.5 mg/ml BSA in LuMPIS buffer. Prepared freshly before use (see Note 4).
4. Elution buffer: 10 mM maltose, 2.5 mg/ml BSA, 5 μg/ml leupeptin in dH_2O. Prepared freshly before use.
5. Amylose beads from NEB. Stored at 4°C (see Notes 2 and 3).
6. Luciferase Detection Reagent (Promega). Stored at –20°C (see Note 8).
7. 96-Well multiscreen high-throughput screen (HTS) art. MCBVN1250 (Millipore, Germany) (see Note 5).
8. 96-Well flat bottom assay plates, white polystyrene, untreated (Costar, Germany).
9. 96-Well PP microtiter plates (Costar).

3. Methods

3.1. Recombinatorial Cloning for Tagging Interacting Proteins Using the Gateway™ Technology

Recombinatorial cloning using the GATEWAY™ system (Invitrogen) is based on the site-specific integration system of phage λ and replaces multistep cloning approaches using restriction digest, purification, and ligation procedures by a single recombination. The first two steps of GATEWAY cloning involve the consecutive amplification of the target gene by PCR, thereby introducing attB sites on each fragment end (Fig. 2). The following BP-clonase reaction requires the attB sites at the ends of the PCR-fragment and attP sites on the so-called donor vector. The resulting vector (entry vector) encompasses the target gene flanked by new formed attL sites (Fig. 2). The LR-clonase reaction targets the attL sites of the entry vector and attR sites on the destination vector. Recombination results in the insertion of the target gene flanked by new formed attB sites.

3.1.1. Nested PCR for Amplification of Jun, Fos, and ΔFos

In the first PCR, ORFs with *attB*-sites are amplified using an ORF-specific primer set: *ORF-forward*: 5′-AAAAAGCAGGCTCC-GCC(18–22 ORF-sequence specific nucleotides including start codon)-3′ and *ORF-reverse*: 5′-AGAAAGCTGGGTC(18–22 ORF-sequence specific nucleotides including stop codon)-3′ and a proof-reading polymerase (*Pfu*) according to the following protocol:

Reaction Mix

1 μl: Template (VZV genomic DNA).
5 μl: 10× PFU reaction buffer.
1 μl: 10 μM ORF-forward.
1 μl: 10 μM ORF-reverse.
1 μl: 10 mM dNTP mix.

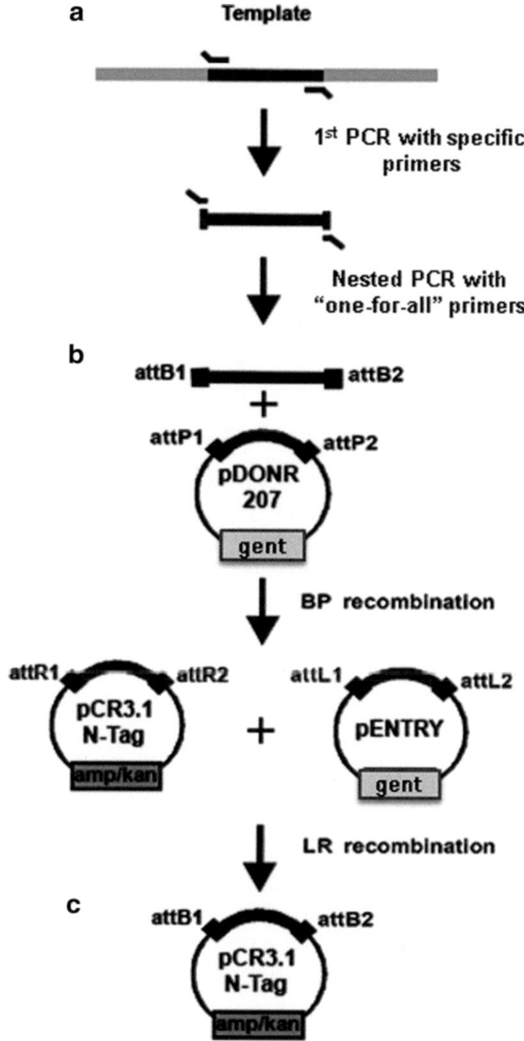

Fig. 2. Recombinatorial cloning. (**a**) Nested PCR for the amplification of ORFs with attB sites, (**b**) BP reaction into pDONR207, (**c**) LR reaction to insert the cloned fragment flanked by attL sites into the customized mammalian expression vectors pCR3.1-N-MBP-[rfB] or pCR3.1-N-eGFP-Luc-[rfB].

1 μl: *Pfu* DNA polymerase 3 U/μl.

40 μl: dH_2O.

Thermocyclcer parameters

Step	Temperature (°C)	Time	Cycle
Initial denaturation	95	2 min	
Denaturation	95	30 s	
Annealing	53.5	30 s	20 Cycles
Elongation	72	2 min/kbp	
Final elongation	72	10 min	

In the second PCR, the *attB*-sites are completed by using a (one-for-all) primer set: *One-for-all-forward*: 5′-GGGGACAAGTTTGTACAAAAAAGCAGGCT-3′ and *One-for-all-reverse*: 5′-GGGGACCACTTTGTACAAGAAAGCTGGGTC-3′. The reaction mix is set as follows:

Reaction Mix
5 μl: Template (first PCR mix).
5 μl: 10× PFU reaction buffer.
1 μl: 10 μM One-for-all-forward.
1 μl: 10 μM One-for-all-reverse.
1 μl: 10 mM dNTP mix.
1 μl: *Pfu* DNA polymerase 3 U/μl.
35 μl: dH_2O.

The thermocycler parameters are the same as shown above.

PCR products are separated on a 1% agarose gel, and the corresponding band is purified using the QIAquick Gel Extraction Kit following the manufacturers' instructions.

3.1.2. BP Clonase Reaction

PCR products encoding the respective ORFs and functional *attB*-sites are then recombinatorially cloned into the *attP*-sites of pDONR207 using BP-clonase II enzyme mix as follows:

BP Clonase Reaction
PCR product: 3 μl.
pDONR207 (Invitrogen): 1 μl.
BP-clonase (Invitrogen): 1 μl.
Room temperature, overnight.

BP reactions are transformed into chemically competent *E. coli* DH5α using standard protocols and plated onto LB-agar plates supplemented with 12.5 μg/ml gentamycin. Individual colonies are grown in LB-broth supplemented with gentamycin at 37°C overnight. Plasmid DNA is isolated using the QIAprep Spin Miniprep Kit (Qiagen). The integrity of the resulting pENTR207 vectors and the recombinatorially inserted ORFs can be verified by *Ban*II (NEB) restriction analysis and forward sequencing (LGC Genomics, Germany).

Restriction Digest with *Ban*II
3 μl: plasmid preparation.
3 μl: 10× Buffer 4 (NEB).
0.3 μl: *Ban*II 10 U/μl.
23.7 μl: dH_2O.
37°C, 3 h.

The restriction digest is supplemented with gel loading dye, and separated by electrophoresis on a 1% agarose gel, 100 V, 1 h.

3.1.3 LR Clonase Reaction

LR recombination reactions using LR-clonase II enzyme mix (Invitrogen) are performed according to the manufacturers' instructions. Briefly, pENTR207 vectors containing ORFs flanked by *attL*-sites were recombinatorially cloned into the *attR*-sites of the customized vectors pCR3-eGFPLuc-N-[rfB] and pCR3-MBP-N-[rfB] to tag the proteins with eGFP-luciferase or MBP, respectively. The latter vectors have been constructed by insertion of a customized cassette consisting of 5′-*Hin*dIII-ATG-[eGFPLuc]-*Eco*RV-[ccdB/CmR(rfB)]-*Eco*RV-*Xba*I-3′ or 5′-*Hin*dIII-ATG-[MBP]-*Eco*RV-[ccdB/CmR(rfB)]-*Eco*RV-*Xba*I-3′ into the backbone of the mammalian expression vector pCR3.1 (Invitrogen), respectively.

LR Clonase Reaction

1 μl: Entry vector (pENTR207).

1 μl: (Customized) Destination vector.

1 μl: LR clonase.

2 μl: dH_2O.

37°C, 2 h.

LR clonase reactions are subsequently transformed into chemically competent *E. coli* DH5α and plated onto LB plates supplemented with 50 μg/ml kanamycin and 100 μg/ml ampicillin (Sigma-Aldrich). Plasmid DNA of individual colonies is isolated as indicated before. The integrity of the resulting pCR3-eGFPLuc-N and pCR3-MBP-N constructs and the recombinatorially inserted ORFs are verified by *Eco*RV (NEB) restriction analysis (see Note 1).

Restriction Digest with *Eco*RV

3 μl: plasmid preparation.

3 μl: 10× buffer 3 (NEB).

0.3 μl: BSA.

0.3 μl *Eco*RV 10 U/μl.

23.7 μl: dH_2O.

37°C, 3 h.

The restriction digest is supplemented with gel loading dye, and separated by electrophoresis on a 1% agarose gel, 100 V, 1 h. DNA concentration is measured photometrically at 240 nm.

3.2. Cell Culture and Transfection

1. For a typical experiment HEK 293 T cells are trypsinized when approaching confluency, resuspended in DMEM medium and counted. Each 12-well plate is prepared using 2.5×10^5 cells/well and incubated at 37°C and 5% CO_2 for approximately 6 h until the cells attach to the surface.
2. Half an hour before transfection, DMEM medium is replaced by OPTIMEM (250 μl/well). Transfection solutions are prepared in a 100-μl scale, sufficient for the transfection of two (12-) wells. For this purpose, 500 ng pCR3-eGFPLuc-N-ORF and 500 ng pCR3-MBP-N-ORF (or 500 ng pCR3-MBP-N-[rfB] as negative control) are diluted within 50 μl OPTIMEM within 1.5 ml reaction tubes. In a second step, 3 μl FUGENE 6™ is added to a reaction tube containing 50 μl OPTIMEM, vortexed for 3 s, and incubated at RT for 5 min. After incubation,

the DNA mixture is pipetted into the second reaction tube containing FUGENE 6™, vortexed and incubated at RT for another 15 min. Finally, 50 μl of the latter DNA/FUGENE mixture is added to each (12-) well drop-wise. Cells are incubated at 37°C and 5% CO_2 overnight.

3. OPTIMEM is replaced by DMEM and the cells are further incubated for 24 h.

3.3. LuMPIS

1. Optional: before starting the assay, transfection can be qualitatively checked for green fluorescence using an inverse fluorescence microscope.
2. 96-Well HTS plates are prepared by adding 100 μl slurry amylose beads with a multistepper pipette. Take care of resuspending the beads thoroughly and use bore tips for pipetting. It is very important that the amount of beads in each well is equal, for that reason it is necessary to continuously resuspend the beads during the procedure. Amylose beads are equilibrated by resuspending them with LuMPIS buffer and applying vacuum at least three times. It is convenient to use multichannel pipettes or a 96-well pipetting device (Liquidator96™, Steinbrenner Systems, Germany).
3. At this point the transfected cells are placed on ice and rinsed once with 500 μl PBS.
4. Cell lysates are prepared by resuspending the cells in 500 μl LuMPIS lysis buffer and pipetting up and down several times. The lysates are transferred to centrifuge tubes and sonicated (five bursts of 15 s) at 4°C using an ultrasound bath (Sonorex). Lysates are cleared from membranes and cellular debris by centrifuging at 17,000 × g for 10 min at 4°C.
5. Cleared lysates are diluted 1:20 in LuMPIS buffer in 96 deep-well plates using a multichannel pipetting device. The plate is kept on ice until the end of the assay. The buffer in which the beads have been equilibrated is removed by aspirating it using the vacuum device. 150 μl/well of each dilution are added to the respective wells of the HTS plate (see Note 7). Beads and the diluted lysate are mixed by pipetting up and down 20 times avoiding foaming as far as possible.
6. HTS plates are placed into the vacuum device and the buffer is aspirated. The flow-through is discarded.
7. HTS plates are washed four times with 200 μl LuMPIS buffer by suspending the beads using the multichannel device (tips should be changed each time) and applying vacuum.
8. Dilute the PPI complexes by adding 150 μl elution buffer on to the HTS plates. The eluates are collected on a 96 deep-well plate.

9. Fifty microliters of eluate per well are pipetted onto a white 96-well plate. Luciferase measurements are carried out using an Optima FLUOstar Luminometer system (BMG LABTech) after the automatically addition of 50 μl luciferase detection reagent per well (see Note 6).
10. Fifty microliters of the diluted lysates are also measured to calculate the LuMPIS interaction ratio (LIR) and to normalize the transfection efficiencies. The LIR is calculated using the following equation:

$$\text{LIR} = \frac{\text{Sample A (eluate) LU/Sample A (diluted lysate) LU}}{\text{Neg. control (eluate) LU/Neg. control (diluted lysate) LU}}.$$

LU light units, *Neg. control* N-MBP versus N-EGFP-Luc-ORF.

The result of a typical LuMPIS experiment testing for viral protein–protein interactions is shown in Fig. 3.

4. Notes

1. Recombinatorial cloning is highly efficient. It is usually enough to pick two colonies to get the right clone. Checking for reading-frame shifts is mandatory.
2. Care should be taken, that the amylose beads do not suffer from changes in temperature, as this favors bubble building, which should be avoided. All steps should be carried out on ice and solutions should be ice-cold before starting.
3. When pipetting the amylose bead slurry, it is useful to use bore tips.
4. Using other protease inhibitors than leupeptin results in a significant reduction of luciferase activity.
5. The 96-well white assay plates should be used only once for the measurements of the eluates, as we have observed a significant increase in background when reusing them.
6. Although the eGFP-luciferase is very stable, it is best to work quickly and on ice. Especially, after eluting the PPI complex, it is necessary to immediately measure luciferase activity, as the enzyme activity decreases rapidly in the presence of elution buffer.
7. This protocol can be adapted for batch procedures, using centrifuge tubes instead of plates.
8. Dyer et al. (6) have developed a noncommercial luciferase assay system which may also be used for this protocol.

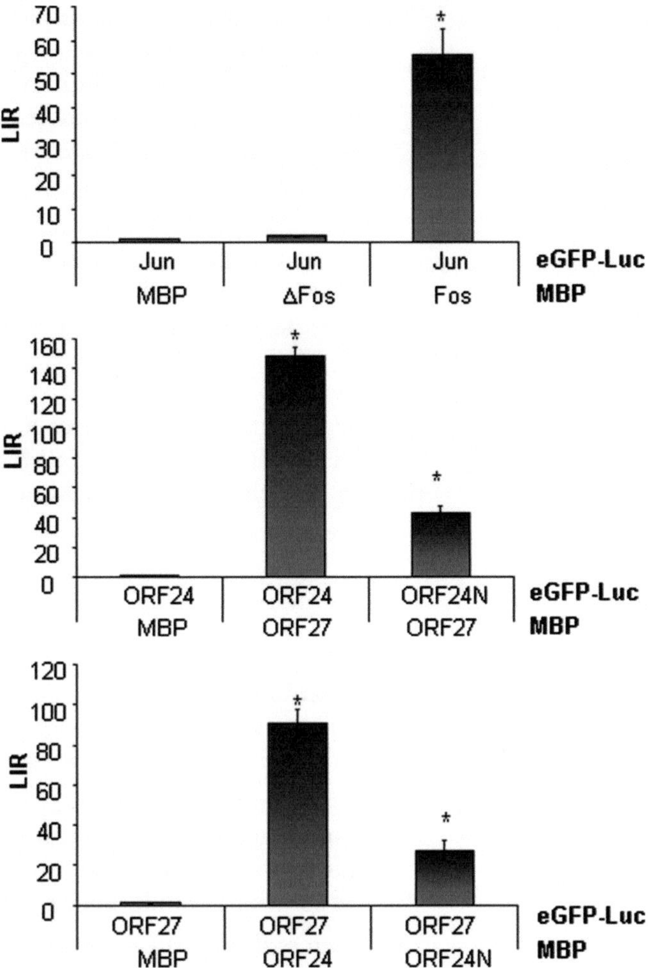

Fig. 3. LuMPIS for testing viral protein–protein interactions. HEK-293 T cells were transfected with maltose binding protein (MBP)- and eGFP-Luciferase (eGFP-Luc)-tagged proteins. Protein–protein interactions were determined by measuring luciferase activity on maltose eluates. (**a**) Two known interactors, the transcription factors Jun and Fos, were used as positive controls, whereas ΔFos, whose ZIP interaction domain had been deleted was used as negative control. (**b**) Putative VZV nuclear egress complex components: VZV proteins ORF24 and ORF27 interact with each other. In the mutant ORF24N, the transmembrane domain that anchors the protein to the nuclear membrane has been deleted. Results are plotted as LIR (LuMPIS interaction ratio) ± SE; $n = 6$. Kruskal–Wallis test and Dunn's test for multiple comparisons with a control (eGFP-Luc-tagged bait against MBP) (*asterisk* indicates $P < 0.05$).

References

1. Stellberger, T., Hauser, R., Baiker, A., Pothineni, V. R., Haas, J., and Uetz, P. (2010) Improving the yeast two-hybrid system with permutated fusions proteins: the Varicella Zoster Virus interactome, *Proteome Sci 8*, 8.

2. Barrios-Rodiles, M., Brown, K. R., Ozdamar, B., Bose, R., Liu, Z., Donovan, R. S., Shinjo, F., Liu, Y., Dembowy, J., Taylor, I. W., Luga, V., Przulj, N., Robinson, M., Suzuki, H., Hayashizaki, Y., Jurisica, I., and Wrana, J. L. (2005) High-throughput mapping of a dynamic signaling network in mammalian cells, *Science 307*, 1621–1625.

3. Vizoso Pinto, M. G., Villegas, J. M., Peter, J., Haase, R., Haas, J., Lotz, A. S., Muntau, A. C., and Baiker, A. (2009) LuMPIS – a modified luminescence-based mammalian interactome mapping pull-down assay for the investigation of protein–protein interactions encoded by GC-low ORFs, *Proteomics 9*, 5303–5308.

4. Seldeen, K. L., McDonald, C. B., Deegan, B. J., and Farooq, A. (2008) Evidence that the bZIP domains of the Jun transcription factor bind to DNA as monomers prior to folding and homodimerization, *Arch Biochem Biophys 480*, 75–84.

5. Hu, C. D., Chinenov, Y., and Kerppola, T. K. (2002) Visualization of interactions among bZIP and Rel family proteins in living cells using bimolecular fluorescence complementation, *Mol Cell 9*, 789–798.

6. Dyer, B. W., Ferrer, F. A., Klinedinst, D. K., and Rodriguez, R. (2000) A noncommercial dual luciferase enzyme assay system for reporter gene analysis, *Anal Biochem 282*, 158–161.

Chapter 21

Yeast Two-Hybrid Screens: Improvement of Array-Based Screening Results by N- and C-terminally Tagged Fusion Proteins

Thorsten Stellberger, Roman Häuser, Peter Uetz, and Albrecht von Brunn

Abstract

Matrix-based yeast two-hybrid screens are an alternative to library-based screens. Recent improvements of matrix screens (also called array screens), use various pooling strategies as well as novel vectors that increase its efficiency while decreasing the false-negative rate, thus increasing reliability. In this chapter, we describe a screening strategy that systematically combines N- and C-terminal fusion proteins using a recently developed vector system.

Key words: Yeast two-hybrid, Protein–protein interactions, Permutated fusion tags

1. Introduction

1.1. The Yeast Two-Hybrid System

The yeast two-hybrid (Y2H) method was originally developed by Stanley Fields and is a genetic method to detect binary protein–protein interactions (PPIs) (1). It exploits the modularity of eukaryotic transcription factors and the ease of genetic engineering of the yeast, *Saccharomyces cerevisiae*, to monitor PPIs. A bait protein is fused to the DNA-binding domain (DBD) and a prey protein is fused to the activation domain (AD) of a transcriptional activator, often the yeast Gal4 protein. The term "two-hybrid" is based on these two chimeric proteins. The bait and prey fusions are co-expressed in yeast and upon physical interaction between the bait and prey protein, the functional transcription factor (TF) is reconstituted. This results in the activation of a reporter gene, which allows either growth under selective conditions or produces a color or fluorescence signal (auxotrophic yeast strain, lacZ, or GFP reporter gene).

Michael Kaufmann and Claudia Klinger (eds.), *Functional Genomics: Methods and Protocols*,
Methods in Molecular Biology, vol. 815, DOI 10.1007/978-1-61779-424-7_21, © Springer Science+Business Media, LLC 2012

1.2. Matrix-Based Yeast Two-Hybrid Screens

In a matrix-screen, the possible combinations of open reading frames (ORFs) are systematically examined by performing direct mating of a set of baits with a set of preys expressed in opposite yeast mating types. This has two major advantages compared to the library screen approach:

1. Each prey is arrayed on an individual position. Thus, the interacting prey can be simply identified by the matrix position and additional identification steps of the interacting prey by a colony PCR and sequencing reactions are obsolete.
2. Array screens can be automated by using a replication robot.

1.3. Combining N- and C-terminally Fused Test Domains

Yeast two-hybrid screens do not generate complete protein interactomes. As for any other detection method, it is almost impossible to detect all physiologically occurring interactants of every screened bait protein. Apart from effects that originate in the heterogenic yeast expression system, e.g., due to a lack of posttranslational modifications, false-negative interactions can be partly traced back to steric hindrance effects due to the used fusion tags. They can prevent physical interactions by covering the respective interaction sites or preventing subsequent transcriptional activation.

Most Y2H vector systems use N-terminally fused test domains, but this can avoid any interactions which involve regions around the N-terminus of these proteins. Thus, we developed C-terminal fusions of the DNA-binding and activation domains and also tested pairwise combinations of N- and C-terminal fusions (2, 3). Stellberger et al. (2) tested all pairwise interactions among the ~70 ORFs of the Varicella Zoster Virus using both N- and C-terminal vectors as well as combinations thereof (Fig. 1). About ~20,000 individual Y2H tests resulted in 182 NN, 90 NC, 151 CN, and 146 CC interactions (Fig. 2). Overlaps between screens ranged from 17% (NC–CN) to 43% (CN–CC). Performing four screens (i.e., permutations) instead of one resulted in about twice as many PPIs, and thus fewer false-negatives. Different vector combinations show unique, as well as overlapping PPI-data, supporting the impact of steric hindrance and the need of free termini for a subfraction of PPIs (Fig. 3).

2. Materials

2.1. Yeast-Rich Media

1. 1 YEPD liquid medium: 10 g yeast extract, 20 g peptone, and 20 g glucose. Make up to 1 L with sterile water and autoclave.
2. YEPD solid medium: 10 g yeast extract, 20 g peptone, 20 g glucose, and 16 g agar. Make up to 1 L with sterile water and autoclave. After autoclaving cool media to ~60°C, add 4 ml of 1% adenine solution [1% in 0.1 M NaOH (see Note 1)]. Pour

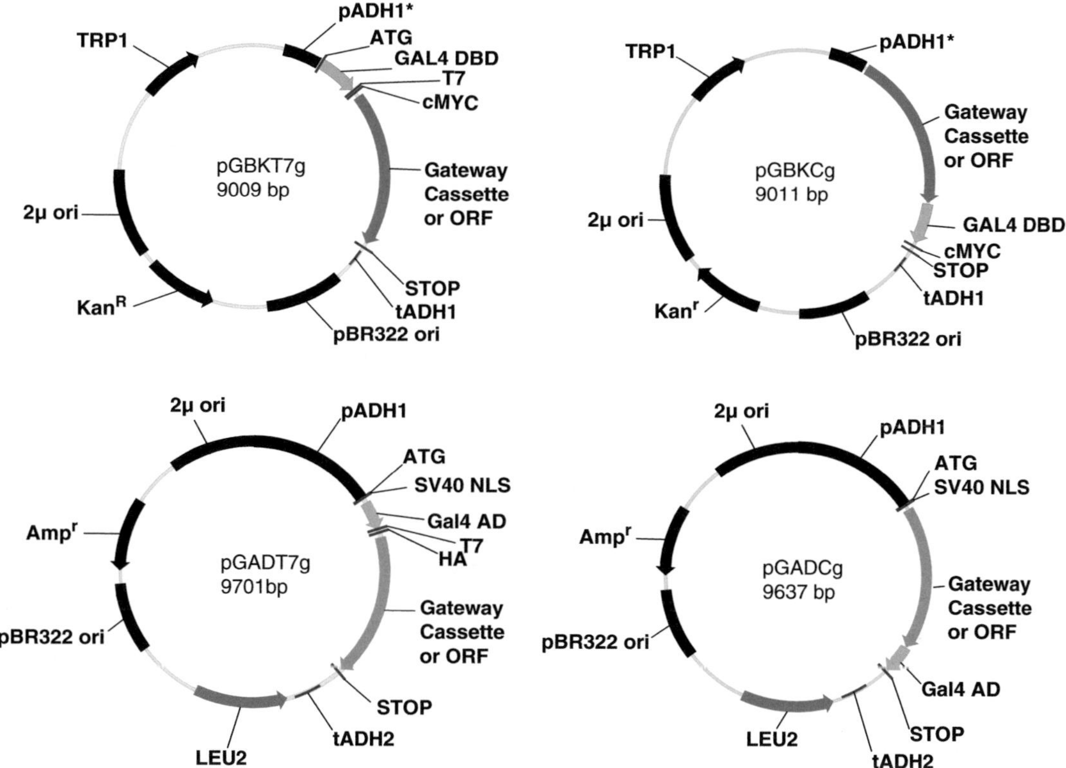

Fig. 1. Vectors described in this study including their parental vectors. pGBKT7g and pGADT7g generate N-terminal fusions of DNA-binding (DBD) and activation domain (AD) fusions, respectively. The new vectors pGBKCg and pGADCg fuse DBD and AD at the C-terminus of inserted ORFs. Note that both pGBK-vectors use a truncated version of the ADH promoter (indicated by *asterisk*) which may reduce expression levels and thus interaction signals (8, 9).

about 40 ml into sterile 1-well plates in a clean bench and let them solidify (see Note 2).

2.2. Yeast-Selective Media

1. Dropout mix (-His, -Leu, and -Trp): 1 g methionine, 1 g arginine, 2.5 g phenylalanine, 3 g lysine, 3 g tyrosine, 4 g isoleucine, 5 g glutamic acid, 5 g aspartic acid, 7.5 g valine, 10 g threonine, 20 g serine, 1 g adenine, and 1 g uracil. Mix all components and store under dry conditions at room temperature.
2. Medium concentrate (5×): 8.5 g yeast nitrogen base, 25 g ammonium sulfate, 100 g glucose, and 7 g dropout mix. Make up to 1 L with water and sterile filter. Store at 4°C (see Notes 3 and 4).
3. Amino acid stock solutions (see Note 1): 4 g/L histidine, 7.2 g/L leucine, and 4.8 g/L tryptophan. Each amino acid dissolved in water and sterile filtrated.
4. 3-amino-triazole (3-AT) stock solution: 0.5 M. Sterile filtrate (see Note 1).
5. For 1 L of minimal medium autoclave 16 g of agar in 800 ml of water, cool the medium to ~60°C, and then add 200 ml 5×

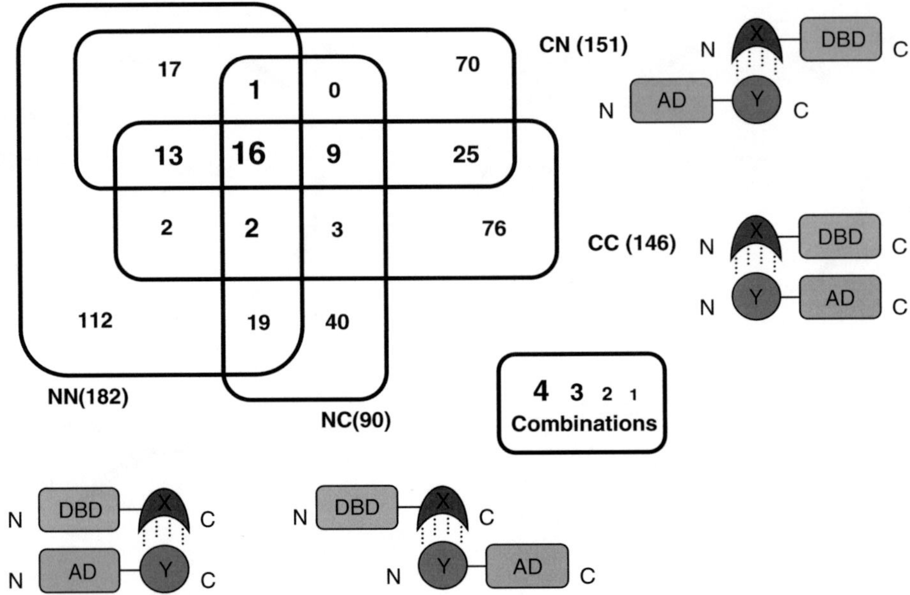

Fig. 2. Overlaps between tag-topology combinations. This *Venn diagram* shows the distribution of 405 VZV interactions found in either one or more tag-topologies. One hundred and eighty two interactions were detected in the traditional screen (NN); 90, 151 and 146 PPIs were additionally identified in the new combinatorial screens (NC, CN, and CC, respectively). The differences in the tag topologies are sketched next to the corresponding topology abbreviations. The number of overlaps between different screens is indicated by the thickness of numbers decreasing from 4 to 1. Modified after ref. 2.

medium concentrate and mix. Pour ca. 40 ml into each sterile Omnitray plate under sterile hood and let them solidify (see Note 2). Depending on the required selective plates you have to add the missing amino acids or 3-AT. Liquid minimal media can be prepared without adding agar. Corresponding amino acids are added from the amino acid stock solutions as follows (see Note 5).

6. Selection of baits (-Trp plates): 8.3 ml leucine and 8.3 ml histidine.
7. Selection of preys (-Leu plates): 8.3 ml tryptophan and 8.3 ml histidine.
8. Selection of diploids (-Leu-Trp plates): 8.3 ml histidine.
9. Readout medium (-Leu-Trp-His plates): add 3-AT from 0.5 M stock solution as needed for screening self-activating baits.

2.3. Yeast Transformation

1. Carrier DNA (salmon sperm DNA): dissolve 7.75 mg/ml salmon sperm DNA in water and store at –20°C following a 15 min 121°C autoclave cycle.
2. 96 PEG solution (100 ml): mix 45.6 g PEG, 6.1 ml of 2 M LiOAc (lithium acetate), 1.14 ml of 1 M Tris–Hcl pH 7.5, and

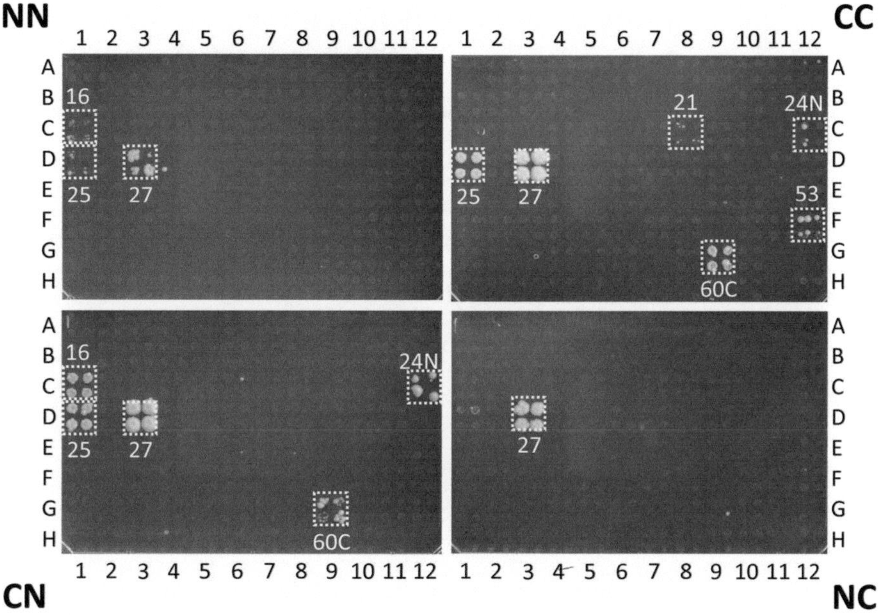

Fig. 3. Different vector combinations detect common, as well as different PPIs. Y2H screens of the four different vector combinations showing the differences on 25 mM 3AT. The same bait, ORF24N (Uniprot accession P09280, 238 N-terminal amino acids) was used as bait with N- and C-terminally fused DNA-binding and activation domains and screened against a whole-genome array of Varicella Zoster Virus (VZV). The N-terminal bait and prey constructs (in pGBKT7g, pGADT7g, NN) show different interaction patterns compared to the C-terminal constructs cloned into pGBKCg and pGADCg (CC) as well crosswise combinations thereof (NC and CN). Preys are indicated by their *ORF number*, e.g., the prey ORF27 is a subunit of the VZV nuclear egress complex together with ORF24 and was described in HSV-1 (10). Note that *N* and *C* labels near yeast colonies indicate N- and C-terminal protein fragments, not AD or DBD fusions (e.g., 60 C is a C-terminal domain of ORF60).

232 μl 0.5 M EDTA; make up to 100 ml with sterile water and autoclave. Store 96 PEG solution at room temperature.

3. CT110: mix 20.73 ml 96PEG, 0.58 ml boiled salmon sperm DNA (boil frozen salmon sperm DNA at 95°C for 5 min) and 2.62 ml DMSO. Add DMSO last and mix quickly after adding by shaking vigorously and vortex for 30 s (see Note 6).

2.4. Screen Procedure, Retests, and Bait Self-activation Test

1. 96-Well microtiter plates, round bottom.
2. 1-Well plates.
3. Bleach solution (20%): dilute a 12% sodium hypochlorite solution 1:5 with water (see Note 7).
4. 95% ethanol solution, industrial.
5. Autoclaved water.
6. Replication tool or robot, 96- and 384-pinning tool.
7. 1% (*w*/*v*) adenine solution (1% in 0.1 M NaOH), sterile filtrate.
8. YEPD and selective media as liquids and agar plates as described (see Subheading 2.1).

2.5. Vectors

1. Bait plasmids: pGBKT7g (4) and pGBKCg (2).
2. Prey plasmids: pGADT7g (4) and pGADCg (2).

Any other vectors can be used as long as they are compatible with each other and the yeast strains.

2.6. Yeast Strains

1. AH109: genotype (MAT a, trp 1-901, leu2-3, 112, ura3-52, his3-200, Δgal4, Δgal80, LYS2: $GAL1_{UAS}$-$GAL1_{TATA}$-HIS3, $GAL2_{UAS}$-$GAL2_{TATA}$-ADE2, URA3: $MEL1_{UAS}$-$MEL1_{TATA}$-lacZ) (5, 6).
2. Y187: genotype (MAT α, ura3-52, his3-200, ade2-101, trp1-901, leu2-3, 112, Δgal4, met, Δgal80, URA3: $GAL1_{UAS}$-$GAL1_{TATA}$-lacZ) (7).

3. Methods

The following protocols describe the Y2H assay with the HIS3 reporter and the pGBKT7g/pGADT7g vector system. The protocols are applicable for the combinations with the pGBKCg/pGADCg system and others as well. However, if using other vector systems, different yeast and *E. coli* selection markers have to be considered during the selection steps and the selection media have to be exchanged.

Each Y2H screen may be repeated four times using the two different bait- and two different prey arrays:

1. pGBKT7g–pGADT7g (NN).
2. pGBKT7g–pGADCg (NC).
3. pGBKCg–pGADT7g (CN).
4. pGBKCg–pGADCg (CC).

3.1. Yeast Transformation for Bait and Prey Construction

This protocol is suitable for 100 yeast transformations and may be scaled up or down as needed. Selection of the transformed yeast cells requires leucine or tryptophan-free media ("-Leu" or "-Trp," depending on the selective marker on the plasmid). Moreover, at least one of the haploid strains must contain a two-hybrid reporter gene under GAL4 control.

1. Inoculate 50 ml YEPD liquid medium with ~200 μl liquid stock of yeast strains (e.g., AH109, Y187 or any other appropriate yeast strain; use Y187 strains for preys and AH109 for baits or vice versa) in a 250-ml flask and grow overnight with shaking at 30°C (minimum 15 h and max. 24 h).
2. Spin down cells in 50-ml conical tube (3,000×*g*, 5 min at room temperature), pour off supernatant and dissolve the pellet by adding 2 ml LiOAc (0.1 M) and transfer resuspended yeast to two 1.5-ml microfuge tubes. Spin out yeast and resuspend in a total volume of 1.8 ml LiOAc (0.1 M).

3. Prepare CT110 solution.
4. Add all the competent yeast cells prepared above and mix vigorously by hand or by vortexing for 1 min. Immediately pipette 245 μl into each of 96 wells of a 96-well plate.
5. Add 50–100 ng of plasmid and positive control (e.g., empty vector) and negative control (only CT110). Seal the 96-well plate with plastic or aluminum tape and vortex for 4 min.
6. Incubate at 42°C for 30 min.
7. Spin the 96-well plate for 10 min at 1,500 g; discard the supernatant and aspirate with eight channel wand or by tapping on cotton napkin for couple of times. Add 150 μl of sterile water to all 96 wells, resuspend and plate cells on selective agar plates (e.g., standard Petri dishes) with -Leu for pGADT7g/pGADCg or -Trp for pGBKT7g/pGBKCg.
8. Incubate plates at 30°C for 3 days. After 2 days, the colonies start to appear; pick colonies after 3 days.
9. Rearray baits and preys in 96-well plates. Grow them up again for 1–2 days in -Leu- or -Trp-liquid minimal medium at 30°C (see Note 8).
10. The bait and prey plate can now be used to make a couple of copies on selective agar medium, to backup the arrays as glycerol (25%) stocks for −80°C long-term storage, and to use the baits directly for the self-activation test (see below). For plate storage at 4°C, it is recommended to have haploids rather on minimal agar medium than on YEPD medium since loss of plasmids can occur on nonselective medium.

3.2. Bait Self-activation Test

The aim of this test is to measure the background reporter activity (here: HIS3) of bait proteins in the absence of an interacting prey protein. This measurement is used for choosing the selection conditions used during the interaction screen and can be achieved by mating individual bait strain with a single prey strain that carries the empty prey plasmid. Ninety-six individual bait activation tests can be carried out on one plate simultaneously.

1. Load a 96-well plate (round bottom) with ~200 μl YEPD liquid medium.
2. Inoculate plate with baits by replicating the 96-format bait plate from solid medium into the destination plate by using a sterile 96-pinning tool (see Subheading 3.3.1 for sterilization details).
3. Inoculate the yeast strain Y187 which carries the empty prey vector in 30–50 ml YEPD liquid medium.
4. Grow yeast for ~18 h at 30°C (it is not necessary to shake the 96-well plate, whereas shaking of the prey strain in a flask is recommended).

5. Pellet yeast by centrifugation for 10 min at 1,500 × *g*; discard the supernatant; and aspirate with eight channel wand or by tapping on cotton napkin for a couple of times.
6. Use 96-replication tool to pin baits from 96-well source plate onto a YEPD single-well agar plate as quadruplicates.
7. Pour the yeast strain with the empty prey vector into a single-well plate.
8. Use 384-replication tool to pin yeast onto the YEPD single-well agar plate that harbors the baits already.
9. Mating occurs at 30°C for 1 to max. 2 days.
10. Replicate from mating plate on -Leu-Trp agar single-well plates to select diploids.
11. Incubate for 2–3 days at 30°C.
12. Pin diploids on -Leu-Trp-His agar medium in single-well plates with different concentrations of 3-AT (e.g., 0, 1, 2, 4, 8, …, 128 mM).
13. Select yeast for about 7 days at 30°C.
14. Determine minimal-inhibitory concentration of 3-AT which is needed for a single bait to suppress self-activation growth for use in the interaction screen.

3.3. Yeast Two-Hybrid Screen

3.3.1. Preparations

1. Sterilization steps: sterilize the pinning tool by dipping the pins into a 20% bleach solution for 20 s, sterile water for 1 s, 95% ethanol for 20 s, and sterile water again for 1 s. Repeat this sterilization after each transfer (see Note 9).
2. Prepare prey array for screening: use the sterile replicator to transfer the yeast prey array (e.g., 384 format) from selective plates to single-well plates containing solid YEPD medium and grow the array overnight in a 30°C incubator (max. 24 h) (see Note 10). Ideally, the template prey array should be kept on selective plates.
3. Prepare bait liquid culture (DBD fusion-expressing yeast strain): inoculate 20–30 ml of liquid YEPD medium in a 50-ml conical flask with a bait strain from plates with selective medium and grow in a 30°C shaker for 18–22 h.

3.3.2. Mating Procedure

1. Add a corresponding volume adenine from a 1% adenine stock solution to a final concentration of 0.004% into the bait liquid culture. This step is recommended to obtain a higher mating efficiency (see Note 11).
2. Pour the overnight liquid bait culture into a sterile 1-well plate. Dip the sterilized pins of the pin-replicator [thick pins (diameter >1 mm) should be used to pin baits] into the bait liquid culture and place directly onto a fresh 1-well plate containing solid YEPD media. Repeat with the required number of plates

and allow the yeast to dry onto the plates for ca. 10 min (see Note 12).

3. Pick up the prey array yeast colonies with sterilized pins [thin pins (≤1 mm diameter) should be used, see Note 13] and transfer them directly onto the baits pinned onto the YEPD plate so that each of the 384 bait spots per plate receives different prey yeast cells (i.e., a different AD fusion protein).
4. Incubate 1–2 days at 30°C to allow mating. Mating will take place in <15 h, but a longer period is recommended (max. 2 days) because some baits strains show poor mating efficiency.

3.3.3. Selection of Diploids

1. Transfer the colonies from YEPD mating plates to single-well plates containing -Leu-Trp medium using the sterilized pinning tool (thin pins should be used in this step).
2. Grow for 2–3 days at 30°C until the colonies are >1 mm in diameter (see Note 14).

3.3.4. Interaction Selection

1. Transfer the colonies from -Leu-Trp plates to a single-well plate containing solid -His-Leu-Trp agar, using the sterilized pinning tool. If the baits are self-activating, they have to be transferred to -His-Leu-Trp with the specific concentration of 3-AT which was determined in the self-activation assay (see Subheading 3.2). Incubate at 30°C for 6–10 days (see Note 15).
2. Score the interactions by looking for growing colonies that are significantly above background by size and that are present as duplicate (or quadruplicate) colonies. Scoring can be done manually or using automated image analysis procedures. When using image analysis, care must be taken not to score contaminated colonies as positives.

3.4. Retests

Testing for reproducibility of interactions greatly increases the reliability of the interaction data. This protocol is used for retesting interaction pairs detected in a Yeast two-hybrid screen.

1. Rearray bait and prey strains or positively tested prey pool of each interaction pair to be tested in 96-well microtiter plates. Use an individual 96-well plate for the baits, as well as for the preys. For each retested interaction, fill one well of the bait plate and one corresponding well of the prey plate with ~200 μl YEPD.
2. For each retested interaction, inoculate the bait strain into a well of the 96-well bait plate and the prey strain at the corresponding position of the 96-well prey plate. For example, bait "X" is transferred at positions B1, B2, and B3 of the bait plate.

The preys to be tested are arrayed into B1 (prey 1), B2 (prey 2), and the prey strain that carries the empty prey vector into B3 of the prey plate. The B3 test position is the control that helps to verify the background/self-activation.

3. Incubate the plates over night at 30°C.
4. Spin the bait and prey plates for 10 min at 1,500 × *g*.
5. Discard the supernatant and aspirate with eight channel wand or by tapping on cotton napkin a couple of times.
6. Pin baits with a sterile 96-pinning tool on -Trp and preys on -Leu selective agar medium as quadruplicates.
7. Allow baits and preys to grow at 30°C for 2–3 days.
8. Mating: first, transfer baits with a sterile 384-pinning tool on YEPD mating plates and second, transfer preys onto baits.

The rest of the procedure can be done according to the screening protocol. For interaction retesting diploids are pinned on -Leu-Trp-His selective media plates with different concentration of 3-AT. The control test position has to be compared to bait self-activation background signals. Reproducible interactions should show up on different concentrations of 3-AT, whereas the activation control test position indicates clearly no colony growth (see Note 16).

4. Notes

1. Stock solutions can be stored up to 6 months at 4°C. Alternatively, the stock solutions can be frozen as aliquots at −20°C for long time storage.
2. 1-Well plates are available from NUNC (Thermo Fisher Scientific). Prepared agar plates should be stored for 1–2 days with closed lid under a sterile hood before use. Fresh solidified media is often wet and cannot be used directly.
3. Medium concentrate can be stored at 4°C up to 6 months.
4. Some components of the medium concentrate (e.g., amino acids) are not well soluble in water. The solution has to be stirred before the filtration step for up to 5 h until all components are dissolved. Heating is not recommended because of the heat sensitivity of amino acids.
5. Selection media may differ due to the used Y2H expression vector system and have to be adapted. For instance, in the pDEST32/pDEST22 system the selection markers for baits and preys are interchanged (baits are selected on -Leu and preys on -Trp) while selection of pGBKT7g/pGBKCg baits is

done on -Trp. pGADT7g/pGADCg preys must be selected on -Leu medium.

6. CT110 has to be prepared freshly before yeast transfection and should not be stored.

7. Sodium hypochlorite solution is not very stable and has to be freshly prepared. Alternatively, other disinfection solutions with a bleaching effect can be used. We do not recommend to use a final concentrations higher than 2.4% since the steel pins of the replication tool might stain.

8. Yeast on agar medium can be stored for ~2 months at 4°C. The plates should be sealed with a sealing film to avoid drying-out. Baits and preys should be stored on the corresponding selective media since loss of plasmids can occur on nonselective medium.

9. Sterilization steps have to be established for the robotic system and sterilization solutions that are used. For instance, the minimal time required for sterilization should be tested in advance since this will speed up the whole screen. However, it must be ensured that no cross-contamination occurs.

10. The needed baits and prey arrays can also be used for the mating procedure when grown on/in selective medium. To our knowledge this does not influence the mating efficiency much but we recommend using YEPD medium since yeast grows faster and higher cell numbers can be achieved.

11. Adenine achieves a higher mating efficiency. Many yeast strains (e.g., AH109 and Y187) are deficient in synthesizing adenine since they can carry an additional adenine selection marker.

12. After transfer from the liquid culture allow the plates to dry for 10–30 min. The positions should be dry when the preys are copied onto the bait spots. Also the plate should be checked if enough bait cells were transferred. Reasonable amounts were transferred when each spot occurs cloudy. This is critical for a good mating efficiency.

13. Thick pins can be used as well. We use thin pins since more replication steps can be done from a single source plate. If only a replication tool with thick pins is available more prey array plates have to be prepared since only a couple of transfer steps can be done because of source plate depletion.

14. This step is an essential control step because only diploid cells containing the Leu2 and Trp1 marker on the prey and bait vectors will grow on this medium. It also leads to an amplification of diploid cells, which increases the efficiency of the next selection step.

15. We normally score interactions after 7 days. But the plates should be examined every day. Most two-hybrid positive

colonies appear within 3–5 days, but occasionally positive interactions can be observed later. Very small colonies are usually designated as background; however, there is no absolute measure to distinguish between the background and real positives. When there are many (i.e., >30) large colonies per array of 6,000 positions, we consider these baits as "random" activators. In this case, the screen should be repeated to ensure that these positives are reproducible (unless the screen is done already in duplicate or quadruplicate).

16. Pinning the retest onto readout medium with various concentrations 3-AT can be used to semi-quantify interactions. This helps, e.g., to distinguish between "strong" and "weak" signals and might also help to separate spurious ones.

Acknowledgments

This work was supported by the German Government via BMBF (Zoonosis Network, Consortium on ecology and pathogenesis of SARS, project code 01KIO701) to A.v.B, by a grant of the Baden-Württemberg Stiftung (Germany) to P.U., R.H. and T.S., and by a grant from the European Union (HEALTH-F3-2009-223101) to P.U.

References

1. Fields, S., and Song, O. (1989) A novel genetic system to detect protein–protein interactions. *Nature 340*, 245–6.
2. Stellberger, T., Hauser, R., Baiker, A., Pothineni, V. R., Haas, J., and Uetz, P. (2010) Improving the yeast two-hybrid system with permutated fusions proteins: the Varicella Zoster Virus interactome. *Proteome Sci 8*, 8.
3. Chen, Y. C., Rajagopala, S. V., Stellberger, T., and Uetz, P. (2010) Exhaustive benchmarking of the yeast two-hybrid system. *Nat Methods 7*, 667–8.
4. Uetz, P., Dong, Y. A., Zeretzke, C., Atzler, C., Baiker, A., Berger, B., Rajagopala, S. V., Roupelieva, M., Rose, D., Fossum, E., and Haas, J. (2006) Herpesviral protein networks and their interaction with the human proteome. *Science 311*, 239–42.
5. James, P. (2001) Yeast two-hybrid vectors and strains. *Methods Mol Biol 177*, 41–84.
6. James, P., Halladay, J., and Craig, E. A. (1996) Genomic libraries and a host strain designed for highly efficient two-hybrid selection in yeast. *Genetics 144*, 1425–36.
7. Harper, J. W., Adami, G. R., Wei, N., Keyomarsi, K., and Elledge, S. J. (1993) The p21 Cdk-interacting protein Cip1 is a potent inhibitor of G1 cyclin-dependent kinases. *Cell 75*, 805–16.
8. Bartel, P., Chien, C. T., Sternglanz, R., and Fields, S. (1993) Elimination of false positives that arise in using the two-hybrid system. *Biotechniques 14*, 920–4.
9. Legrain, P., Dokhelar, M. C., and Transy, C. (1994) Detection of protein–protein interactions using different vectors in the two-hybrid system. *Nucleic Acids Res 22*, 3241–2.
10. Ye, G. J., Vaughan, K. T., Vallee, R. B., and Roizman, B. (2000) The herpes simplex virus 1 U(L)34 protein interacts with a cytoplasmic dynein intermediate chain and targets nuclear membrane. *J Virol 74*, 1355–63.

Chapter 22

Inducible microRNA-Mediated Knockdown of the Endogenous Human Lamin A/C Gene

Ina Weidenfeld

Abstract

RNA interference (RNAi) enables the suppression, and hence the functional analysis, of individual genes. The use of the tetracycline (tet)-controlled transcription activation system for RNAi has become a valuable tool for conditional gene inactivation both in vitro and in vivo. Here, the generation of a conditional RNAi cell line for microRNA (miRNA)-mediated downregulation of the endogenous lamin A/C gene is described. A tet-responsive transcription unit, encoding a designed miRNA against human lamin A/C, is directly placed into a predefined genomic site of our previously developed cell line HeLa-EM2-11ht. This chromosomal locus permits the stringent control of miRNA expression, which results in the precise adjustment of lamin A/C protein concentrations. The utilization of this conditional RNAi system for the controlled inactivation of any gene of interest may significantly contribute to the study of gene functions under highly defined conditions.

Key words: Conditional RNA interference, Rationally designed microRNA, Tet-On system, Flp recombinase-mediated cassette exchange, Clonal cell lines, Lamin A/C, Western blotting, Immunofluorescence, Confocal laser scan microscopy

1. Introduction

Our understanding of gene function greatly benefits from techniques that permit predictable activation or inactivation of the expression of an individual gene and analysis of subsequent phenotypic changes. As such, the combination of RNA interference (RNAi) with the tetracycline (tet)-controlled transcription activation system (1, 2) promises to be a powerful tool to precisely control gene knockdown. This implies both the quantitative control of inhibitory RNA expression in incremental steps, as well as the temporal restriction of target gene inactivation.

Michael Kaufmann and Claudia Klinger (eds.), *Functional Genomics: Methods and Protocols*, Methods in Molecular Biology, vol. 815, DOI 10.1007/978-1-61779-424-7_22, © Springer Science+Business Media, LLC 2012

Presently, several lentivirus- and retrovirus-based delivery systems are available to stably introduce conditional RNAi systems into target cells (3). Unfortunately, these approaches often result in a considerable clone-to-clone variability in absolute expression levels and in expression characteristics of the inhibitory RNAs. Responsible for these effects are copy number and spatial arrangement of the transgenes, as well as epigenetic modifications.

To circumvent these problems, we previously identified a genomic locus in a HeLa cell line that is an optimal recipient of tet-responsive transcription units, in the sense that this locus is transcriptionally silent in the absence of doxycycline (dox), but highly active in its presence (4). This chromosomal site can directly be retargeted utilizing Flp recombinase-mediated cassette exchange (RMCE) (5) to efficiently insert a single copy of any gene of interest.

This chapter describes how to employ the retargetable genomic site of HeLa-EM2-11ht cells for the generation of a conditional RNAi system to efficiently knockdown the endogenous lamin A/C gene. Upon exchange of a tet-controlled transcription unit, encoding a miR 30-based miRNA against the human lamin A/C gene, the resulting stable cell line allows for the precise adjustment of lamin A/C protein concentrations by fine-tuning miRNA expression in a dox-dependent manner. This is monitored by Western blotting as well as by immunofluoresecence against endogenous lamin A/C, whereas the latter method additionally verifies the uniformity of the miRNA-driven knockdown throughout this clonal cell population.

2. Materials

2.1. Conditional Mammalian Gene Expression System

2.1.1. Master Cell Line HeLa-EM2-11ht (4)

This HeLa-based clonal cell line contains a preselected genomic integration site that can directly be targeted with a gene of interest upon applying Flp RMCE. The integration site contains a positive/negative selection cassette consisting of the hygromycin resistance gene and the thymidine kinase gene (HygTK). The selection cassette is flanked by a pair of heterospecific recognition sites for Flp recombinase, F and F3 (5). For optimal function of tet-induced gene regulation, this cell line produces the tet-controlled transcription activator, rtTA (Tet-On system) (6), constitutively and uniformly throughout its population.

2.1.2. Recombination Plasmid p.d1gfp.Ptet.miR (4)

This vector encodes the DNA cassette that will be inserted into the predefined genomic locus of HeLa-EM2-11ht cells. It comprises the bidirectional tet-inducible promoter that, upon induction, drives transcription of both the reporter gene for d1gfp (destabilized gfp) (7) and the sequence encoding a rationally designed miRNA against the endogenous human lamin A/C gene (see Subheading 2.2).

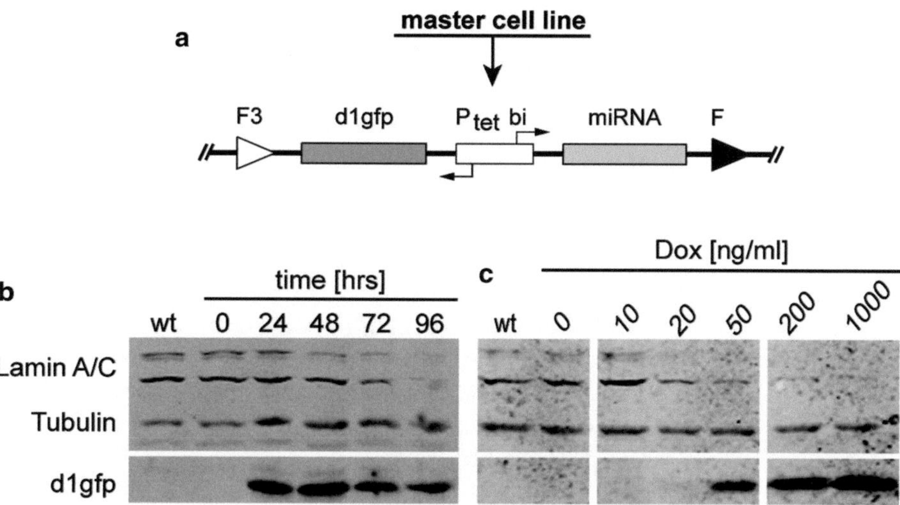

Fig. 1. Conditional knockdown of endogenous lamin A/C by tetracycline-controlled expression of miRNA. (**a**) Schematic representation of the tet-regulated miRNA construct after site-directed genomic integration into the master cell line HeLa-EM2-11ht by RMCE. The bidirectional tet-induced promoter (Ptet.bi) controls the simultaneous expression of a designed lamin A/C-specific miRNA and a destabilized green fluorescent protein (d1gfp). (**b**) Western Blot analysis of lamin A/C downregulation after miRNA expression over time. Clonal cell line miR.L was induced with 200 ng/ml dox for the time indicated. After 96 h, lamin A/C expression is undetectable. Tubulin served as a loading control. (**c**) Dose–response analysis of lamin A/C knockdown in clonal cell line miR.L. Cells were incubated with increasing concentrations of dox for 96 h. A significant reduction of lamin A/C expression is visible with 20 ng/ml dox and reaches a maximum with 200–1,000 ng/ml dox. Lamin A/C-specific miRNA expression is mirrored by the appearance of the coexpressed reporter protein d1gfp (from ref. 4 with permission of Oxford University Press).

The bidirectional expression cassette is flanked by the heterospecific recognition sites for Flp recombinase, F and F3 (see Fig. 1a).

2.1.3. Plasmid pCAGGS-IRES-Puro

This plasmid constitutively expresses a mutant of the site-specific Flp recombinase with improved properties, such as enhanced thermostability at 37°C. The novel mutant is termed Flp<u>e</u> (<u>e</u>nhanced; Gene Bridges) (8) and is the recombinase of choice throughout this study. Coexpression of a puromycin resistance gene allows for positive selection of those cells transfected with the plasmid.

2.2. Rationally Designed miRNAs Against Lamin A/C (4, 9)

For RNAi, an optimized miRNA design is used (see ref. 9 for miRNA3-design). A synthetic double-stranded oligonucleotide is created whose upper strand contains a 21 nucleotide (nt) *sense* sequence of the human lamin A/C gene in 5′–3′ orientation, connected to the loop sequence derived from the endogenous miRNA miR-30 (10), and followed by the 21 nt reverse complement *antisense* sequence. Further elements originating from miR-30, such as the Drosha cleavage site and the nuclear export signal (11) in addition to the sequences necessary for Dicer cleavage (12), are added to constitute the 5′ and 3′ flanking regions of the primary miRNA transcript. Here, the *sense* sequence directly corresponds to positions

608–628 relative to the start codon of lamin A/C (GenBank acc. # X03444) (13).

Upon processing of the primary miRNA transcript, the perfect stem loop structure formed by the present miRNA is excised from the transcript by Drosha and exported out of the nucleus by Exportin 5 (14). After the removal of the terminal loop by Dicer, the strand of the resulting miRNA duplex whose 5′ end is more weakly base-paired (*antisense* strand) is incorporated into the RNA-induced silencing complex (RISC) (15, 16) to then bind lamin A/C mRNAs. Gene silencing occurs either by translational repression or mRNA degradation, or a combination of both (17).

The entire miRNA-encoding sequence described above constitutes one of the two transcription units of the recombination plasmid p.d1gfp.Ptet.miR.

2.3. Cell Culture and Cell Lysis

1. Dulbecco's modified Eagle's medium (DMEM, Gibco) with high glucose (4,500 mg/l glucose) and 2 mM L-glutamine supplemented with 10% (v/v) fetal bovine serum (FBS), 1 mM sodium pyruvate solution as well as 100 U/ml penicillin G sodium salt and 100 μg/ml streptomycin sulfate. Store at 4°C.
2. Opti-MEM I reduced serum medium (Gibco). Store at 4°C.
3. Phosphate-buffered saline (PBS) (1×, Gibco).
4. Solution of 0.25% Trypsin with ethylenediaminetetraacetic acid (EDTA) 4Na (1×). Store in aliquots at –20°C.
5. Dox hydrochloride is dissolved at 1 mg/ml in distilled water, filter sterilized (0.2 μm), and stored in aliquots at –20°C (see Note 1).
6. Puromycin dihydrochloride is dissolved at 10 mg/ml in distilled water, filter sterilized (0.2 μm), and stored in aliquots at –20°C.
7. Ganciclovir is dissolved in distilled water at 10 mM, filter sterilized (0.2 μm), and stored in aliquots at –20°C.
8. Fugene 6 transfection reagent (Roche).
9. Cell lysis buffer (1×), a modification of Laemmli (18) buffer, contains 140 mM Tris base pH 8.0, 2 mM $MgCl_2$, 4% (w/v) sodium dodecyl sulfate (SDS), 50 mM dithiothreitol (DTT), 5 M urea, and 0.03% (w/v) bromophenol blue. Store in aliquots at –20°C.
10. Benzonase® nuclease is added fresh to cell lysis buffer at 5 U/100 μl.
11. Cell culture dishes (60 and 100 mm diameter) and multi-well plates (6-, 12-, and 24-wells).

2.4. SDS–Polyacrylamide Gel Electrophoresis (SDS–PAGE)

1. A 30% (w/v) acrylamide/bis-acrylamide solution (37.5:1). Unpolymerized acrylamide is a neurotoxin and must be handled with extreme caution. Working in a chemical fume hood is recommended especially when weighing solids.

2. N,N,N',N'-Tetramethylethylenediamine (TEMED). TEMED should be stored at room temperature (RT) in a well-ventilated place (see Note 2).
3. Ammonium persulfate (APS) is dissolved in distilled water to obtain a 10% (w/v) solution. It is stored in single use aliquots of 150 μl at −20°C.
4. Resolving gel buffer: 1.5 M Tris–HCl pH 8.8 and 0.4% (w/v) SDS (dissolved in water). Stored at room temperature.
5. Stacking gel buffer: 0.5 M Tris–HCl pH 6.8 and 0.4% (w/v) SDS (dissolved in water). Stored at room temperature.
6. Overlay solution for resolving gel: 0.1% (w/v) SDS.
7. Electrophoresis buffer (10× stock): 250 mM Tris base, 2 M glycine, and 35 mM SDS. Stored at room temperature and diluted to a 1× buffer with water for use.
8. Unstained molecular weight marker or Odyssey two-color protein molecular weight marker (LI-COR Biosciences) (see Note 3).

2.5. Western Blotting for Lamin A/C

1. Transfer buffer: Electrophoresis buffer (1×) plus 20% (v/v) methanol.
2. Nitrocellulose membrane with a pore size of 0.2 μm.
3. 3 mm Chr Whatman chromatography paper.
4. Ponceau S sodium salt is dissolved in 5% acetic acid to obtain a 0.2% (w/v) protein dye solution.
5. Tris-buffered saline (TBS) (10× stock): 1.5 M NaCl, 160 mM Tris–HCl, and 60 mM Tris base. Stored at room temperature and diluted to 1× with water for use (see Note 4).
6. Blocking buffer: 5% (w/v) skimmed milk powder dissolved in 1× TBS buffer. This should be prepared fresh for use.
7. Sodium azide (NaN_3) stock solution of 10% (w/v) in water. NaN_3 is extremely toxic and precautions must be taken (eyeshields, mask, and gloves). Solids must be weighed out under a chemical fume hood.
8. Primary antibodies: Mouse monoclonal anti-lamin A/C (19), mouse monoclonal anti-α-tubulin, and rabbit polyclonal anti-GFP (Sigma-Aldrich).
9. Secondary antibodies labeled with infrared dyes: Goat anti-mouse IgG IRDye 800CW and goat anti-rabbit IgG IRDye 800CW (LI-COR Biosciences).
10. Odyssey® Imaging System (LI-COR Biosciences) for direct infrared fluorescence detection of proteins.
11. Alternatively, membrane-bound proteins can be detected using enhanced chemiluminescence (ECL). Therefore, horseradish peroxidase-conjugated secondary antibodies are needed as well as an ECL reagent and Kodak film.

2.6. Confocal Immunofluorescence for Lamin A/C

1. Glass coverslips of 15 mm diameter and a thickness of 0.13–0.17 mm (No. 1) and glass slides of $25 \times 75 \times 1$ mm in size.
2. Multi-well cell culture plates (12-well).
3. PBS (10× stock): 1.36 M NaCl, 23 mM KCl, 42 mM Na_2HPO_4, and 14 mM KH_2PO_4. Adjust pH to 7.4 with HCl. The stock solution should be autoclaved before it is stored at room temperature.
4. A 40% (w/v) glucose solution prepared with distilled water while gently heated on a hot plate, filter sterilized after cooling, and stored at room temperature.
5. Paraformaldehyde (PFA) ready-to-use solution of 40% (w/v) from Electron Microscopy Sciences is diluted to 3% and 2% with 1× PBS, aliquoted (15 ml) and stored at –20°C (see Note 5a). If only PFA powder is available, all weighing and handling must be done using a chemical fume hood. The solid is dissolved to 3% (w/v) in 1× PBS while slowly stirring on a hot plate (<60°C) (see Note 5b). Adding a few drops (5–10) of 10 M NaOH will rapidly clear the solution. Adjust pH to 7.4 after the solution has cooled to room temperature.
6. Quench buffer: 50 mM NH_4Cl in 1× PBS.
7. Permeabilization buffer: 0.1% (v/v) Triton X 100 in 1× PBS.
8. Blocking solution: 1% (w/v) bovine serum albumin (BSA) and 10% (v/v) goat serum in permeabilization buffer (see Note 6).
9. Primary antibodies: Mouse monoclonal anti-lamin A/C (19) and guinea pig polyclonal anti-Nup-107 (20).
10. Secondary antibodies: Goat anti-mouse Alexa 568 and goat anti-guinea pig Alexa 647.
11. Vectashield® mounting medium (Vector Laboratories).
12. Clear nail polish to seal mounted coverslips.

3. Methods

3.1. Generation of a Stable HeLa Cell Line for Conditional Lamin A/C Knockdown

The generation of stable cell lines is a fast and easy method when utilizing HeLa-EM2-11ht cells (4). The cells are transiently transfected with both a suitably engineered recombination plasmid for genomic integration of the gene of interest (p.d1gfp.Ptet.miR) and the plasmid that will constitutively express Flpe recombinase (pCAGGS-IRES-Puro). Flpe recombinase mediates the replacement of the HygTK selection cassette that is present in the genomic targeting site of HeLa-EM2-11ht cells, by a single copy of the bidirectional tet-controlled transcription unit d1gfp/miR. Cells with correctly recombined genomic loci are enriched by negative selection with ganciclovir. After 8–10 days of selection, stable isogenic cell clones for controlled d1gfp/miR expression can be manually isolated and functionally analyzed.

3.1.1. RMCE in HeLa-EM2-11ht Cells

1. The master cell line HeLa-EM2-11ht is cultured in 100-mm cell culture dishes with supplemented DMEM until the cells reach ~80% cell density. To maintain the culture, cells are quantitatively detached using 0.25% Trypsin/EDTA and reseeded into new culture dishes at 5–10% cell density.
2. To prepare the experimental culture, seed HeLa-EM2-11ht cells into a single well of a 6-well plate at a density of 3×10^5 cells 24 h prior to transfection.
3. For transient cell transfection, place 100 μl of prewarmed (37°C) Opti-MEM serum free medium into a sterile Eppendorf tube. Add 6 μl of Fugene transfection reagent directly into the Opti-MEM, mix gently by tapping the tube and incubate for 5 min at RT. Add 1 μg of the recombination plasmid p.d1gfp. Ptet.miR and 1 μg of the Flpe recombinase encoding plasmid pCAGGS-IRES-Puro, tap to mix the contents and incubate for 30 min at RT (see Note 7).
4. Transfect HeLa-EM2-11ht by directly adding the complete transfection mix to the cells seeded in the 6-well. By applying the mixture in a drop-wise fashion, a uniform distribution across the well is guaranteed. Return the cells to the incubator.
5. Twelve hours posttransfection, briefly wash the cells with prewarmed 1× PBS, detach them with 0.25% Trypsin/EDTA, transfer them to a 100-mm cell culture dish, and culture them in fresh DMEM supplemented with 5 μg/ml puromycin for 24–36 h. This positive selection step enriches cells that express Flpe recombinase.
6. Subsequently remove the puromycin-DMEM along with all the dead cells (see Note 8).
7. Briefly wash the remaining cells attached with prewarmed 1× PBS and add fresh DMEM supplemented with 80–100 μM ganciclovir for negative selection of correctly recombined cells.
8. Renew the supplemented medium every 3 days for the duration of 8–10 days until single resistant cell clones become visible (see Note 9).

3.1.2. Manual Isolation of Single Stable Cell Clones

1. Prepare a 24-well plate with prewarmed (37°C) DMEM.
2. Place the 100-mm culture dish with the stable cell clones on the stage of a light microscope and use a 4× objective to focus on a single cell colony. Choose a colony that is completely isolated from adjacent colonies.
3. Lift the culture dish lid and use a P20 pipette to gently scrape off the selected cell clone while simultaneously aspirating the

cells into the sterile pipette tip. Monitor the entire procedure by looking through the microscope. Place the lid back on the culture dish (see Note 10).

4. Transfer the isolated cells into a well of the 24-well plate, resuspend them gently and immediately label the well.
5. Using a new sterile pipette tip for each clone, repeat steps 3 and 4 until 24 clones are collected.
6. Place the 24-well plate into the incubator for further propagation of the clones.
7. Once the cells have reached ~50% cell density they are detached as described above and transferred into larger wells of 12- or 6-well plates.
8. A fluorescent screen of the clones should be implemented to verify the DNA cassette exchange from HygTK to d1gfp.Ptet.miR (see Note 11):
   (a) Seed 1×10^5 cells of each clone into two wells of a 24-well plate keeping a separate larger backup of each clone in culture.
   (b) Induce gene expression in only one of the two wells by adding DMEM supplemented with 200 ng/ml dox for 24 h. Culture the uninduced cells in DMEM alone.
   (c) Screen for green fluorescent clones after induction.
   (d) As all the resulting positive clones are isogenic, it is sufficient to choose only 5–10 clones for further propagation and for the generation of frozen stocks. One clone is then used to carry out experiments. Here, this clone is referred to as the conditional miR.L cell line.

3.2. Western Blotting for Lamin A/C knockdown

The downregulation of endogenous lamin A/C by tet-controlled expression of miRNAs is analyzed in a time- and dose-dependent manner.

3.2.1. Preparation of Cell Lysates

1. The conditional miR.L cell line and the master cell line HeLa-EM2-11ht are propagated as described in Subheading 3.1.1 *RMCE in HeLa-EM2-11ht cells.*
2. For the preparation of experimental cultures, the conditional miR.L cell line is seeded into 6-well plates at 7×10^4 cells per well.
   (a) To study the effect of miRNA expression over time, five time points are chosen: 0, 24, 48, 72, and 96 h of induction. Starting now and continued for the next 96 h, 200 ng/ml of dox are added to one well every 24 h to induce expression. Two wells are left uninduced: One will serve as the 0 h time point while the other is used to determine the cell number after 96 h of cell growth using a hemocytometer.

(b) For the dose–response study, cells are cultured with different concentration of dox for 96 h: 0, 10, 20, 50, 200, and 1,000 ng/ml.

3. In addition, 7×10^4 cells of HeLa-EM2-11ht are seeded into one 6-well to serve as a control.
4. After 96 h the cell lysates are prepared by briefly washing the cells with prewarmed (37°C) 1× PBS and directly applying cell lysis buffer plus Benzonase® nuclease (5 U/100 μl) onto the cells. The volume of lysis buffer added is defined after determining the cell number by counting the cells of one of the uninduced control wells. A final concentration of 5×10^4 cells per μl of lysate is normally a good choice.
5. Incubate cells with cell lysis buffer for 10 min at room temperature before transferring each sample into a 1.5-ml Eppendorf tube.
6. Heat samples at 95°C for 1–2 min and spin down briefly before loading them onto a polyacrylamide gel or storing them at −20°C.

3.2.2. SDS–PAGE

SDS–PAGE is applied for the separation of proteins in a polyacrylamide gel. The following instructions imply the usage of the Criterion protein electrophoresis system from Bio-Rad, but can be transferred to any other available gel electrophoresis system.

1. Insert a Criterion gel casting cassette (133×87×1 mm) into the appropriate Criterion electrophoresis tank. Remove the comb out of the casting cassette and start preparing 15 ml of a 10% resolving gel by mixing 5 ml of the 30% (w/v) acrylamide/bis-acrylamide solution with 3.75 ml of resolving gel buffer and 6.16 ml distilled water. After addition of 7.5 μl of TEMED and 75 μl of APS, mix briefly and immediately start filling the mixture into the gel casting cassette by using a 10-ml pipette. Pour up to 1 in. below the gel pockets (see marking on cassette). Overlay the gel with 0.1% (w/v) SDS to generate an even surface. Polymerization should be completed after 20–30 min.
2. Pour off the overlay solution and rinse once with distilled water. Use thin filter paper to dry the area above the polymerized gel.
3. Reinsert the cassette into the Criterion tank and prepare 10 ml of the 4% stacking gel by mixing 1.4 ml of the 30% (w/v) acrylamide/bis-acrylamide solution with 2.5 ml of stacking gel buffer and 6.1 ml distilled water. Add 10 μl TEMED and 54 μl APS, mix briefly, pour the stacking gel, and insert the comb. If a bit of the gel solution is spilled on the outside of the cassette or in the tank this can easily be cleaned once the solution has polymerized after 20–30 min.
4. Remove the gel cassette from the tank, peel the plastic strip of the bottom of the cassette, and reinsert it into the tank.

5. Dilute the electrophoresis buffer stock (10×) to 1× with distilled water. To avoid foam formation mix gently by inverting, then fill the Criterion electrophoresis tank with buffer.
6. Carefully remove the comb and wash each well with buffer using a syringe with a needle.
7. Start by loading the molecular weight marker in one well and 10 μl (corresponding to 5×10^4 cells) of each cell sample in the following wells. Use the wells close to the middle of the gel and load the remaining wells to both sides of your samples with lysis buffer only. This will help to generate an even dye front during electrophoresis.
8. Assemble the lid of the Criterion unit and connect the unit to a power supply. The gel can be run at 30–50 mA until the blue dye front is run off. To prevent overheating, electrophoresis should take place in the cold room at 4°C or by placing the unit in an ice bath.

3.2.3. Protein Transfer and Immunodetection

The separated proteins are transferred from the polyacrylamide gel onto a nitrocellulose membrane. This procedure is described assuming that the experimenter is using the Criterion wet electroblotting system from Bio-Rad.

1. To avoid the contamination of the nitrocellulose membrane, it is crucial to wear clean gloves throughout this procedure and to ensure that all assembly trays are clean and free of dyes (e.g. Coomassie brilliant blue, bromophenol blue, or amidoblack) (see Note 3).
2. Prepare 2 L of transfer buffer in a large beaker.
3. Cut a piece of nitrocellulose membrane that is larger than the separating gel as well as 2–3 sheets of Whatman chromatography paper that will fit the size of the transfer cassette.
4. Fill an assembly tray with transfer buffer and equilibrate the Whatman paper and two foam pads. If foam pads are not available, they can be replaced by Whatman paper.
5. The nitrocellulose membrane is equilibrated in transfer buffer in a separate tray for about 5 min.
6. Remove the gel cassette from the electrophoresis unit and disassemble it. The stacking gel can easily be removed from the separating gel by pressing a sheet of lab tissue paper directly on the stacking gel and, in a peeling motion, separating it from the bottom gel. For your orientation, cut off a top corner of the gel that is closest to the lane of sample #1.
7. After briefly equilibrating the separating gel in transfer buffer, the "sandwich" is assembled in the transfer cassette from bottom to top (the top facing you) in the following manner: one foam pad, one sheet of Whatman paper, the membrane, the gel, one

sheet of Whatman paper, and the second foam pad. In order to remove bubbles, a 5-ml glass pipette is used to gently roll over the sandwich 2–3 times before the transfer cassette is closed. The layers of the sandwich must be firmly held together for optimal transfer of proteins.

8. The transfer cassette is inserted into the Criterion blotter such that the top (gel) of the sandwich is facing the cathode (−) and the bottom (membrane) is facing the anode (+).
9. The blotter tank is filled with transfer buffer and a magnetic stir bar is placed into the tank before closing the lid and connecting the unit to a power supply. The blotter is placed in an ice bath on a magnetic stirrer and transfer is carried out for 2 h at 100 V (see Note 12).
10. After electrotransfer, disassemble the sandwich and remove the nitrocellulose membrane together with the gel. Place them on a clean glass plate and use a new razor blade to trim the sides of the membrane to fit the gel and to mark a corner of the membrane for orientation. Optional: stain the gel with Coomassie brilliant blue to verify the transfer.
11. Incubate the membrane in a 0.2% Ponceau S solution for 5 min while gently rocking the membrane on a platform. Note: all of the membrane incubations or washes described here are to be carried out on a rocking platform at room temperature.
12. Recycle the Ponceau S solution and rinse the membrane once with distilled water before incubating it in 3% acetic acid for several min to remove background staining. If the transfer was successful, immobilized protein bands will become visible on the membrane. Document the stained membrane using a conventional scanner (see Note 13).
13. Wash the membrane in 1× TBS for 10 min to remove the Ponceau S staining. Continue with the next step even if remnants of the dye are still visible.
14. Incubate the membrane in blocking buffer for 30 min.
15. Discard the blocking buffer and add the following primary antibodies diluted in blocking buffer: Mouse monoclonal anti-lamin A/C (1:100), mouse monoclonal anti-α-tubulin (1:10,000), and rabbit polyclonal anti-GFP (1:4,000). Incubate for 2 h at room temperature or over night at 4°C. Add 0.01% NaN_3 to prevent contamination of the milk-based blocking buffer.
16. Collect the antibody mixture and store at 4 or −20°C for repeated use.
17. Wash the membrane three times for 10 min each with 1× TBS.
18. Prepare a 1:40,000 diluted mixture of the secondary antibodies labeled with infrared dyes using blocking buffer: Goat anti-mouse IgG IRDye 800CW and goat anti-rabbit IgG IRDye 800CW.

Incubate the membrane with the mixture for 45 min and cover the tray with aluminum foil to keep dark.

19. Collect the secondary antibody solution for it can be stored at 4°C in the dark or at –20°C for repeated application.
20. Wash the membrane three times for 10 min each with 1× TBS.
21. The Odyssey infrared imaging system allows you to scan the membrane while it is still wet or after it has been dried. For drying, place the membrane between two sheets of Whatman filter paper until dry (see Note 14).
22. Scan the membrane using the Odyssey imager according to the manufacturer's instructions. An example is shown in Fig. 1b, c.

3.3. Immunofluorescence Analysis of Lamin A/C Knockdown

The shutdown of lamin A/C synthesis and the induction of the reporter d1gfp are tracked at single cell level by monitoring the cells with immunofluorescence.

3.3.1. Preparation of Cell Samples

1. The conditional miR.L cell line and the master cell line HeLa-EM2-11ht are cultured as described in Subheading 3.1.1 *RMCE in HeLa-EM2-11ht cells*.
2. Prepare a 12-well plate by carefully placing one 15-mm glass coverslip into each well of the plate (see Note 15).
3. Add growth medium such that the coverslips are not floating in the medium but remain at the bottom of the well.
4. The conditional miR.L cell line is seeded in two wells at 2×10^4 cells per well, whereby cells in only one well are treated with 200 ng/ml dox to induce expression for the duration of 96 h. HeLa-EM2-11ht cells are seeded at the same density in one well to serve as a negative control. For the experimenter who is still learning to carry out immunofluorescence staining, the preparation of duplicates of each sample is recommended.

3.3.2. Immunofluorescence Staining

1. The cells growing on glass coverslips in a 12-well plate are briefly washed with 1× PBS plus 0.4% glucose prewarmed to 37°C (see Note 16).
2. Immediately, the cells are fixed for 2 min with 3% prewarmed (37°C) PFA (see Note 17).
3. Discard the PFA solution into a hazardous waste container and wash the cells twice for 5 min each with 1× PBS.
4. Quench PFA by incubating the cells in 50 mM NH_4Cl for 5 min (see Note 18).
5. Wash the cells twice for 5 min each with 1× PBS. The cells can also be stored at this stage over night at 4°C.
6. The cells are now permeabilized by incubation with Triton X100-based permeabilization buffer for 5 min.

7. Aspirate the permeabilization buffer and keep the cells in 1× PBS until continuing with step 8.
8. Incubate the cells in blocking solution for 30 min. For economic reasons, the coverslips are directly provided with a drop of blocking solution (~100 μl per coverslip). Therefore, the coverslips are sequentially removed from their wells using Dumont forceps and placed on an even surface in a humid chamber (see Note 19).
9. Remove the blocking solution by sequentially blotting each coverslip on tissue paper (Kimwipes). Hold the coverslip vertically only allowing the rim of the glass to touch the tissue. Place the coverslips back into the humid chamber and incubate them with a drop of the primary antibody diluted in blocking solution for 1 h: Mouse monoclonal anti-lamin A/C (1:500) and guinea pig polyclonal anti-Nup-107 (1:5,000).
10. Remove the primary antibody dilution by blotting each coverslip as described in step 9. Place each coverslip into a well of a multi-well plate and wash three times for 5 min each with 1× PBS.
11. After returning the coverslips to the humid chamber they are incubated with a mixture of fluorescently labeled secondary antibodies for 30 min (see Note 20). Goat anti-mouse Alexa 568 and goat anti-guinea pig Alexa 647 each diluted 1:500 in blocking solution.
12. The coverslips are now moved back into the multi-well plate and washed three times for 5 min each with 1× PBS.
13. Postfixation is carried out for 3 min using 2% PFA before the cells are washed twice in 1× PBS for 5 min each and once with distilled water.
14. The coverslips are now ready to be mounted on glass slides. Each coverslip is carefully inverted onto a droplet of Vectashield mounting medium on the glass slide. Avoid generating air bubbles in the mounting medium. Clear nail polish is gently applied around the rim of the coverslip sealing it off and preventing the mounting medium to dry out. Once the nail polish has hardened, the slides can immediately be imaged or be stored at 4°C in the dark (if imaged within the following days) or at –20°C (for long-term storage).

3.3.3. Confocal Microscopy

Confocal microscopy is performed with a Leica TCS SP5 laser-scanning microscope. The laser lines 488-, 561-, and 633-nm are used for excitation of d1gfp, Alexa 568 (for the detection of lamin A/C), and Alexa 647 (for the detection of Nup-107), respectively. All images are taken using a Leica TCX PL APO 63× NA 1.4 oil objective. The resulting images are assembled in Fig. 2.

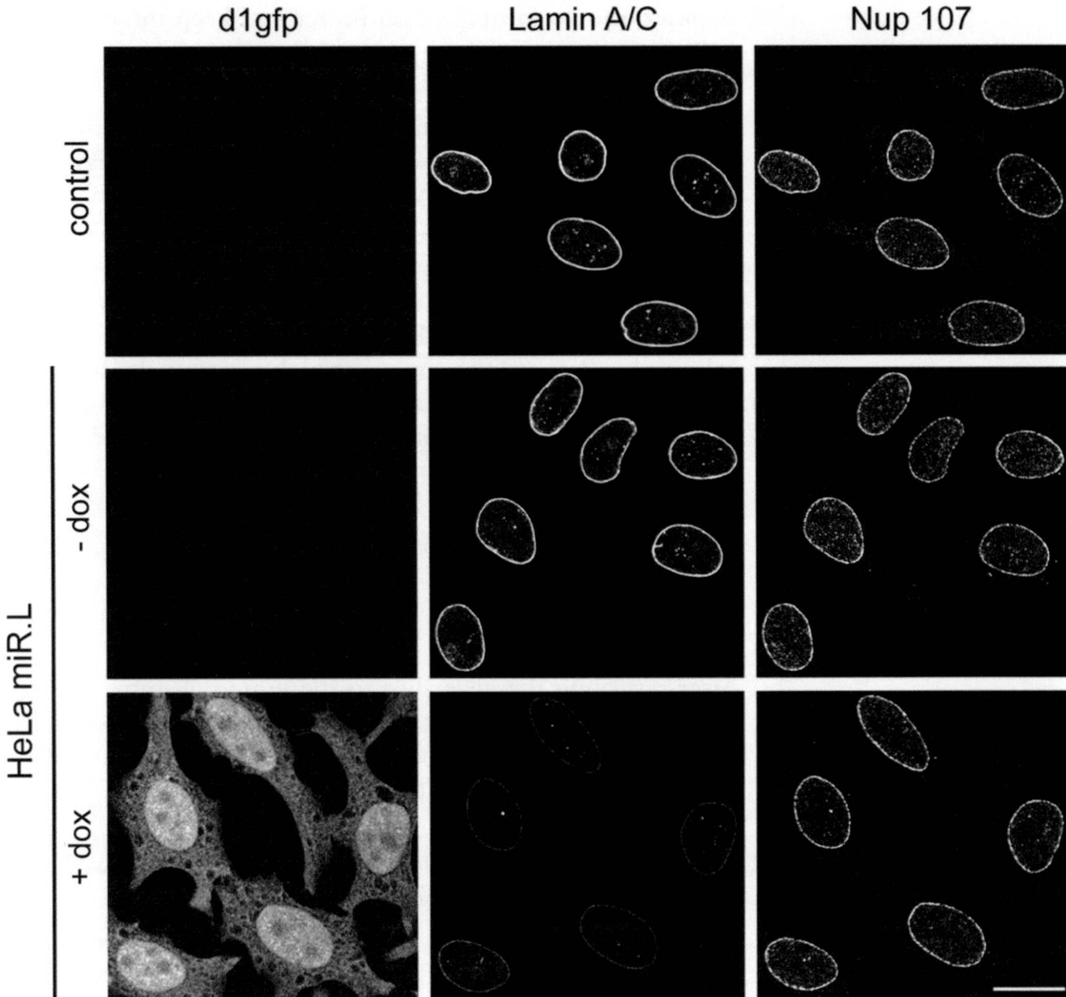

Fig. 2. Efficient and uniform knockdown of endogenous lamin A/C in the conditional miR.L cell line demonstrated by immunofluorescence staining. Cells were seeded on coverslips and cultured in the presence and the absence of 200 ng/ml of dox for the duration of 96 h before stained with anti-lamin A/C antibody. The downregulation of lamin A/C is accompanied by the appearance of the reporter protein d1gfp. Staining of HeLa-EM2-11ht cells served as a negative knockdown control. Antinucleoporin Nup107 staining served as an internal control visualizing nuclear pore complexes within the nuclear envelope. Scale bar 20 μm (from ref. 4 with permission of Oxford University Press).

4. Notes

1. Dox aliquots can also be stored at 4°C for up to a month but are preferentially kept at −20°C for long-term storage. Both the powder and the solutions are light sensitive and should be kept in the dark.
2. TEMED catalyzes the polymerization of polyacrylamide gels. Its quality declines after opening which results in the slower

polymerization of gels. The purchase of small quantities is therefore recommended.

3. Several dyes, even including those commonly used in prestained protein standards as well as in lab markers and pens, can leave permanent traces on nitrocellulose membranes. Even after washing them off, they remain highly visible under infrared light when using the Odyssey® Imaging System for the direct fluorescence detection of proteins. Therefore, most experimenters prefer to use an unstained molecular weight marker to a prestained one when working with the Odyssey® Imaging System.
4. Here, TBS *without* Tween-20 shows best results for immunodetection. The detergent Tween-20 is known to often cause unspecific binding of antibodies to protein lysates when added to blocking buffer.
5. (a) Many protocols give instructions to prepare the PFA solution fresh before each application. I have found no difference in the quality of fixation between freshly prepared or frozen PFA solutions. (b) Exercise caution when heating the PFA solution as it emits formaldehyde gas (pungent odor) when temperatures exceed 60°C.
6. The serum added to the blocking solution is derived from the species in which the secondary antibody was raised. If serum is not available, the BSA concentration should be increased from 1 to 3% (w/v).
7. The transfection mixture is prepared under sterile conditions in a cell culture hood.
8. The dead cells are cells that were not transfected with pCAGGS-IRES-puro and could therefore not undergo Flpe-RMCE. Begin negative selection although the number of viable cells might be small.
9. During mitosis cells detach from the bottom of the culture dish and float in the medium before reattaching. To prevent cross-contamination of adjacent cell clones, rapid movement of the culture dish should be avoided during the selection period.
10. As the manual isolation of clones is performed outside of the cell culture hood this procedure is not sterile. Nonetheless, one can avoid unnecessary exposure to contaminants by reducing the time that the culture dish lid is removed, by thoroughly cleaning the P20 pipette with 70% ethanol in water, by using sterile filter tips that are only opened under the hood and by leaving the 24-well plate, in which the clones are collected, in the hood throughout the isolation procedure.

11. At times, a resistance to ganciclovir can be observed in nonrecombined clones and it is probably caused by spontaneous mutations within the TK gene.
12. The circulation of the buffer ensures uniform temperature and conductivity during protein transfer.
13. This image documents the total amount of protein of each sample that was transferred and serves as a useful control. In addition, you can now highlight the bands of the unstained molecular weight marker on the membrane using only a pencil (see Note 3). To avoid the contamination of the membrane while scanning, place a transparency between the glass of the scanner and the membrane.
14. When stored at 4°C in the dark the antibody signals will remain stable and can still be detected after weeks.
15. The glass coverslips should be autoclaved sterile in a Pyrex glass Petri dish before use. Alternatively, they can be UV sterilized for 15 min in the tissue culture hood after they have been placed in the wells of the multi-well plate.
16. The time between removing the cells from the incubator and the addition of the fixative should be kept short in order to avoid morphological changes or relocalization of proteins.
17. Fixation time and temperature are critical parameters and must be optimized for different antibodies and even different cell and tissue types. For many applications, PFA is the fixative of choice. In some cases however, fixation with organic solvents or fixation with a mixture of PFA and glutaraldehyde gives better results.
18. This step will decrease autofluorescence. Alternatively, 50–100 mM glycine or a $NaBH_4$ solution (1 mg/ml in 1× PBS) can be used for quenching.
19. To make a humid chamber, take a round cell culture dish (100 or 150 mm depending on the number of coverslips), cut out a sheet of Whatman filter paper to fit the size of the dish, place inside and moisten with water. Afterward, place a sheet of Parafilm over the filter paper covering the whole area. With the cells facing upward, lay the coverslips on the Parafilm and gently pipette a drop of solution on the cells. Ensure that the drop is evenly distributed across the cover-slip. Close the lid of the chamber during the incubation period. To be able to discriminate between different coverslips, the Parafilm can easily be labeled.
20. From now on, it is recommended to cover the fluorescent samples to keep them dark.

Acknowledgments

I would like to thank the Professors Hermann Bujard and Dirk Görlich as well as Dr. Kai Schönig for their support and encouragement.

References

1. Gossen M and Bujard H (1992) Tight control of gene expression in mammalian cells by tetracycline-responsive promoters. *Proc Natl Acad Sci USA*, 89(12): 5547–5551
2. Gossen M, Freundlieb S, Bender G, Müller G, Hillen W and Bujard H (1995) Transcriptional activation by tetracyclines in mammalian cells. *Science*, 268(5218): 1766–1769
3. Wiznerowicz M, Szulc J and Trono D (2006) Tuning silence: conditional systems for RNA interference. *Nature Methods*, 3(9): 682–688
4. Weidenfeld I, Gossen M, Löw R, Kentner D, Berger S, Görlich D, Bartsch D, Bujard H and Schönig K (2009) Inducible expression of coding and inhibitory RNAs from retargetable genomic loci. *Nucl Acids Res*, 37(7): e50
5. Schlake T and Bode J (1994) Use of mutated FLP recognition target (FRT) sites for the exchange of expression cassettes at defined chromosomal loci. *Biochemistry*, 33(43): 12746–12751
6. Urlinger S, Baron U, Thellmann M, Hasan MT, Bujard H and Hillen W (2000) Exploring the sequence space for tetracycline–dependent transcriptional activators: novel mutations yield expanded range and sensitivity. *Proc Natl Acad Sci USA*, 97(14): 7963–7968
7. Li X, Zhao X, Fang Y, Jiang X, Duong T, Fan C, Huang CC and Kain SR (1998) Generation of destabilized green fluorescent protein as a transcription reporter. *J Biol Chem*, 273(52): 34970–34975
8. Buchholz F, Angrand PO and Stewart AF (1998) Improved properties of flp recombinase evolved by cycling mutagenesis. *Nat Biotechnol*, 16(7): 657–662
9. Berger S, Pesold B, Reber S, Schönig K, Berger A J, Weidenfeld I, Miao J, Berger MR, Gruss OJ and Bartsch D (2010) Quantitative analysis of conditional gene inactivation using rationally designed tetracycline-controlled miRNAs. *Nucl Acids Res*, 38(17): e168
10. Boden D, Pusch O, Silbermann R, Lee F, Tucker L and Ramratnam B (2004) Enhanced gene silencing of HIV-1 specific siRNA using microRNA designed hairpins. *Nucl Acids Res*, 32(3): 1154–1158
11. Lee Y, Ahn C, Han J, Choi H, Kim J, Yim J, Lee J, Provost P, Radmark O and Kim S (2003) The nuclear RNase III Drosha initiates microRNA processing. *Nature*, 425(6956): 415–419
12. Zeng Y, Wagner EJ and Cullen BR (2002) Both natural and designed miRNAs can inhibit the expression of cognate mRNAs when expressed in human cells. *Mol Cell*, 9(6): 1327–1333
13. Elbashir SM, Harborth J, Lendeckel W, Yalcin A, Weber K and Tuschl T (2001) Duplexes of 21-nucleotide RNAs mediate RNA interference in cultured mammalian cells. *Nature*, 411(6836): 494–498
14. Lund E, Güttinger S, Calado A, Dahlberg JE and Kutay U (2004) Nuclear export of microRNA precursors. *Science*, 303(5654): 95–98
15. Schwarz DS, Hutvagner G, Haley B and Zamore PD (2002) Evidence that siRNAs function as guides, not primers, in the Drosophila and human RNAi pathways. *Mol Cell*, 10(3): 537–548
16. Khvorova A, Reynolds A and Jayasena SD (2003) Functional siRNAs and miRNAs exhibit strand bias. *Cell*, 115(2): 209–216
17. Wu L, Fan J and Belasco JG (2008) Importance of translation and nonnucleolytic ago proteins for on-target RNA interference. *Curr Biol*, 18(17): 1327–1332
18. Laemmli UK (1970) Cleavage of structural proteins during the assembly if the head of bacteriophage T4. *Nature*, 227(5259): 680–685
19. Rober RA, Gieseler RK, Peters JH, Weber K and Osborn M (1990) Induction of nuclear lamins a/c inmacrophages in in vitro cultures of rat bone marrow precursor cells and human blood monocytes, and in macrophages elicited in vivo by thioglycollate stimulation. *Exp Cell Res*, 190(2): 185–194
20. Hase ME and Cordes VC (2003) Direct interaction with nup153 mediates binding of tpr to the periphery of the nuclear pore complex. *Mol Biol Cell*, 14(5): 1923–1940

Chapter 23

Multiple-Gene Silencing Using Antisense RNAs in *Escherichia coli*

Nobutaka Nakashima, Shan Goh, Liam Good, and Tomohiro Tamura

Abstract

We have developed four expression vectors to express antisense RNAs (asRNAs) by which genes of interest are silenced in *Escherichia coli*. The vectors are all IPTG-inducible and co-transformable in any combination and target genes are silenced conditionally and concurrently. Furthermore, in order to improve silencing efficacy, the vectors are designed to express uniquely shaped antisense RNAs, named paired termini antisense RNAs (PTasRNAs). The vectors are useful for comprehensive investigation of gene function and are applicable even if the target genes are essential for cell growth. Here, we describe methods to construct PTasRNA-expressing vectors and to evaluate silencing efficacy.

Key words: Gene silencing, Antisense RNA, *Escherichia coli*, Plasmid compatibility, Co-transformation, Metabolic engineering, Essential gene, RNA silencing

1. Introduction

A gene disruption (knock-out) method is frequently employed to investigate gene function in bacteria (1). However, the method is difficult to apply to genes essential for cell growth; it is time-consuming and unfavorable for functional genomics which requires handling multiple genes simultaneously. In such cases, gene silencing (knock-down) by using antisense RNAs (asRNAs) is more suitable because the method is simple and target genes can be conditionally silenced (2–4).

In eukaryotes, RNA interference (RNAi)-mediated gene silencing has become very popular. However, a conserved RNAi mechanism is absent in bacteria and hence, single-stranded asRNAs expressed from expression vectors are used (Fig. 1) (5). Expressed

Michael Kaufmann and Claudia Klinger (eds.), *Functional Genomics: Methods and Protocols*,
Methods in Molecular Biology, vol. 815, DOI 10.1007/978-1-61779-424-7_23, © Springer Science+Business Media, LLC 2012

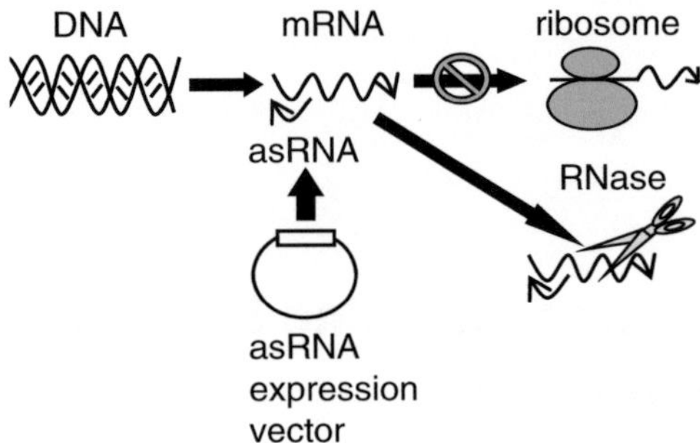

Fig. 1. Action of asRNAs. asRNAs bind target mRNAs and block translation and facilitate degradation.

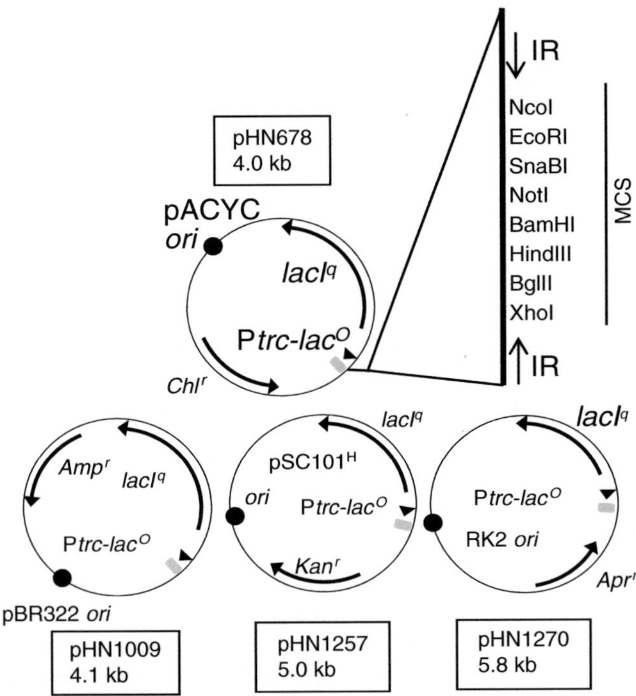

Fig. 2. PTasRNA expression vectors. *Arrows* indicate open reading frames or promoters, and *circles* indicate *oris*. *MCS* multiple cloning site, *IR* inverted repeat, P*trc-lac*^{O} *trc* promoter and lactose operator sequence, *Amp*^{r} ampicillin-resistance gene, *Chl*^{r} chloramphenicol-resistance gene, *Kan*^{r} kanamycin-resistance gene, *Apr*^{r} apramycin-resistance gene. Sequences at *gray squares* (IR and MCS) are identical between four vectors. Restriction enzyme sites shown are *unique* in the vector.

asRNAs are typically designed to hybridize to the ribosome-binding site (RBS)/start codon region of target mRNAs. The asRNAs prevent ribosomes from recognizing the RBS and thus inhibit translation. Furthermore, mRNAs that are not loaded by ribosomes (called naked mRNAs) tend to be degraded rapidly (6, 7). These effects lead to specific asRNA-mediated silencing of target genes (Fig. 1). Despite having obvious advantages, this method had not been frequently used until recently because silencing efficacy was low and silencing multiple genes concurrently had not been reported.

To improve the method, we first constructed an expression vector (pHN678) that expresses asRNAs with paired termini antisense RNAs (PTasRNAs) (Fig. 2) (2). The PTasRNAs have flanking inverted repeats that create paired double-stranded RNA termini (Fig. 3). We found that PTasRNAs had much higher silencing efficacies than asRNAs lacking paired termini probably due to improved RNA stability which increases RNA abundance in cells. Indeed, for several genes, the expected knock-out phenotypes did not appear upon expression of asRNAs lacking paired termini but were clearly observable upon expression of PTasRNAs (2). The vector pHN678 had the *trc* promoter (P*trc*) and the lactose repressor gene (*lacI^q^*) to drive conditional expression of PTasRNAs with the addition of IPTG. These design features enable efficient and conditional gene silencing, which can also be applied to genes essential for cell growth.

Next, we constructed three additional PTasRNA-expressing vectors (pHN1009, pHN1257 and pHN1270) (Fig. 2) to achieve multiple gene silencing (3). The resulting four vectors, including

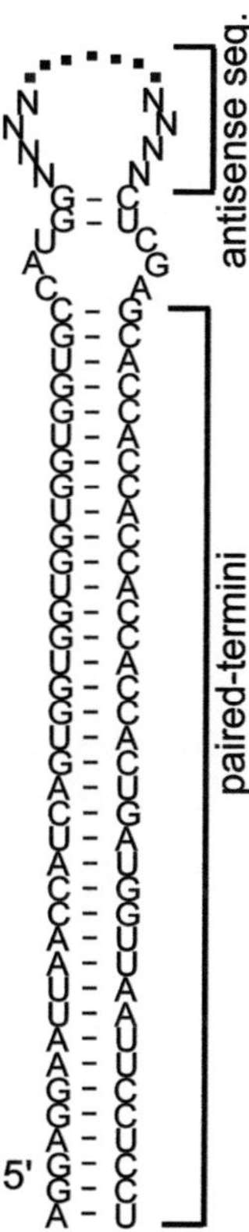

Fig. 3. Structure of a typical PTasRNA. "NNNN….NNNN" indicates an antisense sequence or an MCS control sequence in the case of an empty vector.

pHN678, are compatible and co-transformable in any combination. Each vector has a different and compatible plasmid replication origin [*ori*; pACYC, pBR322, pSC101^{H} (a high-copy mutant of pSC101) and RK2] as well as different antibiotic-resistance markers (chloramphenicol, ampicillin, kanamycin, and apramycin). We have experimentally confirmed that co-transformation with the four vectors and quadruple silencing are possible using *lacZ*, *pepN*, *ackA*, and *pta* as reporters (3).

In Table 1, a list of the genes silenced to date and the results are shown. We believe that this gene silencing system is useful for genome-wide investigations of gene function and genetic interactions.

2. Materials

2.1. Media, Escherichia coli Strains and Plasmids

1. The *E. coli* strain MG1655 was used as a host for expressing PTasRNAs unless otherwise described.
2. Luria Broth (LB): 1% tryptone, 0.5% yeast extract, and 1% NaCl. *E. coli* cells were cultured in LB at 37°C unless otherwise indicated (see Note 1).
3. Antibiotics were included when transformants were grown. Working concentrations of antibiotics in LB: Apramycin, 35 μg/ml, chloramphenicol, 24 μg/ml, kanamycin, 15 μg/ml, ampicillin, 50 μg/ml. All are diluted from stock solutions of 1,000-fold concentrations. The stock solution of chloramphenicol was prepared in 100% ethanol.
4. IPTG, isopropyl β-D-1-thiogalactopyranoside: Prepare stock solutions at 1 M in water and store aliquots at –20°C. Avoid repeated freeze-thaw.
5. PTasRNA vectors and an LacZ- and a DsRed Express-reporter vectors (Figs. 2 and 4).

2.2. Genomic DNA Purification

1. SET buffer: 75 mM NaCl, 25 mM EDTA-2Na, 20 mM Tris–HCl (pH 7.5). Autoclave and store at room temperature.
2. 20 mg/ml proteinase-K, 10% SDS, 5 M NaCl, and phenol/chloroform/isoamyl alcohol (25:24:1).

2.3. RNA Extraction and RT-PCR

1. Mueller Hinton Broth (MHB; Fluka, 70192), autoclave, and store at room temperature.
2. 96-well, flat bottom polystyrene plates.
3. Reagents from Applied Biosystems: RiboPure™ Bacteria Kit (AM1925), MultiScribe Reverse Transcriptase (4311235),

Table 1
Silenced genes to date

Gene name	Gene product	Silencing efficacy[a]	Observed phenotypes upon expression of PTasRNA	Reference
ackA	Acetate kinase	78%	Reduced acetate production, no growth on minimal-acetate media	(2, 3)
ackA-DsRed	Fusion of acetate kinase and red fluorescence protein	84%	Reduced red fluorescence	(2)
pta	Phosphotransacetylase	81%		(3)
lacZ	β-Galactosidase	88%		(3)
pepN	Aminopeptidase N	90%		(3)
aceE	Pyruvate dehydrogenase component		Acetate auxotroph, accumulation of pyruvate	
mutS	Methyl-directed mismatch		Increased mutation rate repair protein	(3)
mutD	DNA polymerase III ε subunit		Increased mutation rate	(3)
ndk	Nucleotide diphosphate kinase		Increased mutation rate	(3)
fabI	Enoyl-acyl carrier protein reductase		severe growth (essential gene)	(4)
acpP	Acyl carrier protein		severe growth (essential gene)	(4)
ftsZ	Tubulin-like protein		severe growth (essential gene), elongated cell	(4)

[a]Evaluated as reduced protein activity upon PTasRNA expression

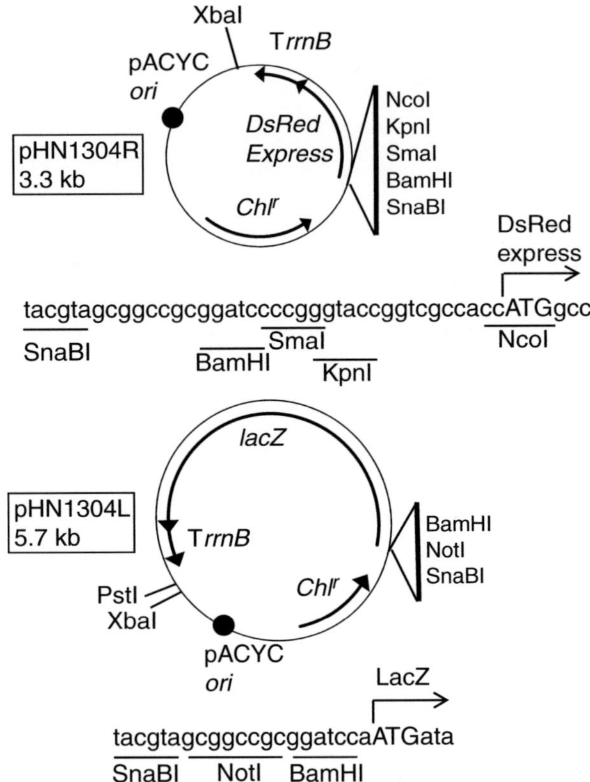

Fig. 4. Maps of reporter vectors. *Arrows* indicate open reading frames, and *circles* indicate *ori*s. *Chlr* chloramphenicol-resistance gene, T*rrnB rrnB* gene terminator. Unique restriction enzyme sites are shown. Sequences around the ATG start codons (shown in *upper cases*) of the reporter genes are shown below the maps. The target genes should be inserted between SnaBI and NcoI sites of pHN1340R or SnaBI and BamHI sites of pHN1340L.

Random hexamers (N8080127), dNTP mix (N8080261), and RNase inhibitor (N8080119).

4. SYBR Green PCR Mastermix (Eurogentec, RT-SN2X-03T).
5. Primers rpoA-F 5′ aagctggtcatcgaaatggaa; rpoA-R 5′ gccgcacgacgaatcg.
6. PowerWave X340 (Bio-Tek) or Safire (Tecan, Switzerland) 96-well plate spectrophotermeter, or equivalents (see Note 2).

2.4. Cell Extract Preparation, DsRed Express Fluorescence, and LacZ Activity

1. 10× Phosphate-buffered saline (PBS): 1.4 M NaCl, 27 mM KCl, 101 mM Na_2HPO_4, 18 mM KH_2PO_4. Autoclave and store at room temperature.
2. Z buffer: 1× PBS, 1 mM $MgSO_4$, 50 mM 2-mercaptoethanol. Store at 4°C.
3. 4 mg/ml *o*-nitrophenyl-β-galactoside (ONPG) solution. Prepare just before use.
4. PowerWave X340 or Safire 96-well plate spectrophotermeter/fluorometer, or equivalents (see Note 2).

5. A black microplate well and a transparent microplate well.
6. Bio-Rad Bradford assay kit (Bio-Rad, Hercules, CA, USA), or equivalents.

3. Methods

3.1. Purification of E. coli Genomic DNA for PCR

1. Culture the MG1655 strain in 5 ml overnight.
2. Spin-down the cells and resuspend the cell pellet with 375 μl of SET buffer in a 1.5 ml microcentrifuge tube.
3. Add 10 μl of 20 mg/ml proteinase-K and 45 μl of 10% SDS, and incubate at 55°C for 1 h.
4. Add 150 μl of 5 M NaCl.
5. Add 600 μl of phenol/chloroform/isoamyl alcohol. Mix by inverting.
6. Centrifuge at top-speed for 3 min.
7. Move the supernatant to a new tube and repeat the steps 5 and 6.
8. Slowly add 1 ml of 100% ethanol to the supernatant and mix by inverting. White DNA clumps will appear.
9. Move the DNA clumps to a new tube by a pipette tip.
10. Add 500 μl of 70% ethanol, vortex, and remove all solutions by pipetting.
11. Dry up and resuspend the DNA clumps with 100 μl of distilled-water.
12. Dilute the DNA solution 1:20–1:100. Typically, 1 μl of the solution is used for subsequent PCR reactions.

3.2. PCR Amplification of Antisense Sequences and Cloning into PTasRNA Vectors

Four PTasRNA vectors (Fig. 2) have different *oris*. Therefore, the plasmid copy number per cell for each vector is different. When the copy number of pHN678 is set to 1, the relative copy number is estimated as following; pHN1009, 2.3; pHN1257, 1.8; and pHN1270, 0.54 (3). Using a high-copy vector typically gives better results (3), but it is important to consider copy number when designing experiments; if moderate silencing is required, using low-copy vectors are useful.

1. Design primers to amplify an antisense fragment of 80–160 bp. Be careful to include RBS and start codon sequences within the fragment and include restriction enzyme recognition sites at both termini. We routinely use NcoI and XhoI which provide efficient cloning. Note that the fragment should be inserted in an inverted orientation relative to P*trc* (see Notes 3 5). An example of a design, where the *acpP* is targeted is shown in Fig. 5.

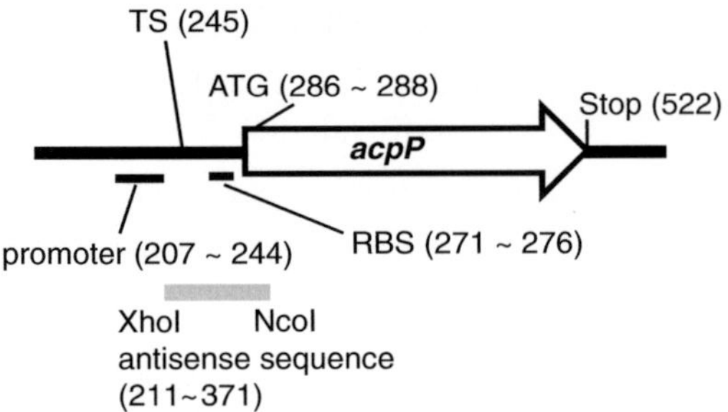

Fig. 5. Structure of the *acpP* gene on a chromosome and an amplified antisense sequence. Positions of genetic elements of *acpP* are shown with their nucleotide numbers. *TS* indicates transcription start site. The *gray bar* is a PCR-amplified fragment to be cloned into the MCS of the PTasRNA vector. In this case, XhoI and NcoI sites were attached to the ends of the fragment.

2. PCR-amplify the fragment from the MG1655 genome.
3. Digest the fragment by restriction enzymes and insert into the MCS of the PTasRNA vector.
4. Transform cells using the resulting PTasRNA vector. If multiple silencing is necessary, mix all vectors beforehand and use the mixture for transformation (see Note 6). Make also transformants with empty PTasRNA vectors to provide negative controls.

3.3. Expression of PTasRNAs

1. Pregrow the transformants overnight.
2. Dilute 1:400 with the medium with or without 1 mM IPTG and grow to logarithmic phase (optical density at 600 nm: OD_{600} = 0.6–0.8) (see Note 7).
3. Harvest cells by centrifugation. The cell pellets may be stored at −80°C until use.
4. If experiments on plates are necessary, prepare 1.8% agarose medium with or without 1 mM IPTG, and streak the transformants (see Note 8).

3.4. Evaluation of Silencing Efficacy by Using Reporter Fusions

For some genes, silencing efficacy can be evaluated by directly measuring enzyme activity of the gene product using the above-mentioned cell pellets. Alternatively, if an antibody is available, Western blotting can be used. If possible, it is helpful to compare the phenotype of silenced strains to that of mutant strains with the same gene disrupted. The same phenotype should appear following PTasRNA induction. However, if these methods are not applicable, as it is the case for essential genes, silencing efficacy need to be evaluated by other methods. Below we describe alternative strategies. In one alternative, a vector containing a "target gene–reporter gene" fusion is constructed, and activity of the reporter gene is measured after silencing. Expression of the fusion gene is driven by the target gene's native promoter (Fig. 4). Here, we describe two examples using Red fluorescent protein (DsRed Express) or LacZ as a reporter.

Note that the LacZ reporter cannot be used in wild-type strains, which have endogenous *lac* operons. Therefore, *lacZ*⁻ strains such as XL1-blue and DH5α need to be used. To our experience, LacZ is a more reliable and sensitive reporter than DeRed Express. DsRed Express is useful because it can be used in any strain and preparation of cell extracts is not necessary. However, we sometimes observed no fluorescence when it is fused to the target genes. The reason is unclear, but the target protein portion may interfere with proper formation of the fluorophore and/or overall protein conformation.

1. Design primers to amplify the open reading frame of the target gene. Be careful to remove the stop codon and adjust the frame to that of the reporter gene (see Notes 9 and 10).
2. Clone the resulting PCR-fragment into the reporter vector (Fig. 4).
3. Co-transform the resulting reporter vector and the PTasRNA vector constructed in Subheading 3.2 (see Note 6). In parallel, establish co-transformants containing the empty PTasRNA vector. Also, if DsRed Express is used, a nontransformant is needed to set fluorescence background.
4. Culture the co-transformants as described in Subheading 3.3 steps 1 and 2.
5. When using DsRed Express, place 50 μl of the culture in a black microplate well.
6. Measure red fluorescence at 532 nm excitation and 580 nm emission wavelengths using the Safire or a similar reader.
7. Measure OD_{600} of the cultures and normalize the fluorescence values with the OD_{600} values. Subtract background fluorescence using a value from a nontransformant.
8. When using LacZ, harvest cells by centrifugation. Cell pellets from 5 ml cultures are sufficient. The cell pellets may be stored at −80°C until use.
9. Resuspend the cell pellets with 500 μl of 1× PBS and disrupt the cells with a beads-beater, by sonication or equivalent methods.
10. Prepare cell extracts by centrifugation at 20,000 ×*g* for 15 min at 4°C and recover supernatant.
11. Determine the protein concentrations in the cell extracts by a Bradford assay.
12. Measure LacZ activity. There are many ways to measure LacZ, but we prefer measurement in a 96-well plate at room temperature. Our reaction mixture is composed of 160 μl of Z buffer, 2 μl of cell extract (typically 0.5–1 μg protein/μl) and 18 μl of 4 mg/ml ONPG. The absorbance is read at 420 nm

using the Safire reader every 30 s, and the slope is used to calculate the specific activity per milligram protein per minute, relative to samples from unsilenced cells. This provides a measure of relative reporter gene expression.

3.5. Evaluation of Silencing Efficacy by Real-Time Quantitative RT-PCR

Measuring the amount of target mRNA by real-time quantitative RT-PCR provides another way to evaluate silencing by PTasRNA constructs. This assay is based on the observation that antisense targeting of mRNA in *E. coli* leads to a decay of the mRNA (see above).

1. Pick a single colony from a freshly streaked LB plate. The plated culture may be stored at 4°C for later use but not longer than 1 week. Resuspend colony in 5 ml MHB and incubate at 37°C, 220 rpm for 16–18 h.
2. Dilute overnight culture 1:10 in MHB and measure OD_{600}. Calculate dilution factor needed to adjust starter culture to ~1×10^6 cfu/ml. We use the formula $OD_{600}/0.003$ = dilution factor for *E. coli* TOP10 and DH5α. Vortex the starter culture well.
3. Prepare a 96-well plate by first adding appropriate amounts of IPTG dilutions to individual wells and adjust the volume in each well to 20 μl. The usual range of IPTG used is 0.1–75 mM; however, this must be optimized for each strain in order to obtain desired levels of growth inhibition or other phenotype changes. Distribute 180 μl of starter culture into each well.
4. Experimental bacterial cultures are grown in a PowerWave X340 spectrophotometer at 37°C with agitation every 5 min in 200 μl volumes in a 96-well plate. Growth is monitored by OD_{600} readings taken every 5 min. Each experimental culture should be grown in triplicate.
5. Harvest cultures at desired phase of growth, e.g., when uninduced (control) culture has increased in OD_{600} by 0.1, by pooling cultures of similar IPTG treatments into one microcentrifuge tube. Spin down cells and use cell pellet immediately or store at −20°C for RNA extraction later. The volume of culture needed to yield a sizeable pellet for RNA extraction depends on the level of growth inhibition. Typically, 8 × 200 μl of uninduced and 24 × 200 μl of highly induced cultures (resulting in very little growth) is sufficient for RNA extraction.
6. Extraction of RNA from bacterial cells, followed by DNase I treatment is carried out using the RiboPure Bacteria Kit according to the manufacturer's protocol (see Note 11).
7. RNA (200–500 ng) is converted to cDNA in a 25 μl reaction consisting of 1× RT reaction buffer, 5.5 mM $MgCl_2$, 0.5 mM of each dNTP, 2.5 μM random hexamers, 0.4 U/μl RNase inhibitor, and 1.25 U/μl MultiScribe Reverse Transcriptase.

8. Relative quantitative PCR is carried out with primers for the target gene of interest and an appropriate reference gene. In our studies of genes essential for growth, we used *rpoA* as the reference gene and primers rpoA-F/R (see Note 12). Each 25 μl of PCR reaction contained 12.5 μl of SYBR Green PCR Mastermix, 100 nM of each primer and 5 μl of cDNA.
9. Target and reference primer pairs should be validated by ensuring that the amplicon is a single product and of the expected size. This can be accomplished by fractionating a small sample of the reaction in a standard agarose gel by electrophoresis.
10. We also optimize PCR reaction efficiency by testing different concentrations of *F* and *R* primers to obtain the lowest C_T value and single dissociation curve peak. Typically, the pairwise concentrations tested are 50, 100, 150 μM of both primers, resulting in 9 *F*/*R* combinations.
11. The optimum *F*/*R* primer concentration determined above is then used to determine primer amplification efficiency. cDNA is serially diluted fivefold at least five times to obtain a standard curve of C_T against log of template quantity, and give an R^2 value indicating linearity and enabling calculation of efficiency using the formula $E = 10^{-1/\text{slope}}$ (8).
12. Quantification of target gene mRNA levels should be normalized against *rpoA* mRNA and calculated relative to the untreated sample. To determine relative qPCR values, the $2^{-\Delta\Delta C_T}$ method can be used for primers with similar amplification efficiencies (9):
    (a) Uninduced (calibrator) sample: Normalize C_T of target gene to C_T of *rpoA* by subtraction: $\Delta C_{T(\text{callibrator})} = C_{T(\text{target})} - C_{T(rpoA)}$.
    (b) Induced (test) sample: Normalize C_T of target gene to C_T of *rpoA* by subtraction: $\Delta C_{T(\text{test})} = C_{T(\text{target})} - C_{T(rpoA)}$.
    (c) Normalize ΔC_T of the test sample to the ΔC_T of the calibrator sample: $\Delta\Delta C_T = \Delta C_{T(\text{test})} - \Delta C_{T(\text{callibrator})}$.
    (d) Calculate the change in target expression relative to *rpoA*: Normalized expression

$$\text{ratio} = 2^{-\Delta\Delta C_T}.$$

13. For primers with different amplification efficiencies, a variation of the above method should be used (10):
    (a). Normalize C_T of target gene in the calibrator to C_T target gene in the test sample by subtraction: $\Delta C_{T(\text{target})} = C_{T(\text{calibrator})} - C_{T(\text{test})}$.
    (b). Normalize C_T of *rpoA* in the calibrator to the C_T of *rpoA* in the test sample by subtraction: $\Delta C_{T(rpoA)} = C_{T(\text{calibrator})} - C_{T(\text{test})}$.

(c). Calculate the change in target expression relative to *rpoA*: Normalized expression $\text{ratio} = \left[(E_{\text{target}})^{\Delta C_{T(\text{target})}}\right] / \left[(E_{rpoA})^{\Delta C_{T(rpoA)}}\right]$, where E_{target} and E_{rpoA} are the primer amplification efficiencies of the target and reference genes determined in steps 9 and 10.

4. Notes

1. M9-glucose media (17 g/L Na_2HPO_4 $12H_2O$, 3 g/L KH_2PO_4, 0.5 g/L NaCl, 1 g/L NH_4Cl, 2 mM MgSO4, 0.1 mM $CaCl_2$, and 1% glucose) can also be used for expressing PTasRNAs because expression from P*trc* is not repressed with glucose.
2. Standard spectrophotometers and fluorometers are useful, or similar 96-well plate spectrophotermeters/fluorometers from other manufactures can be used, provided there is the capacity for incubation, shaking and kinetic analyses or absorbance readings.
3. If one construct does not provide adequate silencing, redesign primers that are shifted by 5–10 bp. The reasons are unclear, but we sometimes experienced great improvement in silencing efficacy by such a sequence shift.
4. If NcoI site is present in the PCR fragment, use compatible restriction enzymes, such as BspHI or PciI. Also, XhoI is compatible with SalI. It is possible to use other restriction enzymes.
5. We usually use fragments of 80–160 bp. This is because fragments less than 80 bp are difficult to handle: Small fragments cannot be purified using commercially available DNA purification kits. Also, we avoid using longer fragments because they may decrease the stability of expressed PTasRNAs.
6. For concurrent transformation of four vectors, competency over about 1×10^6 cfu/μg DNA is required according to our experience. In other words, it is unnecessary to transform vectors sequentially as long as the competency is high. For transformation of two vectors, the competency required is over about 1×10^5 cfu/μg DNA.
7. It is very important at what time both IPTG is added (at the beginning of growth or later) and culturing is stopped. When we evaluate silencing efficacy enzymatically, we always employed this protocol and worked as expected. However, small adjustments should be required depending on the purpose of the experiments or the nature of the target genes.
8. This method is useful when the target gene is essential for cell growth. To silence the gene partially in order to observe

phenotypes associated with intermediate expression levels, vary IPTG concentrations from 0.01 to 1 mM.

9. In some cases, expression of full length target gene is toxic to cells, making it difficult to clone PCR fragments to the reporter vector. In such cases, partial fragments can be used.
10. If pHN678 is used as a PTasRNA vector, it is important to subclone the "target gene–reporter gene" fusion to other vectors to avoid plasmid incompatibility.
11. The RiboPure kit provides relatively high RNA yields from the small volume cultures in 96-well plates. A simple DNaseI inactivation step is also recommended. RNA isolation kits from other manufacturers can be considered provided that they also result in high yields from small volume cultures. RNA integrity, purity, and concentration should be checked by gel electrophoresis and absorbance readings at $OD_{260/280}$.
12. Of the more commonly used reference genes tested (e.g., 16S rRNA, *polA*, and *rpoA*), we found *rpoA* transcript abundance to be the most similar to the genes we were silencing.

Acknowledgments

We thank the members of our research group for their assistance. This work was supported by Kato Memorial Bioscience Foundation.

References

1. Wendland J (2003) PCR-based methods facilitate targeted gene manipulations and cloning procedures. Curr Genet 44:115–123
2. Nakashima N, Tamura T, Good L (2006) Paired termini stabilize antisense RNAs and enhance conditional gene silencing in *Escherichia coli*. Nucleic Acids Res 34:e138
3. Nakashima N, Tamura T (2009) Conditional gene silencing of multiple genes with antisense RNAs and generation of a mutator strain of *Escherichia coli*. Nucleic Acids Res 37:e103
4. Goh S, Boberek JM, Nakashima N, Stach J, Good L (2009) Concurrent growth rate and transcript analyses reveal essential gene stringency in *Escherichia coli*. PLoS One 4:e6061
5. Wagner EGH, Flärdh K (2002) Antisense RNAs everywhere? Trends Genet 18:223–226
6. Dryselius R, Aswasti SK, Rajarao GK, Nielsen PE, Good L (2003) The translation start codon region is sensitive to antisense PNA inhibition in *Escherichia coli*. Oligonucleotides 13:427–433
7. Chen H, Ferbeyre G, Cedergren R (1997) Efficient hammerhead ribozyme and antisense RNA targeting in a slow ribosome *Escherichia coli* mutant. Nat Biotechnol 15:432–435
8. BIO-RAD, Real-Time PCR Applications Guide http://www.gene-quantification.de/real-time-pcr-guide-bio-rad.pdf. Accessed 6 September 2010
9. Livak KJ, Schmittgen TD (2001) Analysis of relative gene expression data using real-time quantitative PCR and the 2(–Delta Delta C(T)) Method. Methods 25:402–408
10. Pfaffl MW (2001) A new mathematical model for relative quantification in real-time RT-PCR. Nucleic Acids Res 29:e45

Chapter 24

Functional Screen of Zebrafish Deubiquitylating Enzymes by Morpholino Knockdown and In Situ Hybridization

William Ka Fai Tse and Yun-Jin Jiang

Abstract

In order to unfold the function of genes, solely performing mRNA over-expression is not enough nowadays. Traditional protein expression experiments, such as Western blotting and immunohistochemical staining, could only provide researchers the changes of expression levels and/or location of their targets. To make a more strong and convincing statement about gene function, it is necessary to perform both "gain-of-function" and "loss-of-function" studies. Both assays can be performed easily by transfecting DNA plasmid and siRNA in cell culture system; while in zebrafish, mRNA and morpholino (MO) microinjection can serve similar purposes. It is common for the zebrafish community to carry out microinjection experiments to explore a gene function. Instead of making a single knockdown/over-expression of a gene, we foresee that more and more large-scale screens on certain protein families will be performed in the future. Here, based on our previous experience in zebrafish "loss-of-function" screening on deubiquitylating enzymes, we describe a general work flow, from morpholino designation, in situ hybridization, to data analysis, as a reference for researchers who may be interested in a similar screen.

Key words: Deubiquitylating enzymes, In situ hybridization, "Loss-of-function" screening, Morpholino, Zebrafish

1. Introduction

Deubiquitylating/deubiquitinating enzymes (DUBs) are involved in numerous biological functions and act as key regulators in different cellular processes (1). They belong to either cysteine or metallo-proteases, which are able to remove ubiquitin (UBQ) from ubiquitin-conjugated proteins. The human genome contains at least 95 DUBs in five major classes (2), while in zebrafish, about 91 DUBs in six major classes are described (3). Over the past 15 years, studies on ubiquitylation have been performed extensively (4, 5); however,

Michael Kaufmann and Claudia Klinger (eds.), *Functional Genomics: Methods and Protocols*,
Methods in Molecular Biology, vol. 815, DOI 10.1007/978-1-61779-424-7_24, © Springer Science+Business Media, LLC 2012

the understanding on how DUB regulates the deubiquitylation process and its cellular function is still in its infant stage (6). We have performed a genome-wide loss-of-function study in zebrafish that forms a basic frame of DUB functions in early zebrafish embryogenesis. Although further studies are required to generate a more complete picture, it provides the first large scale high-throughput knockdown screen in zebrafish. To begin with, we have chosen our target on neuronal development. By in situ hybridization screening, we have classified zebrafish DUBs into five groups based on the neuronal patterning of their morphants and have identified our targets of interest for further studies.

2. Materials

2.1. Microinjection and Zebrafish Maintenance

1. Zebrafish (*Danio rerio*).
2. Egg medium (E3 medium): 5 mM NaCl, 0.17 mM KCl, 0.33 mM $MgSO_4$, 0.33 mM $CaCl_2$.
3. Phosphate-buffered saline, PBS.
4. Mineral oil.
5. Injection materials (MOs from GeneTools).
6. Injection dishes: E3 medium/1% agarose dishes with furrows.

2.2. In Situ Hybridization

1. 0.008% Triciane (w/v): dissolve tricane in deionized water. The solution can be stored at 4°C for months.
2. Hank's saline: 137 mM NaCl, 5.4 mM KCl, 0.25 mM Na_2HPO_4, 0.44 mM KH_2PO_4, 1.3 mM $CaCl_2$, 1 mM $MgSO_4$, and 4.2 mM $NaHCO_3$.
3. 0.003% 1-phenyl-2-thiourea (PTU) (w/v): dissolve PTU in 10% Hank's saline. It remains stable at room temperature for 1 month.
4. PBS with Tween 20 (PBST): 0.1% (w/v) Tween 20 in 1× PBS.
5. 4% PFA (w/v): dissolve paraformaldehyde in 1× PBS solution. Avoid skin/eyes contact and inhalation, since it is toxic.
6. Pronase.
7. Methanol.
8. Proteinase K.
9. 20× SSC.
10. Blocking buffer: 5% goat serum in PBST.
11. RNA probes (prepared by Roche DIG RNA Labeling kit).
12. HYB+: 50% formamide, 5× SSC, 0.1% Tween20, 1 mg/ml yeast torula RNA, 50 μg/ml heparin, adjust pH to 6.0 by 1 M citric acid. Store it at −20°C. For HYB−, leave out heparin.

13. Anti-digoxigenin-AP Fab antibody fragments.
14. AP buffer: 100 mM Tris-HCl (pH 9.5), 50 mM $MgCl_2$, 100 mM NaCl, 0.1% Tween20, 1 mM Levamisol. Store it at 4°C.
15. Staining buffer: 0.5% NBT (nitroblue tetrazolium), 0.375% BCIP (5-bromo-4-chloro-3-indolyl phosphate) in AP Buffer. It should be freshly prepared and kept in dark.

3. Methods

This section describes detailed procedures from morpholino design, in situ hybridization, to data analysis. Researchers are advised to modify the protocol to fit their own experiments. The following procedures provide a general work flow of performing the "loss-of-function" screenings using zebrafish embryos.

3.1. Morpholino Sequence Site Selection and Design

Basically, an morpholino (MO) is a synthetic oligonucleotide of about 25 subunits that contains a morpholine ring instead of a ribose ring. The idea of MO is to introduce an antisense RNA into a cell to inhibit the translation or splicing of its endogenous mRNA. To summarize in brief, there are three kinds of MOs that are generally used for knockdown purposes in zebrafish embryos. They are translation-blocking MOs, which target the ATG translation starting site; 5′-UTR MOs that target the 5′ untranslated region; and splice-inhibiting MOs that target junctions of exons and introns. Genomic information is essentially required to decide a good MO sequence (see Note 1). In addition, designing MOs should follow three criteria: (a) about 40–60% GC content, but less than 37% G content; (b) without any consecutive tri- or tetra-G nucleotide sequences; and (c) minimizing self-pair sequence homology.

1. There are different web-based genome information resources that can help researchers to get the protein/gene sequences for the targeted gene. Identification of translation initiation site (TIS) could be done by using BLAST (http://blast.ncbi.nlm.nih.gov/Blast.cgi) or ENSEMBL (http://www.ensembl.org/index.html). Researchers could delineate their own MOs with the previous criteria stated or allow the company GeneTools to design them instead. GeneTools (http://www.gene-tools.com/) is the company that makes MOs and provides free MO design service. Researchers only need to provide the gene accession number (or input sequence) for the designation. Although the company cannot guarantee the MO effectiveness, we think that it provides an alternative option for designing MO sequences.

2. It is a common practice that you need to have at least two MOs targeting different sites in one gene to show that the knockdown is specific. We recommend generating the first MO that targets on the ATG site; while the second one on the 5′-UTR. Alternatively, making a splicing MO against an intron/exon junction that interferes with the splicing process during maturation of RNA might have the advantage to quantify the efficacy of the MO by normal RT-PCR reaction (If knockdown is successful, two bands could be seen. The efficacy could be determined by quantifying the ratio between splicing-perturbed and -unperturbed PCR products). In addition, a negative control by using a mismatch MO is a general practice in MO knockdown experiments. In the case of zebrafish, making use of internet resources such as Zfin (http://zfin.org/cgi-bin/webdriver?-MIval=aa-ZDB_home.apg) might help researchers to find the published MO sequences for particular genes.

3.2. Zebrafish Mating Setup and Needle Pulling

1. On the evening before the day for microinjection, zebrafish were set up in pairs or groups (consisting of two males and three females) with dividers for crossing purpose (see Note 2).
2. Pull a 0.58-mm ID, 1.0 mm OD glass capillary into two needles by the micropipette puller (Sutter Instrument: P97 Flaming/ Brown Micropipette Puller). We currently use the following parameters: heat = 600, pull = 45, velocity = 75, and time = 90. Pulled needles can be kept in a 150-mm Petri dish with a line of putty ramp on top.

3.3. Loading Injection Material (Morpholino) into the Needle

1. Dissolve the MO in distilled water to make a 5-mM stock and store it at –20°C. Optimize the MO concentration by testing at 0.1/0.5/2 mM final concentrations.
2. Set up the microinjection machine and optimize the injected volume. Adjust the needle opening by using sharp forceps. At least 1.5 μl injection material (mRNA or/and MO) is recommended to be loaded into the needle (see Note 3). Afterward, examine the injection volume under a microscope. The injection volume should be less than 10% of embryo volume (normally 1 nl). The volume can be estimated by using a micrometer. Put a glass slide on top of the micrometer and add a drop of mineral oil on it. Perform a single injection and examine the size of the injected droplet. A diameter of 0.2 mm roughly corresponds to 1 nl. Adjust the volume by either changing the size of needle opening or the injection pressure/time (see Note 4).

3.4. Zebrafish Morpholino Knockdown Injection

1. In the morning, reduce the water amount to about 2 cm above the bottom of the mating cage, then remove the divider for the fish to spawn.
2. Collect the embryos and begin the MO knockdown experiments. More than 100 embryos could be collected from two

pairs of fish. Save 20–30 embryos in the 100 mm Petri dish with egg medium as a control. Transfer the remaining embryos to two injection dishes by using a plastic transfer pipette.

3. Inject embryos at one- to two-cell stage. Lower the needle. Pierce the chorion surface and enter the yolk with one smooth stroke. A larger magnification (1.6×) would work well for that. One dish is for high concentration; while the other is for low concentration (see Note 5). Rinse off the injected embryos to a new 100 mm Petri dish with egg water. Store it in a 28°C incubator (see Note 6).
4. Repeat the same procedure 1–3 for four trials per gene (four tanks).
5. Remove the dead embryos in the evening.

3.5. Embryo Collection, Chorion Removal, and Fixation

1. Collect embryos from different developmental stages. For a typical screening process, we suggest to collect embryos at several critical developmental stages, which are 60–80% epiboly, 12-somite, and prim 5 stages. Researchers can decide different stages depending on their purposes. For example, if researchers are interested in liver development, they need to collect the embryos at later stages when the liver has developed (see Note 7).
2. Collect around 30 embryos by a plastic transfer pipette from each plate and process to chorion removal step. Transfer embryos to a 15-ml plastic tube with egg medium. Add 2.5 mg/ml pronase to the tube for 10 min with general shaking (see Note 8). Remove all the solution and add PBST for washing after 10 min. Repeat the PBST washing step for 2–3 times, every 3–5 min. Transfer embryos to an Eppendorf tube and add 1 ml 4% PFA (see Note 9). After adding PFA, fix overnight with general shaking (see Note 10) at 4°C.

3.6. Dehydration

1. Remove PFA from the Eppendorf tube and add back 1 ml 30% methanol in PBS for 5 min. Repeat with increasing concentration of 50%, 70%, and finally 100% methanol, each for 5 min. Once transferred to 100% methanol, embryos can be stored at –20°C (see Note 11).

3.7. In Situ Hybridization (see Note 12 and Table 1; Rehydration and Proteinase K Digestion)

1. Rehydrate embryos from 100% methanol to 70, 50, and 30% methanol in PBS, each for 5 min. Remove 30% methanol and add PBST. Repeat the PBST washing step for four times, each for 5 min.
2. Add 1 ml PBST with 5 μg/μl proteinase K (see Note 13) for various time periods depending on the developmental stage of embryos (Table 2).
3. Remove the proteinase K/PBST and replaced with PBST. Wash embryos with PBST for two times, 5 min each.

Table 1
In situ hybridization protocol for experienced users

Reagent	Time period
70% Methanol/30% PBST (v/v)	5 min
50% Methanol/50% PBST (v/v)	5 min
30% Methanol/70% PBST (v/v)	5 min
PBST	5 min × four times
Proteinase K	Refer to Table 2
PBST	QR, 5 min × two times
4% PFA (w/v)	20 min
PBST	QR, 5 min × five times
HYB+	1 h (65°C)
RNA probe	Overnight (65°C)
HYB−	QR, 5 min (65°C)
66% HYB−/33% 2× SSC	10 min (65°C)
33% HYB−/66% 2× SSC	10 min (65°C)
2× SSC	10 min (65°C)
0.2× SSC	30 min × two times (65°C)
66% 0.2× SSC/33% PBST	10 min
33% 0.2× SSC/66% PBST	10 min
PBST	10 min
Blocking buffer	1 h
Anti-DIG antibody (1:5,000)	Overnight (4°C)
PBST	QR, 15 min × four times
AP buffer	15 min × two times

QR quick rinse

4. Remove the PBST and add 4% PFA for 20 min. Wash embryos with PBST for five times, each for 5 min (see Note 14).

3.8. In Situ Hybridization (Prehybridization and Hybridization)

1. Remove PBST completely and replaced with 300 μl chilled HYB+.
2. Incubate the Eppendorf tube at 65°C for 1 h in a water bath (see Note 15).

Table 2
Suggested time for proteinase K treatment

Developmental stage	Proteinase K treatment (5 μg/ml)
<10 somite	No digestion
10 somite	1 min
20 somite	5 min
24 hpf	10 min
36 hpf	15 min
48 hpf	40 min
3 days	1 h
4 days	1 h 15 min
5 days	1 h 45 min

3. Transfer at least ten embryos into a new Eppendorf tube for each probe (if 30 embryos, then you could perform three probes tests, see Note 16).
4. Add 80 μl probe to the Eppendorf and incubate it at 65°C overnight. We suggest to use a water bath for the incubation.

3.9. In Situ Hybridization (Posthybridization)

1. Remove the probe and save it for future usage. Quickly rinse and wash embryos with preheated HYB− for 5 min and then replace the solution with preheated 66% HYB−/33% 2× SSC for 10 min at 65°C.
2. Exchange the solution with preheated 33% HYB−/66% 2× SSC for 10 min followed by 10 min 2× SSC wash. All at 65°C.
3. Remove the solution and change to preheated 0.2× SSC. Wash embryos for two times and each for 30 min at 65°C.
4. Remove the solution and replace it with 66% 0.2× SSC/33% PBST, followed by 33% 0.2× SSC/33% PBST, and finally 100% PBST. All for 10 min and at room temperature.

3.10. In Situ Hybridization (Color Development and Long-Term Storage)

1. Add 1 ml blocking buffer for 1 h.
2. Remove the buffer and add 1 ml 1:5,000 anti-DIG at 4°C overnight.
3. Discard the buffer and rinse embryos with PBST. Wash embryos with PBST for four times, each for 15 min.
4. Wash embryos with AP buffer for two times, each for 15 min.
5. Remove the AP buffer, change with staining buffer and then incubate embryos in dark (covered by aluminum foil). Constant checking is suggested to prevent overstaining (see Note 17).

6. Once the pattern appears, remove the solution and rinse embryos with PBST.
7. Add 4% PFA to stop reaction for at least 20 min. Store embryos at 4°C. For long-term storage, remove PFA and wash embryos with PBST for two times, each for 5 min. Add 50% glycerol and store embryos at 4°C in dark.

3.11. Results Anticipation

In situ hybridization screening based on the strength of the staining might give you false indication. It is because the "darker" staining might be due to many reasons, from the background staining to personal subjective feelings. To present a more solid screening, it is recommended to base it on the size and pattern of the stains. It is necessary to have a general hypothesis before starting the screen. Researchers should know their objectives and targets. If researches are interested in small molecules toxicity tests, they should pay more attention to the developmental defects in early stages; while those who are interested in the specific organ development should choose several specific and clear in situ molecular markers that target on their organs. As other screening methods, it is expected that "false positives" will be generated through the screen. However, researches should bear in mind that the "false positive" might be due to different outcomes as well. For example, dosage factor will be important if performing a chemical screen, while nonspecific phenotypes generated by MOs can be observed sometimes. Overall, fruitful screening is based on laboratory experience and a careful set-up before the screen. It is important to have a scoring system for every screen, simply from yes or no, to more advanced scaling like degree of changes, will help researchers to extract a general idea from the screening results.

Here, we use our previous published data as an example to explain briefly how we group our screening from knocking down over 80 genes in zebrafish. Initially, we have focused our target on neural development. Based on that, we chose *huC* as our primary marker because of its strong and easily recognizable staining patterns in early zebrafish development. The distinctive distribution patterns of *huC* allowed us to classify the DUB family into five groups. The classification was based on the strength and the patterns of *huC* in morphants: group I, increase *huC* expression; group II, decreased *huC* expression without changed patterning; group III, decreased *huC* expression with slightly destructive patterning; group IV, decreased *huC* expression together with severely destructed patterning; and group V, no change in both expression and patterning. After the first grouping, researchers could decide to focus on a particular group. In our case, we examined the dorsoventral development of the group IV morphants. Different common in situ hybridization molecular markers had been chosen and tested: ventral markers (*bmp4*, *eve1*, and *gata2*), dorsal markers (*chd*, *gsc*) at 50–60% epiboly; *myoD*, *gata1*, and *pax2a* for 10-somite embryos. From the in situ hybridization results, we found that group IV DUB morphants are dorsalized. To go further, we checked the potential pathway that affect the dorsoventral patterning and performed follow-up experiments. Finally, we proposed that group IV DUBs play roles in the BMP pathway. Details can be found in ref. 2 (see Note 18).

4. Notes

1. Practical problems have been raised from extensive usage of MO. The most common problem is mis-targeting. Thus proper controls are necessary to make good use of the results generated by MOs. Details can be found in ref. 7. If a published MO does not give rise to the expected phenotypes as the paper described, it may be due to the genetic polymorphism found in the target sequence between published genomic data and individual fish.
2. In order to maximize the embryo productivity, we recommend keeping the male and female fish separate.
3. Unload the injection material at the lower part of the needle to reduce the chance of bubble trapping. We use gel loading tips (Eppendorf) for this.
4. Consistency of the injection volume is very critical. Researchers should always pay attention to the injected volume. Be careful not to make a "short" needle as it would damage the embryo and cause lethality. Injection pressure should be less than 20 psi. with a microinjector pressure at around 7 psi, while the injection time should be 0.1 s.
5. This step is only necessary for optimizing the MO concentration for the first trials.
6. Temperature will affect the growth rate. Lower temperature will result in developmental delay.
7. Zebrafish will develop pigment after 24 h postfertilization (hpf). 1× PTU should be added at around tail bud stage to stop pigmentation. Pigmentation will affect in situ hybridization staining outcomes. Embryos older than 16 hpf should be anesthetized by Triciane (Sigma) before fixation to prevent curly tail.
8. There are two ways to remove chorion: manually by a pair of forceps or by pronase treatment. If researchers only handle a small amount of embryos, we suggest using the manual method. Transfer embryos to a new 100 mm Petri dish, half-filled with PBST. Then remove chorion by using a pair of

forceps under a microscope. Once finished the whole process, transfer all embryos to an Eppendorf tube and add 1 ml 4% PFA in it. Young embryos (<prim-5 stage) can go through the fixation process first, and then remove chorion. Chorion can be easily removed after fixation.

9. Fresh PFA always give better results. Do not leave the embryos in PFA for more than 2 days otherwise it will compromise the ISH quality.
10. In this in situ hybridization protocol, most of the processes require shaking. Unless specified, shaking is needed, which means to put the Eppendorf tube on the nutator at a general speed of 40 rpm.
11. Embryos should be treated with 100% methanol for at least 2 h at −20°C.
12. A quick protocol for experienced researchers can be found in Table 1.
13. Beware of excessive proteinase K digestion. Advised time is listed in Table 2 for different developmental stages. Over-digestion will make the embryos fragile and cause progressive damage when going through the in situ hybridization steps. It should be noted that if the probe is expressed in internal organs, such as liver, heart, pancreas, etc., the proteinase K treatment can be longer than usual (plus 30 min, for example, digestion time for 3-dpf embryos can be increased to 90 min). Alternatively, researchers could increase the concentration of proteinase K to reduce digestion time.
14. Be sure that all PFA had been removed.
15. Prolonged washing of HYB+ (>2 h) will decrease the ISH staining.
16. For screening purpose, it is recommended to pool the embryos and undergo the ISH (collect about ten embryos from each plate, if you have four plates, total embryos tested will be forty, which is good enough to provide you with first preliminary data). The advantage of pooling up the sample is to ensure you can collect samples at different developmental stages from the same batch.
17. Once the expected pattern appears, wash the embryos with PBST immediately and fix them. The time required for visualization varies with different probes. Once the signal is developed, the stain will get dark in a short time. It is advised that researchers should check their expected patterns from online resources, such as Zfin (http://zfin.org/cgi-bin/webdriver?MIval=aa-ZDB_home.apg).
18. Details of the study example could be found in the cited reference (2). The general suggestion from authors to perform a

screen is to do it step by step, one by one with clear objectives and targets. Then, the whole process is simple, direct, and easily understandable.

References

1. Nijman SM, Luna-Vargas MP, Velds A et al (2005) A genomic and functional inventory of deubiquitinating enzymes. Cell 123: 773–786
2. Tse WKF, Eisenhaber B, Ho SHK et al (2009) Genome-wide loss-of-function analysis of deubiquitylating enzymes for zebrafish development. BMC Genomics 10: 637
3. Balakirev MY, Tcherniuk SO, Jaquinod M et al (2003) Otubains: a new family of cysteine proteases in the ubiquitin pathway. EMBO Rep 4: 517–522
4. Hershko A, and Ciechanover A (1998) The ubiquitin system. Annu Rev Biochem 67: 425–479
5. Pickart CM, and Eddins MJ. (2004) Ubiquitin: structures, functions, mechanisms. Biochim Biophys Acta 1695: 55–72
6. Ventii KH, and Wilkinson KD. (2008) Protein partners of deubiquitinating enzymes. Biochem J 414: 161–175
7. Judith SE, and James CS. (2008) Controlling morpholino experiments: don't stop making antisense. Development 135: 1735–1743

Chapter 25

Silencing of Gene Expression by Gymnotic Delivery of Antisense Oligonucleotides

Harris S. Soifer, Troels Koch, Johnathan Lai, Bo Hansen, Anja Hoeg, Henrik Oerum, and C.A. Stein

Abstract

Antisense oligodeoxyribonucleotides have been used for decades to achieve sequence-specific silencing of gene expression. However, all early generation oligonucleotides (e.g., those with no other modifications than the phosphorothioate backbone) are inactive in vitro unless administered using a delivery vehicle. These delivery vehicles are usually lipidic but can also be polyamines or some other particulate reagent. We have found that by employing locked nucleic acid (LNA) phosphorothioate gap-mer nucleic acids of 16 mer or less in length, and by carefully controlling the plating conditions of the target cells and duration of the experiment, sequence-specific gene silencing can be achieved at low micromolar concentrations in vitro in the absence of any delivery vehicle. This process of naked oligonucleotide delivery to achieve gene silencing in vivo, which we have termed gymnosis, has been observed in many both adherent and nonadherent cell lines against several different targets genes.

Key words: Gymnosis, Antisense, Silencing, Phosphorothioate, Locked nucleic acid, gap-mer

1. Introduction

For 20 years, it was universally accepted that to achieve gene silencing by oligonucleotides (oligos) to any significant extent in vitro, the oligos had to be transfected into cells, usually by a lipidic or particulate transfection reagent (1). The requirement for oligo transfection in vitro and the corresponding need for a delivery vehicle in vivo creates a series of formidable barriers that have impeded the development of oligo therapeutics: (1) Transfection of some cell types, such as suspension cells and primary cells, is very inefficient; (2) The transfection reagents (e.g., cationic lipids, peptides, and

Michael Kaufmann and Claudia Klinger (eds.), *Functional Genomics: Methods and Protocols*,
Methods in Molecular Biology, vol. 815, DOI 10.1007/978-1-61779-424-7_25, © Springer Science+Business Media, LLC 2012

virus envelope proteins) themselves produce cytotoxicity through interactions with mitochondrial membranes and nonspecific activation of intracellular pathways (2); (3) In addition to systemic and intracellular toxicity, delivery vehicles add cost and regulatory complexity to any potential oligonucleotide therapeutic agent; (4) There is little or no correlation between gene silencing in vitro achieved by lipid/particulate transfection and gene silencing in vivo. This is particularly true when oligos are administered as saline formulations in the absence of a delivery agent; and (5) The in vivo results obtained with particulate formulations are often poorly reproducible and rarely successful in different target organs. Some combination of these barriers exists with all current delivery vehicles, whether lipidic or not. In sum, encapsulation of oligonucleotides does not solve the delivery problem and also interferes with our understanding of the mechanism of antisense oligo activity in vivo. Our discovery of gymnosis, in which the delivery of "naked" LNA phosphorothioate gap-mer oligos to cells can achieve robust gene silencing activity, provides a way around these hitherto intractable problems (3).

Gymnosis represents a far-reaching, paradigm shift in oligo delivery. First, oligos delivered by gymnosis are predominantly found in the cytoplasm, which challenges the long-held notion that endogenous activity only takes place when the oligonucleotides are concentrated in the nucleus (4, 5). Second, we observe that the oligos are concentrated in vacuolar structures in the cytoplasm, including endosomes, but oligos are also found dispersed throughout the cytoplasm (6, 7). Although it is unclear whether endosomal localization is just part of the oligo uptake mechanism, or whether endosomal localization is also important for gene silencing activity (6, 7), gymnosis provides a new way to address the old question on how oligonucleotides are taken up by cells and delivered to cellular locations to function as gene silencing molecules.

Phosphorthioate-LNA gap-mers (PS-LNA gap-mer), which are highly gymnotically active, permit the oligonucleotide "length penalty" to be observed. These PS-LNA gap-mer molecules are stable against nucleases and the high binding affinity provided by the LNA nucleotides forms for the basis for reducing oligo length while retaining knockdown potency. Whereas oligos of 18–20 nts were once considered the optimum length, it is now clear from many in vitro experiments that the optimum LNA gap-mer length under gymnotic delivery is 12–16 mer (8). Indeed, short LNA antisense oligos administered as saline formulations in vivo can be more potent than longer oligos (8).

Gymnotic delivery can define the optimum oligo chemistry for in vivo applications, as in vitro gymnosis appears to correlate with productive in vivo gene silencing. In the absence of transfection reagents, gymnotic delivery exposes cells in culture to a constant concentration of oligo, with concentrations of oligo (1–10 μM) that are similar to the concentrations found when plasma and tissues are exposed to saline-formulated oligos that are dosed in the single digit milli-/kilogram range. In contrast, particulate-mediated transfection creates an artificially high intracellular concentration of oligo that might *only* silence target gene expression in vitro. Such high intracellular oligo concentrations, however, are hard to achieve uniformly in tissues in vivo even with an optimized delivery vehicle. In addition, all transfection-reagent mediated nonspecific effects that may be confused with target-specific gene silencing are eliminated by gymnosis. Thus, gymnotic delivery offers the possibility of reducing the number of "false-positive" oligos that demonstrate silencing in vitro, but fail to reduce target gene expression in vivo. In conclusion, gymnosis strengthens the translational relationship in RNAi drug discovery, since gymnotic delivery of oligos in vitro parallels the delivery of "naked," saline-formulated oligos in vivo, and thereby forms the basis of better understanding the pharmaceutical potential of antisense gene silencing.

Gymnosis is effective in a large number of cell types: cell cultures (adherent/nonadherent), primary cells, and dividing and nondividing cells. Robust gene-silencing activity has been observed against many mRNA targets at concentrations in the low micromolar range (3). A successful gymnotically delivered oligo requires extensive modification of the oligonucleotide backbone. Gymnosis is not effective when all-phosphodiester oligos or oligos that lack nuclease-resistant 3′ and 5′ chemical modifications are employed. Optimum results have been obtained with locked nucleic acid (LNA) phosphorothioate (PS) gap-mers (9), which are highly gymnotically active. Presented herein is the process of gymnosis in both adherent (HeLa and 518A2 melanoma) and nonadherent (Namalwa B-cells). In addition, we have provided detailed protocols for Western blot and quantitative PCR (qPCR) analysis following gymnosis to measure down-regulation of target gene expression.

2. Materials

2.1. Cell Culture for Gymnosis in Adherent Cells

1. Human cervical carcinoma HeLa cells (American Type Culture Collection # CCL-2, Manasas, VA, USA). The 518A2 mycoplasma-free human melanoma cell line was a kind gift of Dr. Voker Wachek (University of Vienna, Austria).
2. Basal Medium for HeLa and 518A2 cells: Dulbecco's modified Eagle's medium (DMEM). Store under sterile conditions at 4°C until expired (see Note 1).
3. Basal medium supplement for HeLa and 518A2 cells: 10% (*v*/*v*) fetal bovine serum (FBS), heat inactivated for 30 min at 56°C.

4. Basal medium supplement for HeLa and 518A2 cells: 100 U/ml penicillin G sodium; 100 mg/ml streptomycin sulfate.
5. Basal medium supplement for HeLa and 518A2 cells: 2 mM L-glutamine.
6. Solution of trypsin (0.25%) and ethylenediaminetetraacetic acid (EDTA) (1 mM).
7. 1× phosphate-buffered saline (PBS), sterile. Store under sterile conditions at 4°C until expired.

2.2. Cell Culture for Gymnosis in Nonadherent Cells Namalwa B-Cells

1. The Burkitt's lymphoma cell line Namalwa (American Type Culture Collection #CRL-1432, Manasas, VA, USA).
2. Basal medium for Namalwa cells: RPMI 1640.
3. Basal medium supplement for Namalwa cells: 10% (*v/v*) FBS, heat-inactivated for 30 min. at 56°C (Biochrom).
4. Basal medium supplement for Namalwa cells: 1 mM sodium pyruvate.
5. Basal medium supplement for Namalwa cells: 10 mM Hepes.
6. Basal medium supplement for Namalwa cells: 25 μg/ml gentamicin.

2.3. Gymnosis of Adherent Cells and Nonadherent Cells Using PS-Modified gap-mer Oligonucleotides

1. Distilled, nuclease-free water. Keep sterile and store at room temperature.
2. 10 mM Tris–HCl, pH 7.4, sterile. Prepare under sterile conditions from 1 M Tris–HCl stock using nuclease free water. Sterile filter using a 0.2-μm filter and store at room temperature.
3. Oligonucleotides: Lyophilized PS-LNA gap-mers were resuspended in 10 mM Tris–HCl, pH 7.4 and their concentration in solution determined by UV spectrophotometry. The PS-LNA gap-mer SPC 2996 targets nucleotides 1–16 of the Bcl-2 mRNA. The PS-LNA gap-mer SPC 3046 is a scrambled control oligo that lacks substantial homology with the human genome (Fig. 1) (see Note 2).

2.4. Western Blot Analysis to Measure Target Down-Regulation

1. Cell lysis buffer: 50 mM Tris–HCl, pH 7.4, 1% (*v/v*) Igepal CA-630 (substitute for Nonident P40, Sigma-Aldrich), 0.25% (*w/v*) sodium deoxycholate, 150 mM sodium chloride, 1 mM ethylene-bis(oxyethylenenitrilo) tetraacetic acid (EGTA), 1 mM sodium orthovanadate (Na_3VO_4), 1 mM sodium fluoride (NaF), 1% (*v/v*) Protease Inhibitor Cocktail (Sigma). Prepare fresh just prior to use (see Note 3).
2. Teflon cell scrapers (Fisher).
3. Liqui-gel acrylamide solution (MP Biomedicals): 40% solution of acrylamide:bisacrylamide (29:1). Store at 4°C until expiration date.

SPC2996: 5′-$^{m}\mathbf{C}_s\mathbf{T}_s c_s c_s c_s a_s a_s c_s g_s t_s g_s c_s g_s {}^{m}\mathbf{C}_s{}^{m}\mathbf{C}_s a$-3′

SPC3046: 5′-$^{m}\mathbf{C}_s\mathbf{G}_s{}^{m}\mathbf{C}_s\mathbf{A}_s g_s a_s t_s t_s a_s t_s a_s a_s \mathbf{A}_s{}^{m}\mathbf{C}_s{}^{m}\mathbf{C}_s t$-3′

Fig. 1. Diagram of PS-LNA gap-mers SPC2996 and SPC3046 (scrambled control). LNA-modified riboses are in *bold capital letters*, whereas *small letters* indicate deoxyriboses. *s* phosphorothioate, *m* C5-methylcytosine.

4. ProtoGel Stacking Buffer (National Diagnostics): 1× solution of 0.5 M Tris–HCl, pH 6.8, 0.4% (*w*/*v*) sodium dodecyl sulfate (SDS). Store at room temperature until expiration date.
5. ProtoGel Resolving Buffer (National Diagnostics): 4× solution of 1.5 M Tris–HCl, pH 8.8, 0.4% (*w*/*v*) SDS. Store at room temperature until expiration date.
6. Bovine serum albumin (BSA), electrophoresis grade.
7. 10% ammonium persulfate, molecular biology grade: Dissolve 1 g of ammonium persulfate in 8 ml of autoclaved dH_2O, dissolve by mixing and heat, and then add autoclaved dH_2O to 10 ml. Filter through a 0.2-μm filter and store at 4°C for 1 week.
8. *N*,*N*,*N′*,*N′*-tetramethylethylenediamine (TEMED), molecular biology grade.
9. 10× TGS running buffer: 250 mM Tris–HCl, pH 8.3, 1.92 M glycine, 1% SDS. To prepare 1× running buffer, add 100 ml of 10× TGS running buffer to 900 ml of autoclaved dH_2O and mix well by inverting. Prepare fresh prior to use.
10. 10× SDS–PAGE transfer buffer: Add the following to 1.5 L of autoclaved dH_2O: 60.52 g Tris, 300 g glycine, 8 g SDS. Mix well by stirring until completely dissolved. Do not adjust the pH. Add autoclaved dH_2O to 2 L and store at room temperature for up to 6 months. To prepare 1× SDS–PAGE transfer buffer, add 100 ml of 10× transfer buffer to 900 ml of autoclaved dH_2O and mix well by inverting. Prepare prior to use and store at 4°C.
11. 5% Stacking gel: 1.5 ml of 1× ProtoGel Stacking Buffer, 0.75 ml of 40% Liqui-gel 29:1, 3.7 ml of autoclaved dH_2O, 60 μl of 10% ammonium persulfate (added as a catalyst for polymerization), 6 μl of TEMED (added last as a catalyst for polymerization). Prepare fresh for each use.
12. 10% Resolving gel: 2.5 ml of 4× ProtoGel Resolving Buffer, 2.5 ml of 40% Liqui-gel 29:1, 3.7 ml of autoclaved dH_2O, 0.1% (*w*/*v*) ammonium persulfate (added as a catalyst for polymerization), 100 μl of 10% ammonium persulfate (added as a catalyst for polymerization), 10 μl of TEMED (added as a catalyst for polymerization). Prepare fresh for each use.
13. 5× SDS–PAGE loading buffer: 250 mM Tris–HCl, pH 6.8, 10% (*w*/*v*) SDS, 0.5% (*w*/*v*) bromophenol blue, 500 mM DL-dithiothreitol, and 50% (*v*/*v*) glycerol.

14. Immobilin-P PVDF membrane, 0.45 μm.
15. 1× PBS + Tween-20 (PBS-T): add 2.5 ml Tween-20 to 500 ml of 1× PBS. Store at room temperature for up to 3 months.
16. 10× Tris-buffered saline (TBS): 200 ml of 1 M Tris–HCl, pH 8.0, 600 ml of 5 M NaCl, 1,200 ml of autoclaved dH_2O. Mix well and store at room temperature for up to 1 year.
17. 1× Tris-buffered saline + Tween-20 (TBS-T): add 50 ml of 10× TBS to 450 ml of autoclaved dH_2O. After mixing, add 500 μl of Tween-20 and mix well using a stir bar. Store at room temperature for up to 3 months.
18. Mouse monoclonal Bcl-2 primary antibody (Dako): Use at 1:400 dilution in PBS-T.
19. Mouse monoclonal α-tubulin primary antibody (Sigma-Aldrich) Use at 1:5,000 dilution in TBS-T.
20. Secondary anti-mouse, HRP-labeled antibodies: Sheep anti-mouse secondary antibody (GE Healthcare). Use at 1:3,000 dilution in either PBS-T (to detect Bcl-2) or TBS-T (to detect α-tubulin). The secondary antibody is fused with horseradish peroxidase (HRP) to facilitate detection of the proteins by enhanced chemiluminescence.
21. Enhanced chemiluminescence kit (GE Healthcare).

2.5. Quantitative PCR Analysis to Measure Target Down-Regulation

1. RNeasy Mini kit.
2. Beta-mercaptoethanol.
3. Microtiter plate for cDNA synthesis. Thermo-Fast 96, semi-skirted (ABgene).
4. RNase and DNase free water.
5. Random decamer primer.
6. dATP, dCTP, dGTP, and dTTP (100 mM).
7. MMLV-RT reverse transcriptase.
8. RNase inhibitor.
9. Microtiter plate for qPCR. MicroAmp Fast Optical 96-well plate (Applied Biosystems).
10. Taqman Fast Universal PCR Master Mix (2×) (Applied Biosystems).
11. Taqman 20× primer-probe mix: Bcl-2 Fast Taqman assay (Applied Biosystems). GAPDH Fast Taqman assay (Applied Biosystems).
12. Applied Biosystems 7500 Real-Time PCR instrument (see Note 3).

3. Methods

For adherent cells, the most critical factor influencing maximum gene silencing by gymnosis is the cell culture density: The cells must never become confluent during the entire course of the experiment. Once the cells become confluent, their rate of adsorptive endocytosis slows, and it is absorptive endocytosis that most likely is responsible for the majority of the oligo that is internalized in the cell and subsequently found in endosomes. For nonadherent cells, on the other hand, the strict requirement for nonconfluency is not relevant. However, nonadherent cells should still be cultured at a density that will permit continued cell growth for up to 2 weeks. Since maximum gene silencing can take from 6 to 10 days after the addition of oligonucleotide in some but not all cell lines (see Note 4), it is possible that each well or plate will contain no more than 1,000 to several tens of thousands of cells at the start of the experiment, depending on the cell line and culture vessel used for gymnosis. Therefore, the plating density must be empirically determined for each cell line before the experiment begins assuming that on average it will take 8 days for maximal gymnotic silencing to occur. On the other hand, it is also important not to plate the cells at so low a density that their growth rate is not sufficient for optimum oligo uptake, or that an insufficient numbers of cells are available to harvest for gene expression when the experiment is terminated. Again, these parameters (i.e., number of cells seeded, time to termination of the experiment, oligonucleotide concentration) must be individually determined for each cell line, but in our experience the tolerances of the method are very generous provided the cells do not achieve confluency and enough time has elapsed before gene silencing is measured.

3.1. Cell Culture for Gymnotic Delivery in Adherent Cells

1. Day 0: HeLa or 518A2 cells were trypisinized from exponentially growing stock cultures, washed once with DMEM-complete medium, and then counted with a hemocytometer.
2. Cells were plated into 6-well tissue culture plates at a density of 2,000 cells/well in 2 ml of complete medium. For both HeLa and 518A2 cells, this plating density will yield subconfluent cultures on day 6. For experiments lasting 8–10 days, 1,000 HeLa (or 518A2) cells should be plated per well of a 6-well plate (see Note 5). 6-Well plates containing cells were returned to the incubator for 18–24 h to allow for cell attachment.

3.2. Cell Culture for Gymnotic Delivery in Nonadherent Namalwa Cells

1. Day 0: The Burkitt's lymphoma cell line Namalwa was maintained in RPMI 1640 (Invitrogen, Carlsbad, CA, USA) supplemented with 10% FBS (heat inactivated for 30 min at 56°C), 10 mM HEPES (Sigma-Aldrich), 1 mM sodium pyruvate (Gibco), and 25 μg/ml Gentamicin (Sigma-Aldrich)

(RMPI complete medium). Approximately 5×10^6 cells were pipetted from exponentially growing cell cultures, washed 1× with fresh RPMI complete medium, and then counted with a hemocytometer.

2. Cells were plated into 6-well tissue culture plates at a density of 250,000 cells/well in 2 ml of RPMI complete medium. 6-Well plates containing cells were returned to the incubator for 18–24 h before the gymnotic treatment of oligonucleotides (see Note 6).

3.3. Gymnosis of Adherent and Nonadherent Cells Using PS-Modified gap-mer Oligonucleotides

1. Working concentrations (1 mM) of PS-LNA gap-mer antisense oligonucleotides were prepared by diluting stock oligos with 10 mM Tris, pH 7.5.
2. On Day 1, the 6-well plates containing cells were retrieved from incubator and 10 μl of 1 mM working stock of PS-LNA gap-mer was added to each well that already contains 2 ml of medium (from day 0 plating). The plates were rocked back and forth gently to mix the PS-LNA gap-mer in the medium before being returned to the incubator for the duration of the 6–13 days experiment. The final concentration of PS-LNA gap-mer is 5 μM, which is a reasonable initial concentration for gene silencing by gymnosis (see Note 7).
3. The cells should be monitored periodically throughout the experiment by microscopy for signs of contamination, cell growth, and phenotypic evaluation for the gene silencing effect.
4. On day 6 (or later depending on the experimental requirements) cells can be collected for protein and RNA analysis. For adherent cells, the cells can be detached from individual wells using a rubber spatula. For nonadherent cells, cells can be removed using a pipette. The cells and culture medium from each well were placed in separate 15-ml centrifuge tubes on ice. Cells were pelleted at $750 \times g$ in a refrigerated centrifuge for 5 min, washed once with ice-cold 1× PBS, and then pelleted again by centrifugation as described. At this point, the cell pellets can be harvested for either protein or RNA analysis. For protein analysis, cell pellets were processed as described in Subheading 3.4. For RNA analysis, cell pellets were processed as described in Subheading 3.5.

3.4. Western Blot Analysis to Measure Target Down-Regulation

1. Cell pellets were resuspended by pipetting in 20–30 μl of fresh cell lysis buffer and incubated on ice for 60 min. Cell lysates were subject to centrifugation at $12{,}000 \times g$ at 4°C and the solubilized protein material was collected with a pipette. The insoluble fraction contained within the pelleted material was discarded.
2. Protein concentrations for each cell lysate were determined using the Bradford assay and a spectrophotometer. Twenty five microgram of protein for each cell lysate were mixed with 3 μl

of 5× loading buffer and the volume of each sample brought up to 15 μl using lysis buffer.

3. SDS/PAGE was performed using a Bio-Rad mini-Protean II system. After assembly of the glass plates using a 0.75-mm spacer, ~3.5 ml of the 10% resolving gel was added using a pipette. Using a 1-ml pipette, a layer of autoclaved dH_2O was added on top of the resolving gel to create a level transition with the stacking gel. After 45 min to allow for polymerization of the resolving gel, the autoclaved dH_2O was poured off and ~1 ml of 5% stacking gel was added on top of the 10% resolving gel before insertion of a 15 lane comb. The gel was left untouched for an additional 25 min to allow for polymerization of the stacking gel, after which the comb was removed, and each well pocket was washed several times with running buffer using a pipette.
4. Prestained protein makers were loaded into the first and last lanes of the gel, followed by the careful loading of each individual sample. Proteins were separated at 130 V for 1 h at room temperature, using the prestained protein markers as a guide for sufficient separation.
5. The running apparatus was disassembled and the gel was soaked for 5 min in transfer buffer. The PVDF membrane was activated with methanol and soaked in autoclaved dH_20 prior to its use. Protein transfer was performed using a Bio-Rad mini Trans-Blot cell according to the manufacturer's instructions. The mini Trans-Blot tank was immersed in an ice bucket and protein transfer was performed at room temperature at 80 V for 1.5 h with constant stirring.
6. Following disassembly of the transfer apparatus, the PVDF membrane was rinsed with transfer buffer to remove any gel material and allowed to air dry. Using the prestained protein marker as a guide, a razor blade was used to cut the membrane horizontally at two places; For Bcl-2: between the 37 and 15 kDa markers; For α-tubulin: between the 75 and the 37 kDa marker.
7. Membranes were blocked at room temperature for 1 h in plastic boxes using separate blocking buffers. For Bcl-2, the membrane was blocked with 5% BSA in PBS-T. For α-tubulin, the membrane was blocked with 5% nonfat dry milk in TBS-T.
8. Membranes were incubated with primary antibodies at room temperature for 1 h in plastic boxes. The Bcl-2 monoclonal antibody was diluted 1:500 in 5% BSA/PBS-T. The α-tubulin monoclonal antibody was diluted 1:5,000 in 5% nonfat dry milk/TBS-T.
9. Membranes were washed three times for 10 min each wash at room temperature using separate wash buffers. For Bcl-2, membranes were washed with PBS-T. For α-tubulin, membranes were washed with TBS-T.

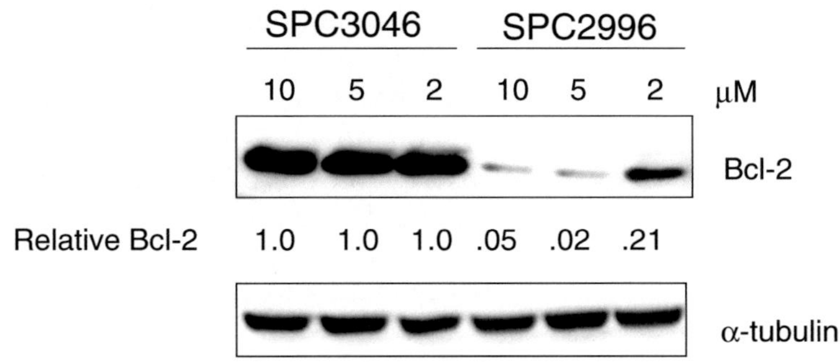

Fig. 2. Western blot analysis following gymnosis of HeLa cells for 6 days with increasing concentrations of SPC2996 or the scrambled control SPC3046. Bcl-2 values were normalized to α-tubulin levels. The relative expression of Bcl-2 is shown.

10. Membranes were incubated with anti-mouse secondary antibodies at room temperature for 1 h in plastic boxes. For Bcl-2, the anti-mouse secondary antibody was diluted 1:3,000 in 5% nonfat dry milk/PBS-T. For α-tubulin, the anti-mouse secondary antibody was diluted 1:3,000 in 5% nonfat dry milk/TBS-T.
11. Membranes were washed three times for 10 min each wash at room temperature using separate wash buffers. For Bcl-2, membranes were washed with PBS-T. For α-tubulin, membranes were washed with TBS-T.
12. Membranes were air dried, placed in the chemiluminescence solution for 1 min, blotted dry, and then wrapped with saran wrap.
13. Membranes were exposed for different times (1 min, 30 s, and 10 s) to chemiluminescent film in a dark room and then developed using an Agfa film processor (Figs. 2 and 4b).
14. The exposed films for Bcl-2 and α-tubulin were scanned into tiff files and the intensity values of the Bcl-2 and α-tubulin chemiluminescent bands were measured using the image processing program ImageJ. To determine the value of normalized Bcl-2, the intensity value for Bcl-2 was divided by the intensity value of the corresponding α-tubulin for that sample. Relative changes in Bcl-2 were determined by dividing the value of normalized Bcl-2 by the value of normalized Bcl-2 of the negative control sample (Fig. 2).

3.5. Quantitative PCR Analysis to Measure Target Down-Regulation

1. Total RNA was extracted using the Qiagen RNeasy kit (Qiagen, The Netherlands) according to the manufacturer's instructions.
2. RNA concentration and quality was analyzed in a plate reader capable of measuring ultraviolet absorbance at 260 and 280 nm. Calculation of the RNA concentration was based on the absorbance at 260 nm; whereas, the RNA purity is judged as the 260/280 nm ratio between 1.8 and 2.0 (see Note 8).

a

Oligo Number:	Sequence
3088:	$5'\text{-}{}^{m}\mathbf{C}_s{}^{o}\mathbf{T}_s{}^{o}c_sc_sc_sa_sa_sc_sg_st_sg_sc_sg_s{}^{m}\mathbf{C}_s{}^{o\,m}\mathbf{C}^{o}{}_{s}a\text{-}3'$
4748:	$5'\text{-}\mathbf{G}_s{}^{o}\mathbf{T}_s{}^{o}t_sc_st_sc_sc_sc_sa_sg_sc_sg_st_sg_sc_sg_sc_sc_sa_s\mathbf{T}_s{}^{o\,m}\mathbf{C}_s{}^{o}\text{-}3'$
4747:	$5'\text{-}\mathbf{T}_s{}^{o}\mathbf{T}_s{}^{o}c_st_sc_sc_sc_sa_sg_sc_sg_st_sg_sc_sg_sc_sc_s\mathbf{A}_s{}^{o}\mathbf{T}_s{}^{o}c\text{-}3'$
2989:	$5'\text{-}\mathbf{T}_s{}^{o\,m}\mathbf{C}_s{}^{o}t_sc_sc_sc_sa_sg_sc_sg_st_sg_sc_sg_sc_s{}^{m}\mathbf{C}_s{}^{o}\mathbf{A}_s{}^{o}t\text{-}3'$
2993:	$5'\text{-}{}^{m}\mathbf{C}_s{}^{o}\mathbf{T}_s{}^{o}c_sc_sc_sa_sg_sc_sg_st_sg_sc_sg_s{}^{m}\mathbf{C}_s{}^{o\,m}\mathbf{C}_s{}^{o}a\text{-}3'$
4746:	$5'\text{-}\mathbf{T}_s{}^{o\,m}\mathbf{C}_s{}^{o}c_sc_sa_sg_sc_sg_st_sg_sc_s\mathbf{G}_s{}^{o\,m}\mathbf{C}_s{}^{o}c_s\text{-}3'$
4745:	$5'\text{-}{}^{m}\mathbf{C}_s{}^{o\,m}\mathbf{C}_s{}^{o}a_sg_sc_sg_st_sg_sc_s\mathbf{G}_s{}^{o\,m}\mathbf{C}_s{}^{o}c\text{-}3'$
4744:	$5'\text{-}\mathbf{T}_s{}^{o\,m}\mathbf{C}_s{}^{o}c_sc_sa_sg_sc_sg_st_sg_s{}^{m}\mathbf{C}_s{}^{o}\mathbf{G}_s{}^{o}\text{-}3'$
4743:	$5'\text{-}{}^{m}\mathbf{C}_s{}^{o\,m}\mathbf{C}_s{}^{o}c_sa_sg_sc_sg_st_sg_sc_s\mathbf{G}_s{}^{o\,m}\mathbf{C}_s{}^{o}\text{-}3'$

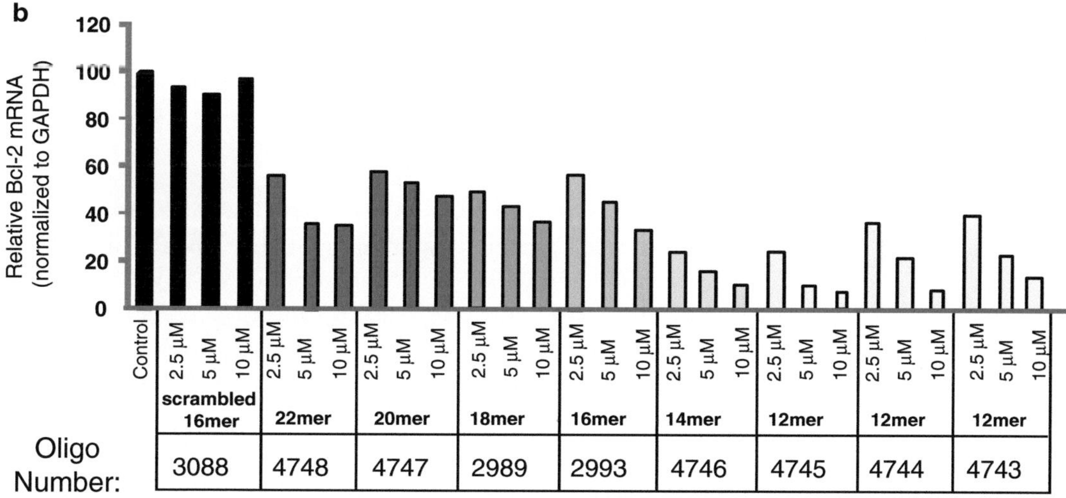

Fig. 3. Effect of different oligo lengths on gymnosis. Quantitative PCR (qPCR) analysis following gymnosis of 518A2 melanoma cells after 5 days with PS-LNA gap-mers of varying lengths (12–22 mer) or a 16 mer scrambled control. (**a**) List of different PS-LNA gap-mer sequences. LNA-modified riboses are in *bold capital letters*, whereas *small letters* indicate deoxyriboses. *o* oxy-LNA, *s* phosphorothioate, *m* C5-methylcytosine. (**b**) qRT-PCR analysis was used to measure Bcl-2 expression after gymnosis. Bcl-2 expression was normalized to the control gene GAPDH.

3. RNA samples were normalized with RNase free H_2O to equal concentration prior to the start of cDNA synthesis by the reverse transcription reaction.
4. 0.5 µg of total RNA were mixed in a 96-well microtiter plate for PCR (ABgene) with random decamer primers (Ambion) and dNTPs mix (Invitrogen) according to the manufacturer's protocol, and heated to 70°C for 3 min in a thermocycler followed by immediate cooling to 0°C.

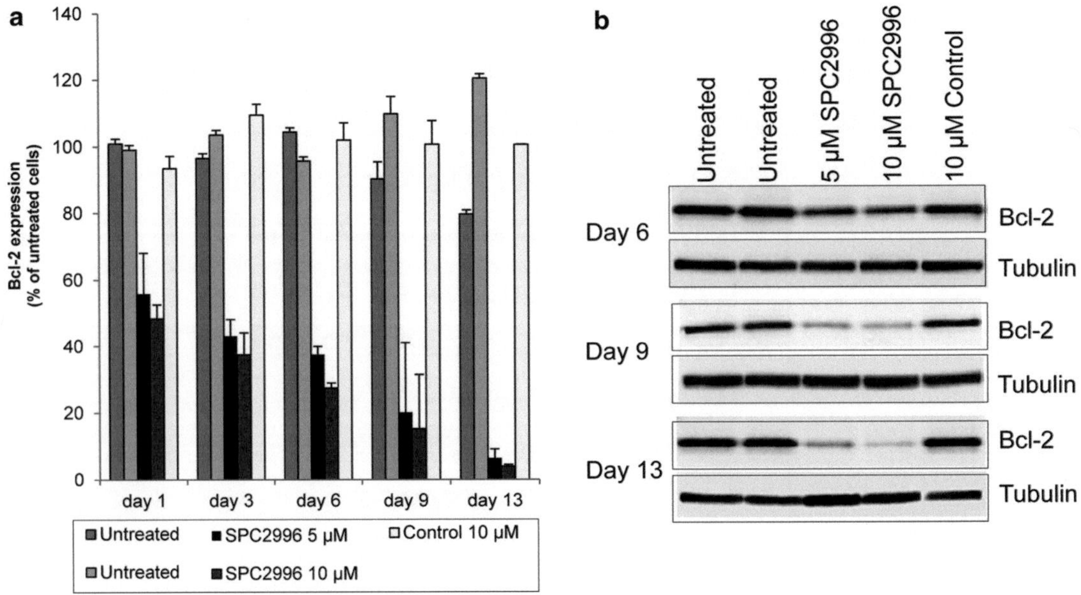

Fig. 4 The effect of PS-LNA gap-mer SPC2996 on Bcl-2 expression over time following gymnosis in nonadherent Namalwa B-cells. (**a**) relative Bcl-2 expression was determined by qPCR and normalized to GAPDH. (**b**) Western blot analysis of Bcl-2 expression as a function of time following gymnosis. The levels of a-tubulin were determined as a loading control.

5. The enzyme reverse transcriptase MMLV-RT (Ambion) was added to each sample, together with the 10× RT first strand buffer (included as a component of the MMLV-RT kit) and RNase inhibitor (Ambion) according to the manufacturer's protocol, and heated for 1 h at 42°C to complete the cDNA synthesis, followed by 10 min at 95°C in order to inactivate/ denature the reverse transcriptase.
6. The reaction was cooled down to 4°C, at which point the samples can be further analyzed or stored at −20°C for long term storage.
7. First strand cDNA was subsequently diluted ten times in nuclease-free water before addition to the real-time PCR reaction mixture, containing 2× Taqman Fast Universal PCR Master mix (Applied Biosystems) in a 96-well MicroAmp Fast Optical reaction plate (Applied Biosystems) in a final volume of 10 µl mRNA quantification of Bcl-2 and GAPDH genes was done using standard TaqMan primer-probe mix (Applied Biosystems). A twofold total RNA dilution series from untreated Namalwa cells served as standard to ensure a linear range [cycle threshold (Ct) versus relative copy number] of the amplification. All samples were measured in duplicate.
8. The Applied Biosystems 7500 Real-Time PCR instrument was used for amplification. A qPCR was run with the following

program: One cycle at 95°C for 20 s, followed by 40 cycles of 95°C for 3 s and 60°C for 30 s.

9. The data were analyzed with Applied Biosystem 7500 Fast System software. Bcl-2 mRNA expression was normalized against GAPDH mRNA expression. Subsequently, the changes in relative Bcl-2 mRNA expression were determined by dividing the value of normalized Bcl-2 of oligonucleotide treated samples by the value of normalized Bcl-2 from the untreated control sample (Figs. 3 and 4a).

4. Notes

1. There are no special cell culture medium, supplements, or additives required for gene silencing by gymnosis. The cell culture medium should be specific to the particular cell line being investigated and/or the experimental goals. To date, gene silencing by gymnotic delivery of PS-LNA gap-mers has been successfully applied to many human cancer cell lines including: 518A2 (melanoma), 333.1 (melanoma), 201.2 (melanoma), 591.8 (melanoma), 1000.36 (melanoma), 1036 (melanoma), PC-3 (prostate), LNCaP (prostate), LAPC-4 (prostate), Huh-7 (liver), HeLa (cervical), CaCo2 (colon), Namalwa (B-cell).
2. All PS-LNA gap-mers must be purified, e.g., by high-pressure liquid chromatography (HPLC). The length of the oligo should be 12–16 mer, but shorter oligos (as short as 10 mer) may be active in some cases. Oligos must have an all-phosphorothioate backbone, i.e., a sulfur atom must replace one nonbridging oxygen atom at each phosphorus atom in the oligo chain. Oligos containing an all-phosphodiester sugar backbone are completely inactive by gymnotic delivery. The 3′ and 5′ termini of the phosphorothioate oligo must also be modified to promote nuclease stability: one or two LNA moieties at each terminus are sufficient to achieve nuclease stability, as well as having the added benefit of increasing the melting temperature (T_m) of the duplex formed between oligo and the target mRNA. LNA-phosphorothioate gap-mer oligos can be purchased from either IDT (Coralville, Iowa) or from Exiqon (Vaedbaek, Denmark).
3. The reagents for Western blot and qPCR analysis can be used for protein and RNA that is harvested from adherent or nonadherent cells.
4. The time required for gene silencing will depend on the half-lives of the mRNA and protein for each target gene. However, in our experience with multiple targets in various cell lines, the optimal time for maximum silencing of mRNA and protein is 3–5 and 6–10 days, respectively.

5. The most critical factor for successful gene knock-down by gymnosis is the plating density of the cells. The cells must never become confluent during the time-course of the experiment because the rate of adsorptive endocytosis of oligo from the medium is reduced. Therefore, preliminary experiments should be performed to determine the best plating density for each individual cell line, keeping in mind the 6–10 days required for gene silencing by gymnosis.
6. For nonadherent cells, the strict requirement for nonconfluency is, of course, not relevant. However, nonadherent cells should be cultured under conditions that maintain cell growth for the 6–10 days required for gymnosis.
7. The concentration of antisense oligonucleotide required for maximum gene silencing by gymnosis must be determined empirically for each cell line and target gene. Preliminary experiments using serial concentrations of oligo (starting at 100 nM and ending with 10 μM) should be performed to determine the lowest oligo concentration that yields the most potent gene knock down.
8. Only high quality nondegraded RNA should be used for qPCR determination of relative gene expression. Therefore, care should be taken to insure that RNA reagents and equipment are free from RNase contamination.

References

1. Lebedeva, I, Benimetskaya, L, Stein, CA, Vilenchik, M (2000) Delivery of oligonucleotides to cells. Euro J Pharm and Biopharm 50:101–119.
2. Akhtar, S, Basu, S, Wickstrom, E, Juliano, R (1991) Interactions of antisense DNA oligonucleotide analogs with phospholipid membranes (liposomes). Nucl Acids Res 19:5551–5559.
3. Stein, CA, Hansen, B, Lai, J, Wu, S, Voskresenskiy, A, Hoeg, A, Worm, J Hedtjarn, M, Souleimanian, N, Miller, P, Soifer, H, Castanotto, D, Benimetskaya, L, Oerum, H, Koch, T (2009) Efficient gene silencing by delivery of locked nucleic acid antisense oligonucleotides unassisted by transfection reagents. Nucl Acids Res 38:e3.
4. Fisher, T, Terhorst, T, Cao, X, Wagner, R (1993) Intracellular disposition and metabolism of fluorescently-labeled unmodified and modified oligonucleotides microinjected into mammalian cells. Nucl Acids Res 21:3857–3865.
5. Moulds, C, Lewis, J, Froehler, B, Grant,D, Huang, T, Milligan, J, Matteuci, M, Wagner, R (1995) Site and mechanism of antisense inhibition of C-5 propyne oligonucleotides. Biochem 34:5044–5053.
6. Gibbings, DJ, Ciaudo, C, Erhardt, M, and Voinnet, O (2009) Multivesicular bodies associate with components of miRNA effector complexes and modulate miRNA activity. Nat Cell Biol 11: 1143–1149.
7. Lee, YS, Pressman, S, Andress, AP, Kim, K, White, JL, Cassidy, JJ, Li, X, Lubell, K, Lim, do H, Cho, IS, Nakahara, K, Preall, JB, Bellare, P, Sontheimer, EJ, Carthew, RJ (2009) Silencing by small RNAs is linked to endosomal trafficking. Nat Cell Biol 11: 1150–1160.
8. Straarup, E, Fisker, N, Hedtjarn, M, Lindholm, M, Rosenbohm, C, Aarup, V, Hansen, H, Oerum, H, Hansen, J, Koch, T (2010) Short locked nucleic acid antisense oligonucleotides potently reduce apolipoprotein B expression and improve the lipoprotein profile of mice and non-human primates. Nucl Acids Res 38:7100–7111.
9. Koch, T, Rosenbohm, C, Hansen, HF, Hansen, B, Straarup, EM, Kauppinen, S (2008) Locked Nucleic Acid: Properties and therapeutic aspects. In: Kurreck, J. (ed) Therapeutic Oligonucleotides. RSC Publishing, Cambridge, p. 103–34.

Chapter 26

Polycistronic Expression of Interfering RNAs from RNA Polymerase III Promoters

Laura F. Steel and Viraj R. Sanghvi

Abstract

In many RNA silencing applications, there is a benefit to expressing multiple interfering RNAs simultaneously. This can be achieved by using a single RNA polymerase II promoter to express multiple micro(mi)RNA-formatted interfering RNAs that are arranged in a polycistronic cluster, mimicking the organization of naturally clustered, endogenous miRNAs. While RNA pol III promoters are often used to express individual short hairpin (sh) RNAs, we have recently shown that pol III promoters can also be used to drive polycistronic expression of miRNA-formatted interfering RNAs. Here, we present methods for the assembly of polycistronic miRNA expression vectors that use pol III promoters. In addition, we present methods for testing the potency and the level of expression of each of the individual miRNAs encoded in the construct.

Key words: miRNA expression vectors, RNA pol III promoters, RNA silencing, Polycistronic miRNA, Dual luciferase reporter plasmid, miRNA northern blot

1. Introduction

The use of technologies based on RNA interference (RNAi) for the regulation of gene expression is becoming widespread both in research applications and in the development of new therapeutic strategies for the treatment of genetic, metabolic, and infectious diseases. It is often advantageous to utilize multiple interfering RNAs, simultaneously targeting either a single or several different transcripts. This is particularly true in the development of antiviral interfering RNAs, where the expression of multiple interfering RNAs is necessary in order to minimize the selection of viral escape mutants and to maximize efficacy across a range of naturally occurring genotypic variants (reviewed in ref. (1)). Several formats have been used for multiplexed expression of interfering RNAs from a

Michael Kaufmann and Claudia Klinger (eds.), *Functional Genomics: Methods and Protocols*,
Methods in Molecular Biology, vol. 815, DOI 10.1007/978-1-61779-424-7_26, © Springer Science+Business Media, LLC 2012

single vector (reviewed in refs. (2, 3)). Polycistronic expression from a single promoter most often utilizes an RNA polymerase II promoter, thereby mimicking the expression of naturally clustered micro(mi)RNAs (4, 5). The use of the stronger pol III promoters is usually reserved for expression of short hairpin (sh) RNAs, where precise transcriptional start and stop positions are important for correct processing and where each promoter can drive only a single interfering RNA. However, we have recently shown that it is possible to use pol III promoters to drive the expression of polycistronic interfering RNAs that are formatted to resemble miRNAs. Functional miRNAs are processed from a primary transcript containing multiple stem-loop structures, using the cellular machinery for endogenous miRNA processing (3). Here, we describe methods for the construction and testing of vectors for the expression of polycistronic, miRNA-formatted interfering RNAs from RNA pol III promoters.

2. Materials

1. TE-4: 10 mM Tris–HCl, 0.1 mM Na_2-EDTA, pH 8.0.
2. 10 mM dNTP stock: 10 mM each dTTP, dATP, dCTP, and dGTP.
3. NuSieve® GTG® agarose, SeaPlaque® (low melting temperature) agarose (Lonza Inc., Allendale, NJ).
4. TAE buffer: 40 mM Tris–acetate, 1 mM EDTA, pH 8.0.
5. psiCHECK-2™ and Dual-Luciferase Reporter Assay System (Promega Corp, Madison, WI).
6. TBE buffer: 45 mM Tris–borate, 1 mM EDTA, pH 8.0.
7. 12% denaturing polyacrylamide gel: For 25 ml gel solution, 10.5 g urea, 10 ml 30% polyacrylamide:bisacrylamide (20:1), 2.5 ml 10× TBE, 100 μl 10% ammonium persulfate, 12.5 μl TEMED.
8. Formamide/EDTA gel loading buffer: 98% formamide, 10 mM EDTA, bromphenol blue.
9. SSPE buffer: 150 mM NaCl, 10 mM NaH_2PO_4, 1 mM Na_2-EDTA.
10. mirVana miRNA Isolation Kit, Decade Marker System, BrightStar Plus nylon membrane, Ultrahyb-Oligo hybridization buffer (Ambion, Austin, TX).
11. Oligonucleotides: see Table 1 for sequence (Integrated DNA Technologies, Coralville, IA).

    F1: Adds Sal I and Xba I sites to the 5′ end of miRNA cassettes based on hsa-miR-30.

    R1: Adds Spe I site, $d(T)_6$, and Kpn I site to 3′ end of miRNA cassettes based on hsa-miR-30.

Table 1
Sequence of oligonucleotides

Oligonucleotide	Sequence[a]
1907-T	AAAGAAGGTATATTGCTGTTGACAGTGAGCGctcggctcctctgc-cgatccatTAGTGAAGCCACAGA
1907-B	CCCTTGAAGTCCGAGGCAGTAGGCAttcggctcctctgccg-atccatTACATCTGTGGC
F1	CTAGGTCGACCACTATTATTTCTATCGTCTAGAAGGCT-AAAGAAGGTAT
R1	CTATGGTACCAAAAAACGGCTGCTGAATCGACTAGTAGCCCCTT-GAAGTCC
F2	CTAGTCTAGAAGGCTAAAGAAGGTATATTGC
R2	CTAGACTAGTAGCCCCTTGAAGTCCGA
1907XN-T	TCGAGTCGGCTCCTCTGCCGATCCAGC
1907XN-B	GGCCGCATGGATCGGCAGAGGAGCCGAC
1907XNmut-T	TCGAGTCGGCTCCTCTGCCGAAGGATGC
1907XNmut-B	GGCCGCATCCTTCGGCAGAGGAGCCGAC

[a] Sequence encoding the sense (in 1907-T) or antisense (in 1907-B) regions of mi-30s/1907A are shown in *lower case*. Restriction enzyme sites are *underlined* in the remaining oligonucleotides

F2: Adds Xba I site to the 5′ end of miRNA cassettes based on hsa-miR-30.

R2: Adds Spe I site to the 3′ end of miRNA cassettes based on hsa-miR-30.

1907XN-Top: Top strand of 1907A target with Xho I compatible 5′ end and Not I compatible 3′ end.

1907XN-Bottom: Bottom strand of 1907A target with Xho I compatible 5′ end and Not I compatible 3′ end.

3. Methods

There can be considerable flexibility in how a polycistronic miRNA expression cassette is designed and assembled. Depending on the goals of the project, expression cassettes can be constructed for either naturally occurring, endogenous miRNAs or for user-designed miRNA-formatted interfering RNAs, or both. The term "user-defined" miRNA indicates a mature silencing RNA that is processed from a transcript that resembles an endogenous primary (pri) miRNA,

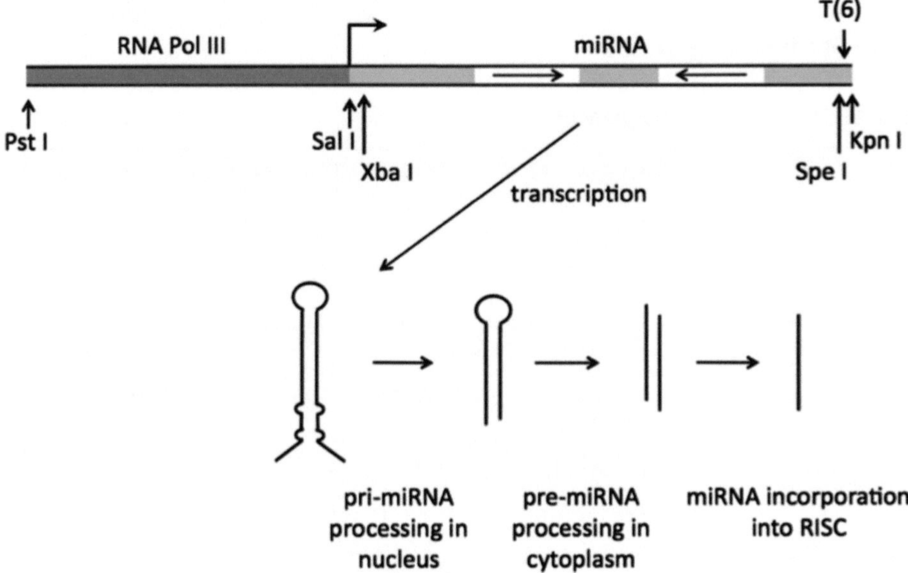

Fig. 1. Organization of a vector for expression of an miRNA-formatted interfering RNA from an RNA pol III promoter. A type 3 pol III promoter (e.g., U6, H1, or 7SK) is placed upstream of Sal I and Kpn I sites that serve as sites for insertion of the first miRNA expression cassette. The miRNA expression cassette is flanked by Xba I and Spe I sites so that additional cassettes can be inserted at either position. A $d(T)_6$ tract downstream of the Spe I site is a transcription termination signal for pol III polymerase. Sequence encoding the sense (passenger) and antisense (guide) strand of the mature miRNA is depicted as *white boxes* with *facing arrows*. Flanking sequence and loop sequence, shown as *light gray boxes*, is derived from an endogenous miRNA gene. The transcriptional start site is shown as a *bent arrow*. The flanking region will provide a pri-miRNA transcript with structure for the binding and processing activities of the Drosha/DGCR8 complex in the nucleus, and the mature miRNA will be produced by further processing in the cytoplasm.

requiring the activity of both nuclear (Drosha/DGCR8, Exportin 5) and cytoplasmic (Dicer/TRBP/Argonaute) RNAi pathway components (see Fig. 1) (miRNA processing is reviewed in ref. 6). To optimize knockdown of a particular RNA, the miRNA should be designed to have complete complementarity to its target sequence, thereby allowing it to act in a catalytic manner to degrade that RNA. Vectors that carry miRNA expression cassettes for the knockdown of mRNAs from humans, mice, rats, and additional organisms are available in plasmid and lentiviral vector formats from numerous commercial suppliers. Similarly, vectors that express endogenous human (and other species) miRNAs are available commercially. Either of these can be a useful source for the individual miRNA components of a polycistronic silencing vector. However, when targeting transcripts from viruses or other organisms that are not available in existing libraries, it is necessary for the user to design the miRNA sequence. Publically available algorithms can aid in the design of either miRNAs or short interfering (si) RNAs (see Note 1). The siRNA sequence can be placed in the context of surrounding sequence derived from an authentic, endogenous

Table 2
miRNAs and miRNA clusters that have been adapted for the construction of miRNA-formatted interfering RNAs

miRNA cluster	miRNA	Ensembl ID	Reference
	mmu-miR-155	ENSMUSG00000065397	(11–13)
	hsa-miR-30	ENSG00000207827	(2, 14, 15)
hsa-miR-17/92	miR-17 miR-18a miR-19a miR-20a miR-19b-1 miR-92a-1	ENSG00000207745 ENSG00000199180 ENSG00000207610 ENSG00000199149 ENSG00000207560 ENSG00000207968	(16)
hsa-miR-106b	miR-106b miR-93 miR-25	ENSG00000208036 ENSG00000207757 ENSG00000207547	(17)

pri-miRNA (7–9) (see Note 2 and Table 2). Nevertheless, the identification of a specific target region that will be most susceptible to interfering RNAs can be challenging and requires empirical testing. Prior to the assembly of a polycistronic miRNA expression vector, the silencing potency of each of the individual miRNA elements should be validated separately.

When individual miRNAs have been identified that show strong and specific silencing of a target transcript, they can be assembled into a polycistronic cassette, as described below. Additional choices in the construction of the expression vector include the vector backbone (for instance, plasmid or viral vectors can be used), the restriction sites that will be used to connect the individual miRNA regions, and the RNA polymerase promoter that will be used to drive expression (see Note 3). The methods outlined below describe the assembly of a polycistronic, user-designed miRNA expression cassette, but they can readily be adapted to accommodate many variations in overall design.

3.1. Assembly of a DNA Cassette that Encodes a Single miRNA for RNA pol III-Driven Expression

The DNA cassette encoding each individual miRNA contains a stem-loop region placed within flanking sequence that is derived from an authentic, endogenous miRNA gene (see Fig. 2a). The example described below is based on the construction of an miRNA (mi-30s/1907A) that targets transcripts of hepatitis B virus (HBV) and is placed in the context of human miR-30 sequence (2).

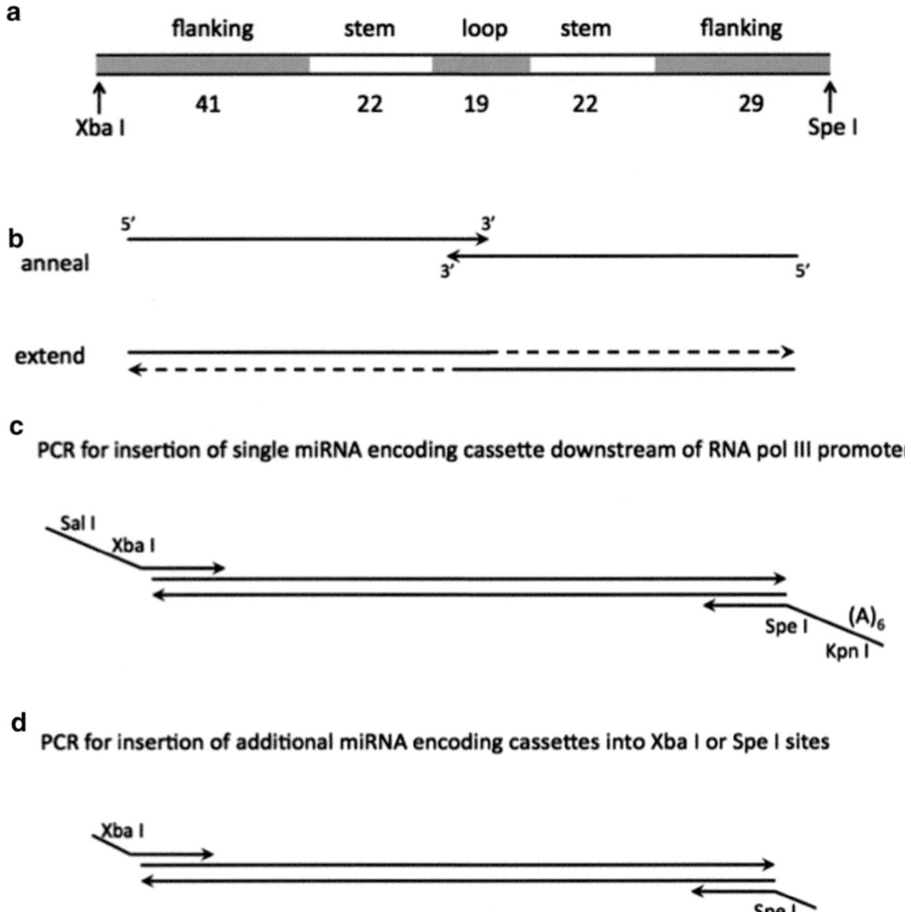

Fig. 2. Construction of individual miRNA expression cassettes. (**a**) The mi-30s/1907A cassette includes segments encoding the stem region of the pri-miRNA (*white boxes*) embedded in sequence encoding flanking and loop regions derived from the hsa-miR-30 gene (*light gray boxes*). The size of each region is indicated in bp and the cassette contains Xba I and Spe I restriction sites at the ends. (**b**) Oligonucleotides with sequence from the top and bottom stands are annealed and then extended by DNA polymerase, Klenow fragment. (**c**) For construction of the first cassette to be inserted downstream of a pol III promoter, the miRNA-encoding cassette is amplified with primers that will add Sal I and Xba I sites at the 5′ end and Kpn I and Spe I sites at the 3′ end. A tract of six T residues is encoded between the 3′ Spe I and Kpn I sites. (**d**) For insertion of additional miRNA cassettes, the PCR amplification uses primers that add a 5′ Xba I site and 3′ Spe I site.

1. Oligonucleotides containing sequence from approximately one-half of the top and bottom strands of the miRNA cassette, respectively, are designed to contain an overlap of 8 nt at their 3′ ends (see Table 1 for the sequence of oligos 1907-T and 1907-B). The overlap should occur in the loop region and the length of the oligos should be approximately 60–70 nt (see Fig. 2b).
2. Anneal and extend oligonucleotides: Resuspend the oligos at 200 μM in TE-4 or H_2O. Combine 1 μl of each oligo, 2 μl 10× Klenow buffer, and 14 μl H_2O. Heat at 85°C for 5 min and

cool to room temperature. Add 1 μl 10 mM dNTPs and 1 μl (2 U) DNA polymerase, Klenow fragment. Final reaction conditions are 200 pmol each oligo, 0.5 mM dNTPs, 1× Klenow buffer, and 2 U Klenow enzyme in a final volume of 20 μl. Incubate at room temperature for 30 min (see Note 4).

3. Resolve the extended products in an agarose gel: Separate the extended reaction products in a 3% NuSieve agarose gel in TAE buffer and stain the gel with ethidium bromide. Using a long wavelength UV light (366 nm) to visualize the DNA, excise a gel slice containing fragments of the length predicted (~120 bp) for the extended oligonucleotide pair (see Note 5).
4. Design two PCR primers to amplify the extended oligos from Subheading 3.1, step 3 for cloning downstream of a pol III promoter, as follows: (1) The forward primer (F1) should contain a restriction site for cloning immediately downstream of a pol III promoter (Sal I in Fig. 1) followed by a spacer region, an Xba I site, and further sequence corresponding to approximately 15 residues at the 5′ end of the top strand of the extended products of Subheading 3.1, step 3. (2) The reverse primer (R1) should contain a restriction site for cloning (Kpn I in Fig. 1), a stretch of six A residues, an Spe I site, and sequence corresponding to approximately 15 residues the 5′ end of the bottom strand of the extended products of Subheading 3.1, step 3. Refer to Table 1 and Fig. 2c.
5. PCR amplify the extended products to add restriction sites for cloning: Melt the excised gel fragment from Subheading 3.1, step 3 at 68°C, cool to 37°C, and add 0.5 μl of the melted gel to a reaction containing 5 μl 10× Taq DNA polymerase buffer, 1 μl 10 mM dNTPs, 1 μl 20 μM PCR primer F1, 1 μl 20 μM PCR primer R1, and 0.5 μl Taq DNA polymerase, in a final volume of 50 μl. The final reaction conditions are 1× Taq polymerase buffer, 0.2 mM dNTPs, 0.4 μM each PCR primer, and 2.5 U Taq DNA polymerase. Cycle the reaction five times at 95°C for 45 s, 37°C for 45 s, and 72°C for 45 s, and then 25 times at 95°C for 45 s, 57°C for 45 s, and 72°C for 45 s. After the last cycle, incubate at 72°C for 10 min and then hold at 4°C. Clean up the PCR products using a commercial kit or, alternatively, spin through a Sephadex G-50 column to remove unincorporated dNTPs.
6. Sequentially digest the amplified DNA products with Kpn I and Sal I and isolate the resulting DNA fragment by electrophoresis in 1% low melting temperature agarose in TAE buffer.
7. Excise the Sal I – Kpn I fragment containing the miRNA cassette and insert it into corresponding sites in a vector with a pol III promoter, using standard ligation reaction conditions.

8. After transformation of bacterial cells and preparation of plasmid DNA, correct insertion of the target region in the resulting reporter plasmid is confirmed by restriction digestion and DNA sequence analysis.

3.2. Assembly of Additional miRNA-Encoding Cassettes for Polycistronic Expression from the pol III Promoter

1. Construct additional miRNA cassettes following Subheading 3.1, steps 1 through 3.
2. Design two primers to amplify the extended oligos for cloning into the Xba I site or the Spe I site of the monocistronic vector constructed above (see Fig. 3), as follows: 1) The forward primer (F2) should contain an Xba I site and further sequence corresponding to approximately 15 residues at the 5′ end of the top strand of the extended products of Subheading 3.2, step 1. (2) The reverse primer (R2) should contain an Spe I site and sequence corresponding to approximately 15 residues the 5′ end of the bottom strand of the extended products of Subheading 3.2, step 1. Refer to Table 1 and Fig. 2d.
3. PCR amplify the extended oligonucleotides as described in Subheading 3.1, step 5.
4. Digest the amplified DNA with Xba I and Spe I, isolate the digested fragment by electrophoresis in 3% NuSieve agarose, and ligate the fragment into a monocistronic vector that has been linearized by digestion with either Xba I or Spe I and treated with alkaline phosphatase to prevent self-ligation (see Note 6).
5. Insert additional cassettes at either the Xba I site or the Spe I site by repeating the steps above (see Fig. 3a).
6. Verify that each new cassette has been inserted in the correct orientation by restriction digestion of the new molecular clone, for instance with Xba I and Kpn I, or with Spe I and Sal I (see Fig. 3b). All constructs are confirmed by DNA sequence analysis.

3.3. Functional Testing of the Silencing Potency of Each Individual miRNA Expressed from the Polycistronic Vector

Individual reporter plasmids are constructed for each of the regions targeted by the component miRNAs in order to test the silencing potency of the individual miRNAs when they are expressed from the polycistronic vector. The procedure below uses the example of a reporter for mi30s/1907A silencing activity (3).

1. Design targets for testing each individual miRNA: For each of the component miRNAs, target sequence is inserted into cloning sites present in the 3′-untranslated region (3′UTR) of the Renilla luciferase gene in the dual luciferase reporter plasmid, psiCHECK-2. Similarly, a target containing 3–4 mutations in the seed region is inserted into the reporter plasmid to serve as a negative control in silencing assays. Complementary oligonucleotides that will encode the target region are designed with additional sequence to produce termini compatible with

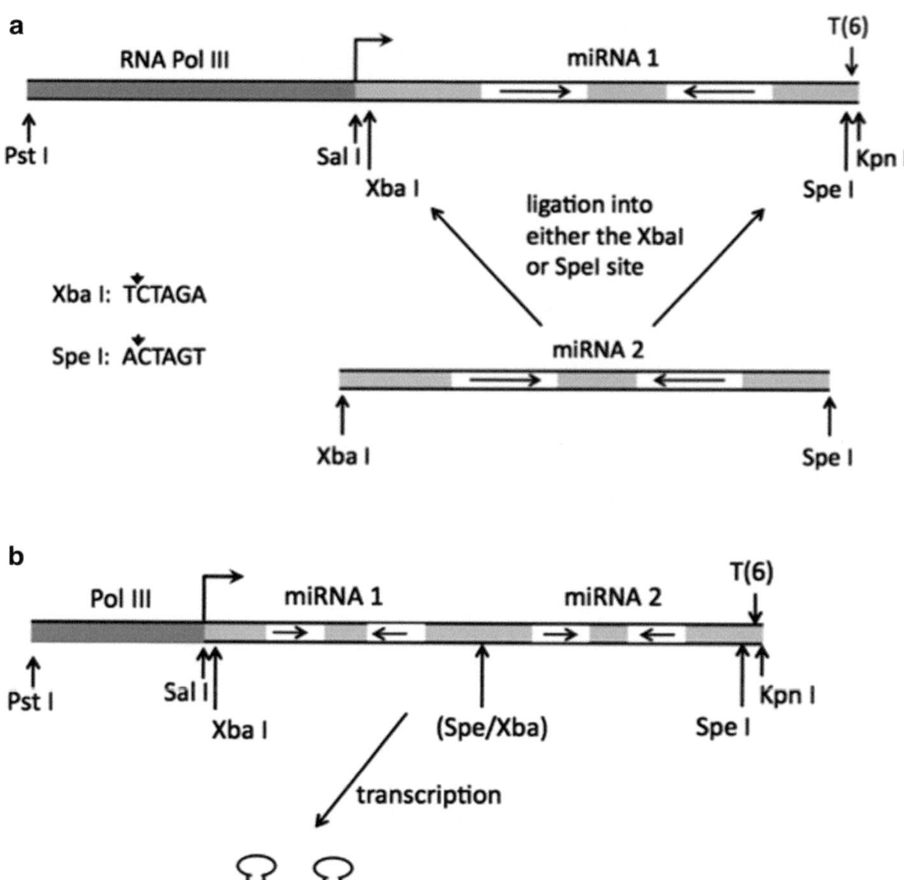

Fig. 3. Insertion of new miRNA expression cassettes to assemble a polycistronic vector. (**a**) Each new miRNA expression cassette can be inserted at either the Xba I site or the Spe I site. (**b**) Transcription from the polycistronic vector results in a pri-miRNA that can be processed into multiple individual miRNAs. The figure shows a bicistronic vector, although up to four miRNAs have been successfully expressed from a single pol III promoter (3).

the vector cloning sites. See Table 1 for the sequence of oligos 1907XN-T and 1907XN-B that are designed to insert target sequence for mi30s/1907A into the Xho I and Not I sites of psiCHECK-2. The oligos 1907XNmut-T and 1907XNmut-B are used to encode a mutant target region in the negative control plasmid (see Note 7).

2. Resuspend each of the oligos at 200 μM in TE-4 or H_2O. Combine 1 μl of each of the complementary oligos in a total volume of 20 μl 1× SSPE and heat to 68°C for 5 min. Allow to cool to room temperature.

3. Linearize psiCHECK-2 by digestion with Xho I and Not I. Combine 500 ng linearized vector with 0.5 μl of the annealed oligos in a reaction with 1 μl 10× ligation buffer, 1 μl (1 U) T4 DNA ligase, and H_2O to a final volume of 10 μl. Incubate at room temperature for 1 h or overnight at 15°C.
4. After transformation of bacterial cells and preparation of plasmid DNA, confirm correct insertion of the target region in the resulting reporter plasmid by DNA sequence analysis.
5. For each reporter plasmid, transfect cultured cells with a constant amount of the reporter plasmid together with increasing doses of the polycistronic miRNA expression vector. Similarly, transfect cells with a corresponding negative control reporter plasmid together with increasing doses of the polycistronic miRNA expression vector. For example, for HEK-293T cells use 100 ng reporter plasmid plus 5, 10, and 20 ng of polycistronic miRNA expression plasmid in a 12-well format. Empty vector DNA (e.g., pUC19) is used to bring the total DNA in each transfection to 1 μg.
6. Two days posttransfection, assay Renilla and firefly luciferase using a dual-luciferase reporter assay system. In the psiCHECK-2 vector, target sequence is inserted into the 3′UTR of a gene encoding Renilla luciferase, and firefly luciferase is expressed from an independent gene on the plasmid to serve as a transfection efficiency control. Silencing activity is reported as a reduction in the ratio of Renilla luciferase activity to firefly luciferase activity. Results from cells transfected with the reporter carrying a target are compared to results from cells transfected with a corresponding reporter carrying a mutant target.

3.4. Determination of the Expression Level of Each Individual miRNA Expressed from the Polycistronic Vector

The amount of mature miRNA produced from each component of the polycistronic expression vector can be determined by northern blotting.

1. Transfect HEK 293T cells with polycistronic miRNA expression vector. In a 6-well format, use 250 ng miRNA expression vector plus 1,750 ng empty vector for a total of 2 μg of DNA in a calcium phosphate-mediated transfection.
2. Two days posttransfection, isolate total RNA using a mirVana miRNA isolation kit (see Note 8).
3. Resolve RNA by electrophoresis in a 12% denaturing polyacrylamide gel in TBE buffer. For each sample, combine 10 μg of RNA in H_2O with an equal volume of formamide/EDTA gel loading buffer. Heat samples 95°C for 5 min and cool briefly prior to loading the gel. Prepare radiolabeled Decade Markers to serve as size markers (10 nt intervals from 10 to 100 nt plus 150 nt). Run the gel until bromphenol blue dye has migrated approximately 3/4 of the length of the gel.

4. Transfer the RNA to BrightStar nylon membrane by electroblotting. Using an immunoblotting apparatus, transfer the RNA in 0.5× TBE at 90 V for 1–2 h. Pack the transfer tank in ice to prevent excessive heating of the buffer during the transfer.
5. Dry the blot by baking at 60°C for 1 h.
6. Prepare radiolabeled probe. Label an oligonucleotide that is completely complementary to the mature miRNA in a reaction that contains 1 μl oligo (20 pmol), 1 μl 10× kinase buffer, 3 μl ^{32}P-γ-ATP (3,000 Ci/mmol, 10 μCi/μl), 4 μl H_2O, and 1 μl (10 U) T4 polynucleotide kinase. Incubate at 37°C for 30 min. and spin through Sephadex G-50 to remove unincorporated ATP.
7. Prehybridize the blot in UltraHyb-Oligo at 37°C for 1 h. Add probe and hybridize overnight at 37°C.
8. Wash the blot in 2× SSPE, 0.1% SDS three times, 10 min each, at room temperature.
9. Detect the hybridized probe using a Storm 820 (GE Healthcare) phosphorimager.

4. Notes

1. Algorithms that are useful in the design of miRNA-formatted interfering RNAs are offered by several companies, including Invitrogen (https://rnaidesigner.invitrogen.com/rnaiexpress/), Applied Biosystems (http://www.ambion.com/techlib/misc/siRNA_finder.html), and Dharmacon (http://www.dharmacon.com/designcenter/designcenterpage.aspx?gclid=CKPL2NuphaUCFQu87Qod4HN7Ow).
2. The regions encoding the stem-loop structure of each miRNA can be embedded in flanking sequence from a single primary miRNA that is repeated in each case, or from different primary miRNAs that will mimic a naturally occurring genomic cluster. see Table 2 for references to primary miRNA sequences that have been used in the construction of miRNA-formatted interfering RNAs.
3. A type 3 RNA pol III promoter should be used, where all sequence necessary for initiation is upstream of the transcriptional start site. In general, the U6 and 7SK promoters are stronger than the H1 promoter. For a detailed discussion of pol III promoters that have been used for the expression of shRNAs, see ref. 10. Plasmid or lentiviral expression vectors that use RNA pol III promoters are widely available. If necessary, the promoters can be obtained by PCR from human DNA, using primers that will attach appropriate cloning sites.

4. For oligos from 60 to 70 nt long, 200 μM will correspond to approximately 4.0–4.6 μg/μl. However, the oligos should be combined in equimolar amounts.
5. It is useful to gel purify the extended DNA to eliminate products that arise from less than full length oligos that may be present in the starting stock solutions. A prominent band should be seen at the predicted length, although it will be surrounded by a significant smear of both longer and shorter DNA fragments.
6. The order in which miRNAs are placed in a polycistronic vector can affect their individual potencies (3), and it may be necessary to test alternative positions for the component miRNAs to optimize overall silencing efficacy. Xba I and Spe I produce compatible ends so that an Xba I/Spe I cassette can be cloned into either site as the polycistronic vector is developed, as shown in Fig. 3a. Many additional pairs of restriction enzymes that produce compatible overhanging ends can be useful in this application. The restriction sites used to join the individual miRNA elements can be chosen for convenience in step-wise assembly of the polycistronic expression vector, or for easy removal or replacement of the individual miRNA-encoding components.
7. Targets can also be generated by PCR or reverse transcriptase-PCR of approximately 100–200 bp of sequence from the authentic targeted RNA.
8. It is not necessary to enrich for RNA that are less than 200 nt. Total RNA can also be isolated using TriReagent (Sigma, St. Louis, MO) or related products.

References

1. Grimm, D., and Kay, M. A. (2007) Combinatorial RNAi: a winning strategy for the race against evolving targets? *Mol Ther* **15**, 878–888.
2. Snyder, L. L., Esser, J. M., Pachuk, C. J., and Steel, L. F. (2008) Vector design for liver specific expression of multiple interfering RNAs that target hepatitiis B virus transcripts, *Anitviral Research* **80**, 36–44.
3. Snyder, L. L., Ahmed, I., and Steel, L. F. (2009) RNA polymerase III can drive polycistronic expression of functional interfering RNAs designed to resemble microRNAs, *Nucleic Acids Res* **37**, e127.
4. Cai, X., Hagedorn, C. H., and Cullen, B. R. (2004) Human microRNAs are processed from capped, polyadenylated transcripts that can also function as mRNAs, *RNA* **10**, 1957–1966.
5. Lee, Y., Kim, M., Han, J., Yeom, K. H., Lee, S., Baek, S. H., and Kim, V. N. (2004) MicroRNA genes are transcribed by RNA polymerase II, *EMBO J* **23**, 4051–4060.
6. Kim, V. N., Han, J., and Siomi, M. C. (2009) Biogenesis of small RNAs in animals, *Nat Rev Mol Cell Biol* **10**, 126–139.
7. Boden, D., Pusch, O., Silbermann, R., Lee, F., Tucker, L., and Ramratnam, B. (2004) Enhanced gene silencing of HIV-1 specific siRNA using microRNA designed hairpins, *Nucl. Acids Res.* **32**, 1154–1158.
8. McManus, M. T., Petersen, C. P., Haines, B.B., Chen, J., and Sharp, P. (2002) Gene silencing using micro-RNA designed hairpins, *RNA* **8**, 842–850.
9. Zeng, Y., Wagner, E. J., and Cullen, B. R. (2002) Both natural and designed micro RNAs

can inhibit the expression of cognate mRNAs when expressed in human cells, *Molecular Cell 9*, 1327–1333.

10. Bos, T. J., De Bruyne, E., Heirman, C., and Vanderkerken, K. (2009) In search of the most suitable lentiviral shRNA system, *Curr Gene Ther 9*, 192–211.
11. Chung, K.-H., Hart, C. C., Al-Bassam, S., Avery, A., Taylor, J., Patel, P. D., Vojtek, A. B., and Turner, D. L. (2006) Polycistronic RNA polymerase II expression vectors for RNA interference based on BIC/miR-155, *Nucleic Acids Research 34*, e53.
12. Du, G., Yonekubo, J., Zeng, Y., Osisami, M., and Frohman, M. A. (2006) Design of expression vectors for RNA interference based on miRNAs and RNA splicing, *FEBS J 273*, 5421–5427.
13. Son, J., Uchil, P. D., Kim, Y. B., Shankar, P., Kumar, P., and Lee, S.-K. (2008) Effective suppression of HIV-1 by artificial bispecific miRNA targeting conserved sequences with tolerance for wobble base-pairing, *Biochem Biophy Res Commun 374*, 214–218.
14. Silva, J. M., Li, M. Z., Chang, K., Ge, W., Golding, M. C., Rickles, R. J., Siolas, D., Hu, G., Paddison, P. J., Schlabach, M. R., Sheth, N., Bradshaw, J., Burchard, J., Kulkarni, A., Cavet, G., Sachidanandam, R., McCombie, W. R., Cleary, M. A., Elledge, S. J., and Hannon, G. J. (2005) Second-generation shRNA libraries covering the mouse and human genomes, *Nat Genet 37*, 1281–1288.
15. Zeng, Y., and Cullen, B. R. (2003) Sequence requirements for micro RNA processing and function in human cells, *RNA 9*, 112–123.
16. Liu, Y. P., Haasnoot, J., Ter Brake, O., Berkhout, B., and Konstantinova, P. (2008) Inhibition of HIV-1 by multiple siRNAs expressed from a single microRNA polycistron, *Nucleic Acids Res 36*, 2811–2824.
17. Aagaard, L. A., Zhang, J., von Eije, K. J., Li, H., Saetrom, P., Amarzguioui, M., and Rossi, J. J. (2008) Engineering and optimization of the miR-106b cluster for ectopic expression of multiplexed anti-HIV RNAs, *Gene Ther 15*, 1536–1549.

Part VI

Metabolite Analysis

Chapter 27

Metabolite Analysis of *Cannabis sativa* L. by NMR Spectroscopy

Isvett Josefina Flores-Sanchez, Young Hae Choi, and Robert Verpoorte

Abstract

NMR-based metabolomics is an analytical platform, which has been used to classify and analyze *Cannabis sativa* L. cell suspension cultures and plants. Diverse groups of primary and secondary metabolites were identified by comparing NMR data with reference compounds and/or by structure elucidation using ^{1}H-NMR, *J*-resolved, ^{1}H-^{1}H COSY, and ^{1}H-^{13}C HMBC spectroscopy. The direct extraction and the extraction by indirect fractionation are two suitable methods for the *C. sativa* sample preparation. Quantitative analyses could be performed without requiring fractionation or isolation procedures.

Key words: *Cannabis sativa*, Metabolomics, Multivariate data analysis, Nuclear magnetic resonance, Principal component analysis

1. Introduction

Cannabis sativa L. is an annual dioecious plant which produces several metabolite groups from plant primary and secondary metabolism. Amino acids, fatty acids, sugars, and steroids are some examples from primary metabolites. More than 247 compounds have been identified as secondary metabolites (1–3). They have been grouped into six groups of secondary metabolism: cannabinoids, flavonoids, stilbenoids, terpenoids, lignans, and alkaloids. Several analytical platforms have been used to analyze, identify, and quantify the different metabolites present in this plant (4–12). From traditional techniques, either spectroscopic or chromatographic as UV, TLC, CPC, HPLC, GC, MS, IR, NMR, to high-throughput techniques as hyphenated methods (LC–MS, GC–MS, Py-GC–MS, GLC). Although NMR is a common tool used for structural elucidation

Michael Kaufmann and Claudia Klinger (eds.), *Functional Genomics: Methods and Protocols*,
Methods in Molecular Biology, vol. 815, DOI 10.1007/978-1-61779-424-7_27, © Springer Science+Business Media, LLC 2012

of compounds new applications are being explored. ^{1}H-NMR spectroscopy is an analytical platform in the field of plant metabolomics (13, 14). It is known that metabolomics has facilitated an improved understanding of cellular responses to environmental changes (15–17). For NMR-based metabolomics, the analysis allows the simultaneous detection of diverse groups of primary and secondary metabolites. The signals in an NMR spectrum are proportional to their molar concentrations, so a direct comparison of concentrations of all compounds is possible. Using two-dimensional NMR measurements, many signals can be identified (18–20). NMR-based metabolomics from Cannabis plants and cell suspension cultures have been reported for classification of *C. sativa* cultivars and cell lines, and for the analysis of its metabolism under stress conditions (21–23).

2. Materials

2.1. Cannabis Plants and Cell Suspension Cultures

1. Air-dried or fresh flowers and leaves of *C. sativa* L. (Stichting Institute for Medicinal Marijuana, Rotterdam, The Netherlands).
2. Seeds of *C. sativa* (The Sensi Seed Bank, Amsterdam, The Netherlands and Dr. D. Watson, HortaPharm, Amsterdam, The Netherlands) for establishing cell culture lines.
3. Cannabis cell suspension cultures.
4. Erlenmeyer (EM) flasks (250 ml) containing 50 ml of Murashige and Skoog (MS) basal medium supplied with 10 mg/L thiamine hydrochloride, 1 mg/L pyridoxine hydrochloride, 1 mg/L nicotinic acid, 1 mg/L 2,4-D (2,4-dichlorophenoxyacetic acid), 1 mg/L kinetin and 30 g/L sucrose.
5. MS basal medium: NH_4NO_3 (1,650 mg/L), KNO_3 (1,900 mg/L), $CaCl_2{\cdot}2H_2O$ (440 mg/L), $MgSO_4{\cdot}7H_2O$ (370 mg/L), KH_2PO_4 (170 mg/L), KI (0.83 mg/L), H_3BO_3 (6.2 mg/L), $MnSO_4{\cdot}4H_2O$ (22.3 mg/L), $ZnSO_4{\cdot}7H_2O$ (8.6 mg/L), $Na_2MoO_4{\cdot}2H_2O$ (0.25 mg/L), $CuSO_4{\cdot}5H_2O$ (0.025 mg/L), $CoCl_2{\cdot}6H_2O$ (0.025 mg/L), $FeSO_4{\cdot}7H_2O$ (27.8 mg/L), $Na_2EDTA{\cdot}2H_2O$ (37.3 mg/L), myo-inositol (100 mg/L) and glycine (2 mg/L).

2.2. Elicitors

1. 10 mg/ml yeast extract (Bacto™ Brunschwig Chemie, Amsterdam, The Netherlands).
2. 8 mg/ml *Pythium aphanidermatum* (Edson) Fitzp. (313.33, Centraalbureau voor Schimmelcultures, Utrecht, The Netherlands).
3. 8 mg/ml *Botrytis cinerea* Pers. (Stichting Institute for Medicinal Marijuana, Rotterdam, The Netherlands).

4. 1 mM salicylic acid.
5. 0.3 mM methyl-jasmonic acid.
6. 100 μM jasmonic acid.
7. 0.1 mg/ml *Citrus fruits* pectin (87% galacturonic acid and 8.7% methoxy groups, Sigma).
8. 84 μg/ml Cannabis pectin extract.
9. 150 μg/ml sodium alginate (Fluka, Buchs, Switzerland).
10. 100 μM $AgNO_3$.
11. 100 μM $CoCl_2$.
12. 100 μM $NiSO_4$.
13. Exposition to UV-irradiation using UV 302 and 366 nm lamps (Vilber Lourmat, France) for 30 s or 30 min.

2.3. Indirect Fractionation Method

1. MeOH:Water (1:1, *v/v*).
2. $CHCl_3$.
3. Ethyl acetate.

2.4. Direct Extraction Method

1. MeOD: 90 mM KH_2PO_4 buffer in D_2O (pH 6.0; adjusted with 1 M NaOD) (1:1, *v/v*) containing 0.1% TMSP-d_4 [3-(trimethylsilyl) propionic-2,2,3,3-d_4 acid sodium salt, 99 atm.% D] (*w/w*).
2. $CDCl_3$ containing 0.01% HMDSO-d_{18} (hexamethyl-d_{18}-disiloxane, 99 atm.% D) (*w/w*).

2.5. Additional Material

1. Liquid nitrogen.
2. Eppendorf tubes, 2 ml.
3. Cap test tubes, 15 ml (plastic and glass).
4. Evaporating flask, 25 ml.
5. Pasteur pipettes.
6. NMR tubes, 5 mm.

2.6. Equipment

1. Freeze-dryer for sample drying.
2. Ultrasonicator.
3. Refrigerated centrifuge.
4. Rotary evaporator.
5. 500 MHz Bruker NMR spectrometer (DMX500, Bruker) equipped with a 5 mm TXI probe and a *z*-gradient system or similar instrument.

2.7. Software

1. Microsoft Excel™.
2. AMIX version 3.7 (Bruker Biospin) for bucketing NMR data.
3. SIMCA-P version 11.0 (Umetrics AB, Umea, Sweden) or comparable software for multivariate analysis.

3. Methods

3.1. Elicitation of Cannabis Cell Suspension Cultures

1. Inoculate 5 g of fresh Cannabis cells into an EM flask containing fresh MS medium.
2. Grow the cultures at 110 rpm and 25°C under a light intensity of 1,000–1,700 lx.
3. Five days after inoculation, add elicitors to Cannabis cell cultures or expose them to UV-irradiation (see Notes 1–3).
4. Collect the Cannabis cells using a Büchner funnel at different time periods (see Note 4).
5. Freeze the cells using liquid nitrogen and store them at −80°C (see Note 5).
6. Keep on freezing the medium of cell cultures if you plan to analyze it too.

3.2. Lyophilization of Plant Material

1. Grind the frozen plant material (Cannabis cells or tissues) using a precooled pestle and mortar under liquid nitrogen (see Note 6).
2. Transfer the grounded plant material into a plastic tube and keep it in dry ice, liquid nitrogen or freezer before lyophilization.
3. Place the samples in a freeze-dryer for 1–3 days (see Notes 7–8).

3.3. Extraction Method by Indirect Fractionation

1. Weigh 1 g (fresh) or 100 mg (dry) Cannabis plant material in a glass cap test tube.
2. Add 4 ml of MeOH:H_2O and 4 ml of $CHCl_3$ (see Note 9).
3. Mix for 30 s using a vortex at room temperature.
4. Ultrasonicate for 10 min at room temperature.
5. Centrifuge at 4°C and 9,000 × g for 20 min to obtain a clear supernatant.
6. Separate $CHCl_3$ fraction (lower phase) and MeOH:Water fraction (upper phase) using Pasteur pipettes and transfer to evaporating flasks.
7. Repeat a second extraction (solvent addition, mixing, sonication, centrifugation, and separation; see Note 10).
8. Evaporate each fraction using a rotary evaporator (see Note 11).
9. Resuspend in 1 ml of $CDCl_3$ (with 0.01% HMDSO-d_{18}) and 1 ml of MeOD:KH_2PO_4 buffer (with 0.1% TMSP-d_4), respectively. Mix for 1–2 min on a vortex (see Note 15).
10. Transfer the MeOD:KH_2PO_4 buffer solution to an Eppendorf tube and centrifuge for 5 min at maximum speed.
11. Load the MeOD:KH_2PO_4 buffer solution to 5-mm NMR tube (see Note 12).

12. Make a filter using a Pasteur pipette and tissue (Kimwipes™ or wool).
13. Load the filter on a 5-mm NMR tube.
14. Filter directly the $CDCl_3$ solution into the NMR tube (see Note 12).

3.4. Direct Extraction Method

1. Weigh 50–150 mg (dry or fresh) Cannabis plant material in a 2.0-ml Eppendorf tube or glass cap test tube.
2. Add 1 ml of MeOD:KH_2PO_4 buffer (with 0.1% TMSP-d_4; see Note 15) and 1 ml of $CDCl_3$ (with 0.01% HMDSO).
3. Mix for 30 s using a vortex at room temperature.
4. Ultrasonicate for 10 min at room temperature.
5. Centrifuge at 4°C and 9,000 × *g* (glass cap test tube) or 13,800 × *g* (Eppendorf tube) for 20 min to obtain a clear supernatants.
6. Separate $CDCl_3$ fraction (lower phase) and MeOD:KH_2PO_4 buffer fraction (upper phase) using Pasteur pipettes and transfer to Eppendorf tubes.
7. Centrifuge for 1 min at 16,200 × *g* at room temperature.
8. Transfer 800 μl of the MeOD:KH_2PO_4 buffer solution to a clean 5-mm NMR tube and 800 μl of the $CDCl_3$ solution to a clean 5-mm NMR tube (see Note 12).

3.5. Extraction Method for Cell Culture Medium

1. Put 10 ml cell culture medium in a separating funnel.
2. Add 10 ml ethyl acetate and shake the separating funnel.
3. Transfer the ethyl acetate phase into a beaker.
4. Repeat a second extraction using 10 ml ethyl acetate.
5. Dry the ethyl acetate phase with Na_2SO_4.
6. Transfer the ethyl acetate phase to an evaporating flask.
7. Evaporate using a rotary evaporator (see Note 11).
8. Resuspend in 1 ml of MeOD and mix for 1–2 min on a vortex.
9. Load 800 μl of MeOD solution to a 5-mm NMR tube (see Note 12).

3.6. NMR Measurement

1. Load the 5-mm NMR tube with your sample into the spectrometer (see Notes 13–14).
2. Set the sample temperature to 25°C and wait for thermal equilibration.
3. Tune and match the NMR tube.
4. Lock the spectrometer frequency to the deuterium resonance arising from the NMR solvents (MeOD or $CDCl_3$).
5. Shim the sample using either a manual or an automated method.

6. Determine the frequency of the water resonance and set the center of the spectrum to this frequency.
7. *For Standard 1H-NMR spectroscopy*: Set up pulse sequence comprising (relaxation delay-90°-acquire), where pulse power is set to achieve a 90° flip angle, 4.0 kHz spectral width and water pre-sat applied during 1.0-s relaxation delay (see Note 23).

   Processing parameters: Zero-fill to 64 k data points, apply exponential line broadening of 0.3 Hz. Free induction decay signals are transformed by Fourier with LB = 1.0 Hz, GB = 0 and PC = 1.0. After Fourier transformation, manually phase spectrum (zero and first phase), correct baseline, and calibrate the spectrum by setting TMPS peak at 0.00 ppm (for methanol:water fraction), MeOD peak at 3.30 ppm (for methanol fraction), $CDCl_3$ peak at 7.26 ppm or HMDSO peak at 0.07 ppm (for chloroform fractions). Record 128 scans for each sample (see Notes 16–21, 24 and 25).
8. *For J-resolved spectroscopy (homonuclear two-dimensional J-resolved NMR spectroscopy)*: Set up *J*-resolved pulse sequence, two-pulse echo sequence (relaxation delay-90°-[t1/2]–180°-[t1/2]-acquire) with water pre-sat during a relaxation delay of 1.5 s. Acquire FID using data matrix of 64 × 4,096 points covering 66 × 6,361 Hz, with 16 scans for each increment. Zero-fill the data to 128 × 4,096 and apply a sine bell-shaped window function in both dimensions before magnitude mode two-dimensional Fourier transformation. Tilt the resulting spectra along the rows by 45° relative to the frequency axis and symmetrize about the central line along F2. Manually correct baseline and calibrate to the internal standard (see Notes 19–21).
9. *For ^1H-^1H COSY (two-dimensional homonuclear ^1H-^1H correlated NMR spectroscopy)*: Use a phase sensitive/magnitude mode standard three pulse sequence with pre-saturation during relaxation delay of 1.0 s. A data matrix of 512 × 4,096 points covering 6,361 × 6,361 Hz, record with eight scans for each increment. Zero fill data to 4,096 × 4,096 points and apply a sine 2 bell-shaped window function shifted by /2 in the F1 and /4 in the F2 dimension before States-TPPI type two-dimensional Fourier transformation. Manually phase all spectra, correct baseline, and calibrate to the internal standard).
10. *For ^1H-^{13}C HMBC (two-dimensional heteronuclear multiple bond correlation NMR spectroscopy)*: Use a data matrix of 254 × 4,096 points covering 27,164 × 6,361 Hz with 256 scans for each increment with a relaxation delay of 1.0 s. The data should be linear to 512 × 4,096 points using 32 coefficients before magnitude type two-dimensional Fourier transformation and apply a sine bell-shaped window function shifted by /2 in

the F1 dimension and /6 in the F2 dimension. Calibrate all spectra according to the internal standard (^{1}H: TMSP = 0 ppm and ^{13}C: CD_3OD = 49.0 ppm).

3.7. Data Processing

1. Convert NMR spectra to an ASCII file using AMIX software. Scale the spectral intensities to HMDSO for the $CHCl_3$ fractions and TMSP for MeOD:Water fractions. For MeOD fractions scale to total intensity.
2. Integrate the peaks into a small bin (bucket) from 0.04 ppm to avoid signal fluctuation by pH or concentration changes.
3. Delete solvent signals (δ3.28–3.34 for MeOD, δ4.6–5.8 for water, and δ7.18–7.30 for $CDCl_3$).
4. Copy the ASCII data to an Excel table to identify and classify your samples according to your requirements.
5. Perform the principal component analysis (PCA) or partial least square discriminant analysis (PLS-DA) using the SIMCA-P software according to the user guides. Select the Pareto scaling for a variance numerically equal to its initial standard deviation.
6. Display the score and loading plots.
7. Look for patterns or clusters in the dataset. Identify the metabolites responsible for those differences or similarities among the datasets, either by comparison with NMR signals to reference compounds or by two-dimensional NMR spectra.

3.8. Quantitative Analysis

1. Identify the proton signals of the target compounds and the internal standard.
2. Determine the integral of the target and standard peaks (see Note 22).
3. Based on the quantity of the internal standard, the concentration (in μmol/100 mg of dry cell material) is calculated using the following equation (see Note 23):

   Concentration (μmol/100 mg dry weight) = [integral of target compound/integral of internal standard] × [number of protons from internal standard/number of protons from target compound] × quantity of internal standard (μmol).
4. Based on the weight of the internal standard, the concentration (in μg/100 mg of cell material) is calculated using the following equation (see Note 23):

   Concentration (μg/100 mg dry weight) = [integral of target compound/integral of internal standard] × [number of protons from internal standard/number of protons from target compound] × [MW of target compound/MW of internal standard] × weight of internal standard (μg).

4. Notes

1. Elicitation should be done during the exponential phase of growth.
2. Elicitors should be sterilized by autoclaving or filtration (0.22-μm filter) before adding to cell suspensions.
3. Methyl-jasmonic acid and jasmonic acid can be dissolved in EtOH or in a 30% EtOH solution (*v/v*).
4. Use identical harvesting times because the metabolite levels from plants vary throughout the day.
5. Liquid nitrogen should be handled carefully. Always use glasses and gloves.
6. Grinding to fine powder plant material has the advantage of improving the efficiency of extraction.
7. In the freeze-dryer, place uncovered tubes with sample or cover with perforated paper.
8. After lyophilization keep your samples in a dry environment because it can absorb moisture.
9. Prepare a new MeOH:Water (1:1) solution every time that you perform extractions. Do not store it because the ratio between MeOH and water may vary over time.
10. After extraction, a fractionation step by solid phase extraction (SPE) may facilitate identification of secondary metabolite signals on removing primary metabolite signals. Use C_{18} or silica gel cartridges (1 or 3 ml).
11. Extracts can be stored at 4°C.
12. Clean NMR tubes using in the following order of solvents: water, ethanol, methanol, dicloromethane, and acetone. Use an NMR tube cleaner.
13. Do not expose NMR tubes to high temperatures during drying because they may lose their properties of uniformity and/or concentricity.
14. On cleaning NMR tubes, be careful not to scratch and not to use reagents that can attack the glass or bear paramagnetic impurities difficult to remove (e.g., chromic mixture).
15. The extracts should be clear, no solid waste with a homogeneous volume.
16. Before NMR measurements the extracts should be placed at room temperature at least half an hour in order to avoid bad shimming owing to the temperature difference in the samples.
17. Before loading the NMR tube into the spectrometer, clean it with a tissue in order to take out grease from your hands.

18. Always clean new NMR tubes before the first use as they may have grease or impurities.
19. Use a suitable buffer, as KH_2PO_4, because the pH of the extracts can have an influence on the chemical shifts of compounds containing acid and basic groups.
20. Interactions with metal ions, hydrogen bonding, and other intermolecular interactions can also cause chemical shift displacements.
21. The chemical shifts of some metabolites can be changed by pH or their concentration (e.g., fumaric acid, citric acid, or malic acid).
22. By visual inspection of an NMR spectrum from Cannabis material, signals of amino acids (δ0.5–2.0), organic acids (δ2.0–3.0), sugars (δ3.0–5.0), and aromatic compounds (δ5–10) from methanol/water fractions, and signals of terpenoids and steroids (δ0.5–3.0), fatty acids (δ0.8–1.9 and δ5.08–5.4), and cannabinoids (δ4.3–8.16) from chloroform fractions can be identified.
23. For identification of metabolites compare chemical shift, kind of peak, and *J* value with those from reported reference compounds or your own NMR spectra database. Be careful that the values were obtained under the same conditions previously reported.
24. According to our conditions, the following metabolites can be identified from Cannabis cell cultures and plants based on chemical shift (δ), kind of peak, and coupling constant (*J*, Hz) from ^{1}H-NMR spectra (21–24).

In MeOD:KH_2PO_4 Buffer

Adenosine δ6.04 (H-1′, *d*, *J*=6.6), δ8.23 (H-8, *s*), δ8.35 (H-2, *s*).

Alanine δ1.48 (H-β, *d*, *J*=7.2), δ3.73 (H-α, *q*, *J*=7.2).

Asparagine δ2.87 (H-3b, *dd*, *J*=16.9, 7.6), δ2.96 (H-3a, *dd*, *J*=16.9, 4.3), δ4.01 (H-2, *dd*, *J*=7.6, 4.3).

Aspartic acid δ2.83 (H-β, *dd*, *J*=17, 7.9), δ2.94 (H-β′, *dd*, *J*=17, 4.0), δ3.95 (H-α, *dd*, *J*=8.1, 4.0).

γ-aminobutyric acid δ1.90 (H-3, *m*, *J*=7.5), δ2.31 (H-2, *t*, *J*=7.5), δ3.00 (H-4, *t*, *J*=7.5).

Choline δ3.21 (H-1′, H-2′, H-3′, *s*).

Cytidine δ5.86 (H-5, *d*, *J*=8.0), δ5.91 (H-1′, *d*, *J*=4.3), δ7.93 (H-6, *d*, *J*=8.0).

Ethanol glucoside δ1.24 (H-2, *t*, *J*=6.9).

Fumaric acid δ6.54 (H-2, H-3, *s*).

α-Glucose δ5.19 (H-1, *d*, *J*=3.8), δ5.24 (H-1, *d*, *J*=3.7).

β-Glucose δ4.58 (H-1, *d*, *J*=7.9), δ4.64 (*d*, *J*=8.0).

Glutamic acid δ2.05 (H-β, *m*), δ2.36 (H-γ, *m*).

Glutamine δ2.13 (H-β, *m*), δ2.49 (H-γ, *m*).

Isoleucine δ0.95 (H-5, *t*, *J*=7.5), δ1.02 (H-6, *d*, *J*=6.8).

Leucine δ0.97 (H-5, *d*, *J*=6.7), δ0.98 (H-6, *d*, *J*=6.7).

Phenylalanine δ3.09 (H-3, *dd*, *J*=14.4, 8.4), δ3.30 (H-3′, *dd*, *J*=14.4, 9.6), δ3.94 (H-2), δ7.36 (H-5, H-6, H-7, H-8, H-9, *m*).

Sucrose δ4.19 (H-1′, *d*, *J*=8.5), δ5.40, 5.42 (H-1, *d*, *J*=3.8).

Threonine δ1.33 (H-γ, *d*, *J*=6.5), δ3.52 (H-α, *d*, 4.9), δ4.24 (H-β, *m*).

Tryptophan δ3.27 (H-3), δ3.50 (H-3′), δ3.98 (H-2), δ7.14 (H-8, *t*, *J*=7.7), δ7.22 (H-7, *t*, *J*=7.7), δ7.29 (H-11, *s*), δ7.47 (H-9, *dt*, *J*=8.0, 1.3), δ7.72 (H-6, *dt*, *J*=8.0, 1.3).

Tyramine δ6.85 (H-3, H-5, *d*, *J*=8.4), δ7.20 (H-2, H-6, *d*, *J*=8.4).

Tyramine glycoside δ7.11 (H-3, H-5, *d*, *J*=8.4), δ7.30 (H-2, H-6, *d*, *J*=8.4).

Tyrosine δ3.01 (H-β), δ3.20 (H-β′), δ3.86 (H-α), δ6.85 (H-3, H-5, *d*, *J*=8.4), δ7.18 (H-2, H-6, *d*, *J*=8.4).

Tyrosol δ6.80 (H-3, H-5, *d*, *J*=8.4), δ7.11 (H-2, H-6, *d*, *J*=8.4).

Trigonelline δ8.86 (H-4, H-6, *m*), δ9.15 (H-2, *s*).

Valine δ1.00 (H-γ, *d*, *J*=7.0), δ1.05 (H-γ′, *d*, *J*=7.0).

In CD_3OD

Phenylalanine δ3.14 (H-3, *dd*, *J*=15.9, 8.9), δ3.86 (H-2, *dd*, *J*=8.0, 4.0), δ7.03 (H-7, *t*, *J*=8.0), δ7.18 (H-11, *s*), δ7.35 (H-9, *d*, *J*=8.0), δ7.68 (H-6, *d*, *J*=8.0).

Gentisic acid δ6.61 (H-3, *d*, *J*=8.2), δ6.99 (H-4, *dd*, *J*=8.2, 2.5), δ7.21 (H-6, *d*, 2.5).

Glutamyl-tyramine δ2.05 (H-3″, *m*), δ2.38 (H-4″, *t*, *J*=7.2), δ3.56 (H-2″, *dd*, 15.0, *J*=7.2), δ3.34 (H-2′, *t*, *J*=8.0), δ2.68 (H-1′, *t*, *J*=8.0), δ6.69 (H-3, *d*, *J*=8.0), δ7.01 (H-2, *d*, *J*=8.0).

Tryptophan δ3.07 (H-3, *dd*, *J*=15.3, 8.0), δ3.91 (H-2, *dd*, *J*=8.0, 4.0), δ7.31 (H-5, *dd*, *J*=8.4, 1.6), δ7.39 (H-6, *t*, *J*=8.4).

Tyramine δ6.69 (H-3, H-5, *d*, *J*=8.4), δ6.78 (H-2, H-6, *d*, *J*=8.4).

Tyrosine δ6.62 (H-3, H-5, *d*, *J*=8.4), δ6.80 (H-2, H-6, *d*, *J*=8.4).

Tyrosol δ6.64 (H-2, H-6, *d*, 8.4), δ6.80 (H-2, H-6, *d*, *J*=8.4).

In $CDCl_3$

Δ^9-Tetrahydrocannabinolic acid (Δ^9-THCA) δ0.90 (H-5′, *t*, *J*=6.9), δ2.49 (H-1′, *m*), δ3.23 (H-10a, *dm*, *J*=7.0), δ6.25 (H-4, *s*), δ6.39 (H-10, *s*), δ12.19 (OH, *s*).

Cannabidiolic acid (CBDA) δ0.89 (H-5″, *t*, *J*=6.9), δ2.10, 2.20 (H-4, *m*), δ4.09 (H-1, *m*), δ5.56 (H-2, *s*), 6.26 (H-5′, *s*), δ6.63 (6′-OH, *s*), δ11.93 (2′-OH, *s*).

Δ^9-Tetrahydrocannabinol (Δ^9-THC) δ0.87 (H-5′, *t*, *J*=7.0), δ2.42 (H-1′, *m*, *J*=7.3, 1.6), δ3.20 (H-10a, *dm*, *J*=10.9), δ4.87 (OH, *s*), δ6.14 (H-2, *d*, *J*=1.6), δ6.26 (H-4, *d*, *J*=1.6), δ6.30 (H-10, *s*).

Δ^8-Tetrahydrocannabinol (Δ^8-THC) δ0.88 (H-5′, *t*, *J*=7.1), δ2.44 (H-1′, *td*, *J*=8.3, 2.1), δ2.70 (H-10a, *td*, 10.8, 4.8), δ3.24 (H-10, *dd*, *J*=16.5, 3.7), δ6.27 (H-4, *d*, *J*=1.5).

Cannabinol (CBN) δ0.89 (H-5′, *t*, *J*=6.8), δ2.5 (H-1′, *t*, *J*=7.5), δ5.13 (OH, *s*), δ6.29 (H-2, *d*, *J*=1.1) δ6.43 (H-4, *d*, *J*=1.1), δ7.07 (H-8, *d*, *J*=7.9), δ7.14 (H-7, *d*, *J*=7.9), δ8.16 (H-10, *s*).

Cannabidiol (CBD) δ0.88 (H-5″, *t*, *J*=6.8), δ1.55 (H-2′, *q*, *J*=7.6), δ2.43 (H-1″, *t*, *J*=7.5), δ3.90 (H-1, *m*, *J*=11.8), δ5.02 (6′-OH, *s*), δ5.57 (H-2, *s*), δ5.99 (2′-OH, *s*), δ6.26 (H-3′, *brs*).

Cannabigerol (CBG) δ0.90 (H-5″, *t*, *J*=6.9), δ1.56 (H-2″, *q*, *J*=7.8), δ2.45 (H-1″, *t*, *J*=7.5), δ3.41 (H-1′, *d*, *J*=7.0), δ5.07 (H-6′, *m*), δ5.29 (H-2′, *m*), δ5.36 (OH, *s*), δ6.0 (H-6, *s*), δ6.26 (H-4, *s*).

In $(CD_3)_2CO$ (Acetone-d_6)

Cannflavin A δ1.54 (H-8″, *s*), δ1.78 (H-9″, *s*), δ1.93 (H-4″, *t*, *J*=10.0), δ2.03 (H-5″, *t*, *J*=7.21), δ3.99 (OMe, *s*), δ5.04 (H-6″, *t*, *J*=7.08), δ7.02 (H-5′, *d*, 8.28), δ7.61 (H-2′, *d*, 1.88), δ13.30 (5-OH, *s*).

Cannflavin B δ1.80 (H-4″, *s*), δ3.36 (H-1″, *d*, *J*=7.12), δ3.99 (OMe, *s*), δ5.29 (H-2″, *tt*, *J*=7.24, 1.52), δ7.00 (H-5′, *d*, *J*=8.28), δ7.60 (H-2′, *d*, *J*=1.88), δ13.30 (5-OH, *s*).

25. For identification of the cannaflavins A and B purification steps by CC over HP-20 resin, silica gel, and Sephadex LH-20 are required.
26. A relative concentration of the intensities from target compounds can also be calculated, where a 100% value is assigned to control samples and increments or decrements are calculated in treated-samples.
27. Insufficient relaxation time gives an underestimation on the concentration of the compounds in the sample.

28. After an NMR measurement, the samples can be used for further analyses by column chromatography (Sephadex LH-20 column chromatography), HPLC or LC–MS. A cannabinoid profiling from chloroform fractions or flavonoid profiling from methanol:water fractions can be obtained.

29. For calibration of the spectrum, the internal standard or the solvent signal can be used.

References

1. Flores-Sanchez IJ, Verpoorte R (2008) Secondary metabolism in cannabis. Phytochem Rev 7: 615–639
2. Radwan MM, Ross SA, Slade D, et al (2008) Isolation and characterization of new cannabis constituents from a high potency variety. Planta Med 74: 267–272
3. Radwan MM, ElSohly MA, Slade D, et al (2008) Non-cannabinoid constituents from a high potency *Cannabis sativa* variety. Phytochemistry 69: 2627–2633
4. Debruyne D, Moulin M, Bigot Mc et al (1981) Identification and differentiation of resinous cannabinoid textile Cannabis: combined use of HPLC and high-resolution GLC. Bull Narc 33: 49–58
5. Ross SA, ElSohly HN, Elkashoury EA, et al (1996) Fatty acids of Cannabis seeds. Phytochem Anal 7: 279–283
6. Hazekamp A, Simons R, Peltenburg-Looman A, et al (2004) Preparative isolation of cannabinoids from *Cannabis sativa* by centrifugal partition chromatography. J Liq Chromatogr Relat Technol 27: 2421–2439
7. Raharjo TJ, Verpoorte R (2004) Methods for the analysis of cannabinoid in biological materials: a review. Phytochem Anal 15: 79–94
8. Choi YH, Hazekamp A, Peltenburg-Looman A, et al (2004) NMR assignments of the major cannabinoids and cannabiflavonoids isolated from flowers of *Cannabis sativa*. Phytochem Anal 15: 345–354
9. Ross SA, ElSohly MA, Sultana GNN, et al (2005) Flavonoid glycosides and cannabinoids from the pollen of *Cannabis sativa* L. Phytochem Anal 16: 45–48
10. Hazekamp A, Peltenburg-Looman A, Verpoorte R, et al (2005) Chromatographic and spectroscopic data of cannabinoids from *Cannabis sativa* L. J Liq Chromatogr Relat Technol 28: 2361–2382
11. Gutierrez A, Rodriguez IM, Del Rio JC (2006) Chemical characterization of lignin and lipid fractions in industrial hemp bast fibers used for manufacturing high-quality paper pulps. J Agric Food Chem 54: 2138–2144
12. Flores-Sanchez IJ, Verpoorte R (2008) PKS activities and biosynthesis of cannabinoids and flavonoids in *Cannabis sativa* L. plants. Plant Cell Physiol 49: 1767–1782
13. Holmes E, Tang H, Wang Y et al (2006) The assessment of plant metabolite profiles by NMR-based methodologies. Planta Med 72: 771–785
14. Ward JL, Beale MH NMR spectroscopy in plant metabolomics. In Saito K, Dixon RA, Willmitzer L (ed) (2006) Plant Metabolomics, Biotechnology in Agriculture and Forestry, Vol. 57. Springer-Verlag, Berlin Heidelberg
15. Fiehn O (2002) Metabolomics: The link between genotypes and phenotypes. Plant Mol Biol 48: 155–171
16. Sumner LW, Mendes P, Dixon RA (2003) Plant metabolomics: large-scale phytochemistry in the functional genomics era. Phytochemistry 62: 817–836
17. Rochfort S (2005) Metabolomics reviewed: A new "omic" platform technology for systems biology and implications for natural products research. J Nat Prod 68: 1813–1820
18. Kim HK, Verpoorte R (2010) Sample preparation for plant metabolomics. Phytochem Anal 21: 4–13
19. Kim HK, Choi YH, Verpoorte R (2010) NMR-based metabolomic analysis of plants. Nat Protoc 5: 536–549
20. Ludwing C, Viant MR (2010) Two-dimensional *J*-resolved NMR spectroscopy: Review of a key methodology in the metabolomics toolbox. Phytochem Anal 21: 22–32
21. Choi YH, Kim HK, Hazekamp A, et al (2004) Metabolomic differentiation of *Cannabis*

sativa cultivars using ^{1}H-NMR spectroscopy and principal component analyses. J Nat Prod 67: 953–957

22. Flores-Sanchez IJ, Pec J, Fei J, et al. (2009) Elicitation studies in cell suspension cultures of *Cannabis sativa* L. J Biotechnol 143: 157–168
23. Pec J, Flores-Sanchez IJ, Choi YH et al (2010) Metabolic analysis of elicited cell suspension cultures of *Cannabis sativa* L. by ^{1}H-NMR spectroscopy. Biotechnol Lett 32: 935–941
24. Choi YH, Hazekamp A, Peltenburg-Looman AMG, et al. (2004) NMR assignments of the major cannabinoids isolated from flowers of *Cannabis sativa*. Phytochem Anal 15: 354–354

Chapter 28

Metabolome Analysis of Gram-Positive Bacteria such as *Staphylococcus aureus* by GC-MS and LC-MS

Manuel Liebeke, Kirsten Dörries, Hanna Meyer, and Michael Lalk

Abstract

The field of metabolomics has become increasingly important in the context of functional genomics. Together with other "omics" data, the investigation of the metabolome is an essential part of systems biology. Beside the analysis of human and animal biofluids, the investigation of the microbial physiology by methods of metabolomics has gained increased attention. For example, the analysis of metabolic processes during growth or virulence factor expression is crucially important to understand pathogenesis of bacteria. Common bioanalytical techniques for metabolome analysis include liquid and gas chromatographic methods coupled to mass spectrometry (LC-MS and GC-MS) and spectroscopic approaches such as NMR. In order to achieve metabolome data representing the physiological status of a microorganism, well-verified protocols for sampling and analysis are necessary. This chapter presents a detailed protocol for metabolome analysis of the Gram-positive bacterium *Staphylococcus aureus*. A detailed manual for cell sampling and metabolite extraction is given, followed by the description of the analytical procedures GC-MS and LC-MS. The advantages and limitations of each experimental setup are discussed. Here, a guideline specified for *S. aureus* metabolomics and information for important protocol steps are presented, to avoid common pitfalls in microbial metabolome analysis.

Key words: Metabolomics, Liquid chromatography, Gas chromatography, Mass spectrometry, Bacteria, *Staphylococcus aureus*, Energy charge, Protocol, Sampling, Ion-pairing

1. Introduction

1.1. Microbial Metabolome Analysis

Metabolomics, the qualitative and quantitative measurement of metabolites within a system, is part of the functional genomics era (1, 2) and added in the last years substantial input into the physiological understanding of organisms (3–6). In particular, microbial metabolomics aims to analyze the metabolome of bacterial cells (7) which consist of approximately 200–2,000 metabolites referring to genome based metabolic network models (8). Today, none of the

Michael Kaufmann and Claudia Klinger (eds.), *Functional Genomics: Methods and Protocols*,
Methods in Molecular Biology, vol. 815, DOI 10.1007/978-1-61779-424-7_28, © Springer Science+Business Media, LLC 2012

established analytical techniques (e.g., ^{1}H-NMR, GC-MS, LC-MS, or CE-MS) are separately able to cover the complete diversity of metabolites of an organism. In principle, these methods can be divided into two groups: (1) spectroscopic analysis and (2) spectrometric approaches coupled to chromatographic separation techniques. ^{1}H-NMR spectroscopy is a powerful method for the investigation of metabolites. The main benefit of NMR is the ability to quantify identified metabolites by integration of proton NMR signals. Major drawbacks of NMR are less sensitivity and separation capacity than LC-MS or GC-MS. In contrast to NMR, chromatographic separation techniques provide the opportunity to analyze subgroups of the chemically diverse metabolome. By hyphenation of high-performance liquid chromatography (HPLC) and gas chromatography (GC) with mass spectrometry, a broad variety of different classes of compounds are accessible with high sensitivity.

In addition to the analytical platform, the method for sample generation should be carefully considered. An inappropriate slow sampling could influence the sample results based on high metabolite turnover rates in the cells. The generation of an appropriate microbial metabolome sample is more complex than routine biofluid sampling and requires an evaluated protocol for each organism (9, 10). No changes in the metabolome should occur during the sampling procedure so that the sample accurately reflects the biological status of interest. Important steps that have to be taken into account for metabolome analysis are (1) separation of cells from the cultivation medium via centrifugation or filtration, (2) washing procedures using appropriate solutions, (3) fast quenching procedures using organic solvent and/or liquid nitrogen (check leakage problems for stated points 1–3), (4) efficient metabolite extraction solutions with low impact on metabolite stability, and (5) cell disruption in a rapid but gentle and effective way. In this chapter, the workflow for sampling and analyzing the intracellular metabolite pool of *S. aureus* is described. The sampling work flow was critically evaluated (10) and applied to different biological approaches (11, 12). For other prokaryotic or eukaryotic cells, the sampling protocol steps must be tested and adapted, if necessary, for the investigated organism. Especially differences in the cell wall structure make it necessary to develop an organism-specific cell disruption protocol. A mechanical cell disruption is expedient for some organisms. For example, the Gram-negative organism *Escherichia coli* needs only an organic solvent cell disruption (13). Further, the variable composition of the metabolome requires a precise testing of the extraction solution to be used. In addition, the broad variety of the cell matrix from different microbes has a strong influence on the metabolome analysis, affecting sampling, cell disruption, extraction procedures, and analytical procedures. A well-established and organism-specific protocol is, therefore, necessary to evaluate. A feasible and established parameter to control the sampling protocol is the adenylate energy charge (see Subheading 3.3.2). Finally, it is recommended to prepare at least 3–5 biological replicates of each condition analyzed, since all steps of the protocol can vary.

1.2. GC-MS Metabolome Analysis

Gas chromatography is based upon the separation of volatile compounds in a gaseous phase. By carriage of suitable derivatized compounds with molecular masses up to 650 Da, a separation of complex mixtures is possible. By using a broadly applicable derivatization procedure such as the combination of methoxylation and *N*-methyl-*N*-trimethylsilylfluoracetamide silylation, a wide coverage of central pathway metabolites is achievable. The detection is made by mass spectrometry. The general procedure is the electron impact ionization (EI) that forms characteristic fragments of metabolites. These mass spectrometric patterns can be used in conjunction with the information of the chromatographic separation to identify the compounds by database comparison as described in detail in Subheading 3.2. This approach facilitates a robust identification and quantification of hundreds of metabolites and is used for a wide range of organisms (14). Nevertheless, the mandatory temperature dependent derivatization procedure and limited capacity to high masses (>650 Da) limits GC-MS to some compound classes and makes additional LC-MS approaches essential.

1.3. LC-MS Metabolome Analysis

The use of LC-MS for untargeted metabolome analyses (metabolomic profile) in diverse fluids or extracts is a common part of metabolomics (15). The wide application range of LC is based on the diversity of stationary and mobile phases (16–18). LC approaches for bacterial metabolome analysis mostly focus on polar metabolites, especially phosphate-containing compounds, of which the majority cannot be analyzed via GC-MS. Therefore HILIC and reversed-phase methods with ion-pairing reagents were used (12, 19, 20), affording access to nucleotides, sugar-phosphates, cell wall precursors, cofactors, vitamins, and others. For untargeted metabolomics, a time-of-flight mass spectrometer (TOF-MS) with an electrospray-ionization technique is sufficient and results in minimal fragmentation and accurate masses of molecular ions. The described IP-LC-MS method is an essential addition to GC-MS to get higher metabolome coverage and is described in detail in Subheading 3.3.

2. Materials

2.1. Cell Sampling and Metabolite Extraction

1. Standard safety laboratory conditions for the cultivation and sampling of *Staphylococcus aureus* with regards to local law demands concerning handling of pathogen bacteria.
2. *Shake flasks*: Caution: the shake flask has to be useful for a fast sampling procedure (e.g., rapidly to open, wide opening to

enable fast sample removal). The volume cell suspension per volume flask should be 1/5 to avoid, for example, oxygen limitations.

3. 200-μl and 1,000-μl pipette.
4. Volumetric glass pipette, size depending on sample and wash volume.
5. 50-ml Tubes (two per sample).
6. 15-ml Tubes (one per sample).
7. Vacuum pump: (~1.7 m^3/h).
8. A filter based system from millipore® (Order No.: XX1004700) was applied and tested for *S. aureus* metabolome sampling (see Fig. 1). This should be tested and if necessary changed for other bacteria. Details of filter system: the filtration area is 9.6 cm^2. The sterile filter paper has a pore diameter of 0.45 μm and a filter size of 47 mm. The inner funnel diameter is 7.6 cm and the inner height is 22.9 cm. As the fitting outlet a No. 8 perforated silicone stopper was used. The funnel and the base were built up by Borosilicate glass.
9. Small tweezer.
10. Millipore filter S-Pak (order no.: HAWG047S6) – see Note 4.
11. Dewar vessel (approx. volume of 1 L), insulated gloves are recommended when working with liquid nitrogen.
12. Liquid nitrogen (volume dependent of the number of samples and the expenditure of time of the experiment).
13. Extraction solution: 60% (*w*/*v*) ethanol HPLC grade or p.a. (purity ≥ 99.8%) ethanol and water in p.a. quality 60% (*w*/*v*). Ethanol is established for a global metabolome analysis of *S. aureus* (see Note 2).
14. Washing solution (0.6% sodium chloride p.a. in water, for chemical defined medium (10)) (see Note 3).
15. Internal standard solution (see Subheading 2.4).
16. Glass beads, diameter: 0.10–0.11 mm (e.g., Sartorius®), 0.5 ml beads per tube with 1 ml quenched cell suspension.
17. Cell homogenizer (e.g., Precellys® Bertin Technologies or fast prep® MP biomedicals).
18. Suitable tubes for the homogenizer.
19. 2-ml Syringe.
20. Sterile filter with pore size 0.45 μm (e.g., Filtropur, S Sarstedt®) – see Note 4.

Different solutions for cell sampling and metabolite extraction should be prepared in advance as followed:

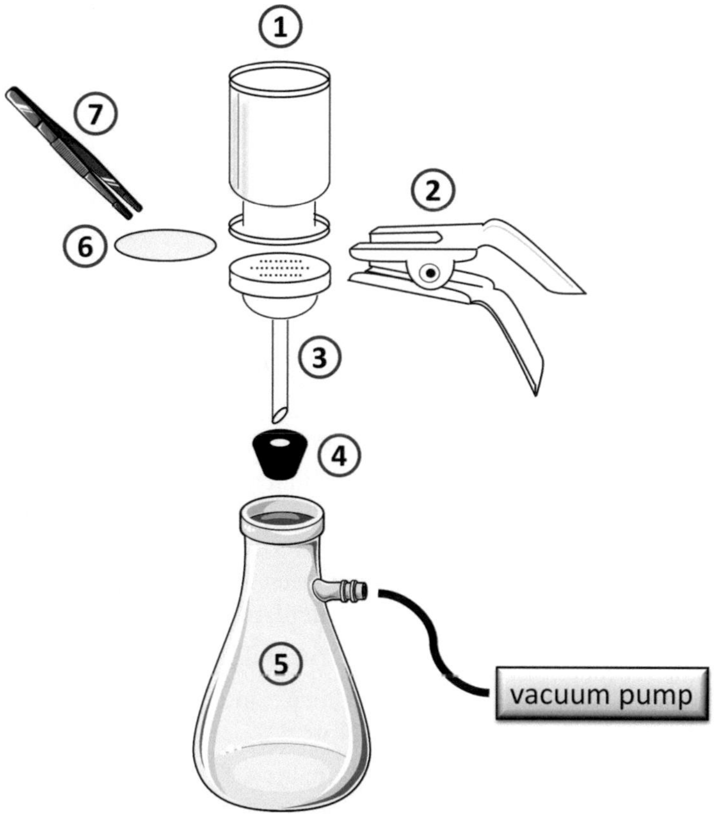

Fig. 1. Filter system for *Staphylococcus aureus* cell sampling consisting of a glass filter holder assembly with funnel (1), clamp (2), fritted base (3), stopper (4), and a vacuum filtering flask 0.5–1.0 L (5). Put the filter (6) with tweezers (7) on the fritted base.

21. Preparation of the extraction solution (see Note 2). Fill 5 ml extraction solution into a 50-ml falcon tube before starting sampling. Further add the internal standard (appropriate to analytic technique applied, see Subheadings 2.4, 3.2 and 3.3) to the extraction solution.

    Caution: The extraction solution with internal standards has to be stored at −20°C before sampling and on ice during sampling.

22. Preparation of the washing solution. Calculate osmolarity of the cultivation medium and prepare a sodium chloride solution in water of same osmolarity and use it as washing solution. Per sample 10 ml of washing solution are needed. Store the washing solution at 4°C before sampling and on ice during sampling.

    Caution: An isotonic washing solution (calculated based on osmolarity of cultivation medium) is essential to avoid cell lysis during washing (see also Note 3).

23. Prearrangement of sampling device. Start the vacuum pump 10 min before the experiment starts, so the vacuum pump has the full power during sampling.
24. Preparation of the filter system. Connect the vacuum pump to the filter flask. The filter plating will be connected to the filter flask by the rubber bung with hole. Place the filter on the filter plating with a tweezers and put the collecting vessel at the top by the metal clamp. (see Fig. 1).

2.2. GC-MS Metabolome Analysis

1. *N*-Methyl-*N*-(trimethylsilyl)trifluoroacetamide (MSTFA) with 1% trimethylchlorosilane (TMCS) as catalyst, store at 4°C and under dry conditions (see Note 5).
2. Methoxyamine hydrochloride (MeOx).
3. Saturated fatty acid methyl esters (FAMEs) of C_8, C_9, C_{10}, C_{12}, C_{14}, C_{16}, C_{18}, C_{20}, C_{22}, C_{24}, C_{26}, C_{28}, and C_{30} linear chain length. FAMEs were common as retention time standards but also homologous alkanes and fatty acids were in use for retention index calculation (21, 22).
4. Myristic acid d_{27}- retention time lock (RTL) locking substance.
5. Ethylacetate and acetone, as washing solutions for GC syringe (see Note 5).
6. High-quality GC-MS vials with microinserts for small volume injections, suitable for autosampler.
7. Helium as carrier gas for gas chromatography.
8. Pipettes and tips suitable for handling organic solvents.
9. Heating block for small 2-ml tubes.
10. PC and software to control GC-MS.
11. Gas chromatograph (e.g., Agilent® 6890GC system with split/splitless injector and electronic pressure control); *settings for Inlet*: Split, initial temp. 250°C, pressure 8.80 psi (0.61 bar), split ratio: 25:1 (depending on accessible biomass), split flow: 25.0 ml/min, total flow: 29.2 ml/min, gas saver: on, saver flow: 20 ml/min at 2.0 min; *settings for the oven*: exact settings used for Agilent® Fiehn Library see manual (see Note 6), settings used for *S. aureus* metabolome samples were as followed, 1 min at 70°C, increasing by 1.5°C/min to 76°C, and then by 5°C/min to 330°C, followed by a 10 min isothermal cool-down to 70°C, transfer line temp: 280°C.
12. Conical single taper split/splitless liner (e.g., Agilent® part no. 5062–3587).
13. DB5-MS column (Agilent® J&W Scientific, Folsom, CA, USA), 30 m long with 0.25 mm inner diameter and 0.25 μm film thickness.

14. Mass spectrometer with, for example, quadrupole mass selective detector (Agilent® 5973 Network MSD) or time-of-flight mass detector (LECO® Instruments Pegasus IV); *settings for MS*: for Agilent® 5973MSD as followed, full-scan modus (45–650 *m/z*) at a rate of 2 scans/s, solvent delay 6.00 min, MS quad temp: 150°C, MS source temp: 250°C.

    Preparation of solutions and standards for GC-MS metabolome analysis:
15. Prepare fresh methoxyamine hydrochloride solution in pyridine (20 mg/L) (see Note 5) and store not longer than 7 days at 4°C.
16. Dissolve retention time standards (FAMEs) in chloroform at a concentration of 0.8 mg/ml (under C_{18}) and 0.4 mg/ml (above C_{18}). Can be stored at −20°C.
17. Dissolve myristic acid d_{27} in water–methanol–isopropanol, 2–5–2 (*v/v/v*). Can be stored at −20°C.
18. Prepare internal standard solution (*see* Subheading 2.4), store in aliquots at −20°C.
19. Prepare QC samples for GC-MS measurements (see Subheading 2.4).

    The following steps should be performed to ensure reliable measurement parameters for GC-MS:
20. Run an air/water check to probe for leakages and quality of gas cleaning cartridges.
21. Clean your ion source frequently, depending on sample matrix and sample throughput.
22. Tune MS in appropriate time intervals, use Agilent® manual procedures for tuning with perfluorotributylamine (PFTBA) and use the Agilent® guidelines for report evaluation (see Note 7).
23. Check retention time locking and retention index calibration in regular periods. Constant values are very important for metabolite analysis. Both used in combination with the EI spectra helps to identify metabolites with databases build up under comparable conditions. If not correct, run new RTL procedure and check system.
24. Prove signal intensities in QC samples to check injection and detection system.

2.3. LC-MS Metabolome Analysis

1. Nitrogen (stream).
2. Tributylamine – ion-pairing reagent (see Notes 5 and 8).
3. HPLC solvent bottles with cap (e.g., Schott, amber Duran® glass bottle with light protection).
4. Mass tune mixture (e.g., Agilent® low tune mix).
5. pH meter.

6. HPLC vials with cap and microinserts for small volume injections.
7. HPLC system equipped with a quaternary pump, an online degasser, and an autosampler (e.g., Agilent® liquid chromatographic System 1100).
8. Column RP-C_{18} Waters® Symmetrie Shield (150×4.6 mm, 3.5 μm) with C_{18} waters® precolumn, flow rate of 0.3 ml/min at a temperature of 25°C. For stable chromatography, a constant temperature is necessary. For this purpose, a column oven is most suitable.
9. Gradient with mobile phases A and B: after a 5 min prerun with 100% mobile phase A the gradient was as followed: mobile phase A 100% for 2 min, 100–80% in 2 min, 80–69% in 11 min, 69–40% in 19 min, 40–0% in 5 min, 0% hold for 15 min, back to 100% A in 6 min, and hold for additional 3 min. At all, the time for a total run was 68 min with the possibility to inject the mass tune mixture within the first 5 min of the chromatography for internal mass calibration of each sample.
10. Mass spectrometer; other than in GC-MS analysis, the variety of mass spectrometers potentially coupled to HPLC systems is quite large. Choose for metabolite analysis optimal source and detector parameters depending on the instrument type; see instructions of the manufacturer. In general, the setup should consist of electrospray ionization in negative ion mode (see Note 8), full scan and a mass range 50–2,500 Da (see Note 9).

    Preparation of solutions and standards for LC-MS metabolome analysis:
11. Prepare mobile phases. Avoid introducing gas in the solutions and, if necessary, use a degasser or an ultrasonic bath. Mobile phase A consists of 5% methanol (gradient grade, more preferable methanol MS grade, see Note 5) and 95% p.a. quality water, containing 10 mM tributylamine as ion-pairing reagent (see Notes 5 and 8) and 15 mM acetic acid. pH adjustment to pH 4.9 is carried out with ammonia. Be aware that an exact pH of the mobile phase is indispensable for stable chromatography. Mobile phase B consists of 100% methanol.
12. Prepare mass tune mix after manufacturer guidelines.
13. Prepare internal standard solution (see Subheading 2.4), store in aliquots at −20°C.

    The following steps should be performed to ensure reliable measurement parameters for LC-MS:
14. It is important to validate the LC-MS system on a daily basis. For this purpose, quality control (QC) samples with a given concentration of representative standards are essential to be analyzed within your batch. For more details about special

sample run order, QC sample use, and more validation guidelines to avoid analytical errors, see Sangster et al. (23).

15. Calibrate the mass spectrometer daily before analyzing samples; see instructions of the manufacturer but be aware that the chosen mass tune mixture should include standard masses which cover the mass range of the method.
16. Flush spray needle of the ion source with water and check it for a straight stream. The ion source and spray shield should be cleaned regularly according to manufacturer guidelines. The frequency depends on your sample throughput.

    Caution: ion-pairing reagents will crystallize under standard conditions in the spray chamber and even in the subsequent glass capillary; additionally complex biological samples lead to a fast contamination of the ion source, especially of the spray shield based on high salt, saccharide, and peptide content.
17. Equilibrate capillaries, precolumn, and analytical column with 100% of the aqueous mobile phase A with which the gradient will start. Observe peak shapes and retention times of QC and biological samples, if chromatography is not convenient, column and guard-column should be changed in a regular manner depending on sample matrix.
18. Check column pressure while running aqueous mobile phase as a control for general HPLC setup quality and leak-proof.
19. In summary, the LC-MS system must be able to provide stable retention time, constant signal intensity, and mass accuracy.

2.4. Internal Standards and Quality Control Samples

An internal standard is a chemical substance that is added in a known and constant amount to the samples, the QC samples, the blanks, and the calibration standards. For metabolome analysis, it is essential to choose compounds not present in the biological sample as internal standards. The concentration of the internal standard must be in the range of the analyzed metabolites in the samples. A well-established and accurate method is the addition of isotopic labeled metabolites (24). In the best case, those isotopic labeled metabolites are gained by a parallel cultivation in, for example, ^{13}C or ^{15}N labeled medium. A defined part of the metabolite extract is added to all relevant samples. Nevertheless, this approach has also limitations including a high cost and the need for large amounts of isotopically labeled standards. The need to generate a new metabolite calibration curve for each experiment is a further time-consuming disadvantage. A related method of equal expense is to purchase a variety of isotopically labeled metabolites and use these as internal standards in defined concentrations (25). This method, using the defined purchasable compounds, could be more reproducible in terms of long-time calibration measurements, because effects from the internal standard sample preparation were diminished. Chemical substances, not occurring in the biological sample were an alternative and a more economical choice as internal standards. If this method is used, it is expedient to choose diverse chemical substances with different and well spaced retention times in the given analytical method (e.g., a mix containing an amino acid, a sugar, a fatty acid and an organic acid). For GC-MS metabolome analysis of *S. aureus* samples, ribitol and norvaline were adopted for the above-introduced sampling and analysis protocol. The applied ion-pairing LC-MS method was used with internal standards such as bromated adenosine triphosphate (Br-ATP) (12) and is further used with camphorsulfonic acid (CSA) and buffer substances such as PIPES and MOPS, as used by others (26).

For *S. aureus*, cultivated in chemically defined medium the following amounts and concentrations of internal standards could be used:

1. For GC-MS samples, add 200 μl internal standard solution of ribitol und norvaline (100 μM in water p.a. quality) into prepared extraction solution (see Subheading 2.2).
2. For LC-MS samples, add 100 μl internal standard solution of CSA, PIPES and MOPS (250 μM in water p.a. quality) into prepared extraction solution (see Subheading 2.3).

The preparation of quality control samples for both GC-MS and LC-MS measurement should be done depending on the sample number of the project. One way to obtain QC samples is to pool small amounts of the biological batch samples to be analyzed and aliquot that to the needed number of QCs. For long-term projects or experiments with huge numbers of samples, this attempt is not feasible and a representative QC sample must be prepared, e.g., extract from a huge amount of biomass. Aliquot such extracts and store them until usage at the lowest available temperature to avoid long-term degradation effects. An additional approach is to use synthetic QCs with a limited number of metabolites adapted to the applied analytical method. Prepare mixtures of 10–20 metabolites in defined concentrations, aliquot and store them appropriate. Use these QCs for the evaluation of your system setup.

3. Methods

3.1. Cell Sampling and Metabolite Extraction

An overview of the work flow for microbial metabolome analysis with fast filtration sampling is given in Fig. 2. When sampling cells from liquid culture as described in Subheading 3.1.1, the common sample size was 20 ml of a cell suspension with OD = 1 (approx. 4 mg cell dry weight). It may be possible to use less volume, depending on the analytical sensitivity. If other bacteria than *S. aureus* are used

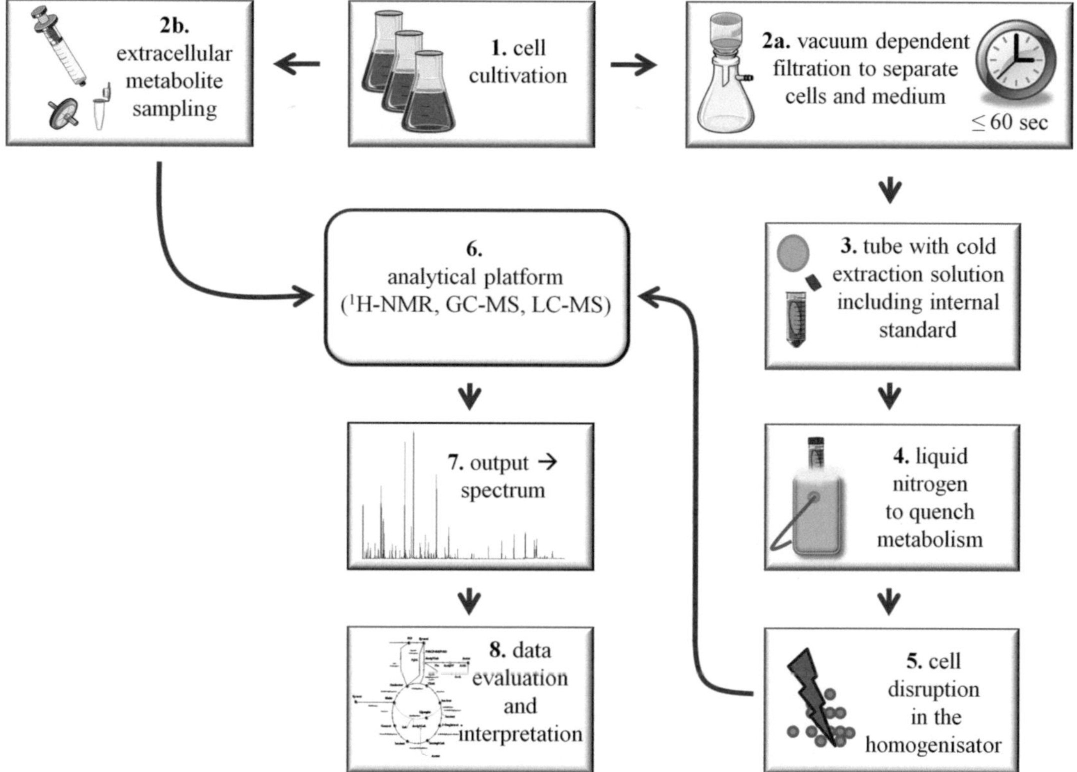

Fig. 2. Work flow for microbial metabolome analysis with fast filtration sampling.

for metabolome analysis, the filter type and procedure as well as the mechanical cell disruption settings of the used homogenizer has to be reevaluated. In Subheading 3.1.2, the cell disruption and metabolite extraction for a global snapshot of the *S. aureus* metabolome is described. After the drying step, samples are ready for further preparation depending on the analytic method used. Most important within the following steps is the required time for cell suspension filtering, avoiding long sampling times to guarantee good metabolome samples.

3.1.1. Cell Sampling

1. Ensure that the vacuum pump is running and the filter system is prepared.
2. 20 OD units of cell culture have to be removed from the shaking flask as fast as possible by a glass pipette. Subsequently fill the 20 OD units cell culture into the filter system (in the funnel and on the fritted base with filter) to separate the cells from the medium (see Note 10).

   Critical step: this step has to be performed as fast as possible, it affects the quality of the sample in a crucial manner (see Note 11).

3. Wash the cells on the filter twice with 5 ml cold and isotonic NaCl solution to guarantee an exact separation of intra- and extracellular metabolites (see Note 3).
4. Quickly and cautiously remove the filter from the filter system with a tweezer and put it, including the cells, into the prepared Falcon tube containing the cold extraction solution and internal standard solution.
5. Close Falcon tube and shake tube strongly for 10 s. This step is used to wash the filter completely with the extraction solution. Subsequently, the Falcon tube must be dropped into liquid nitrogen to quench the metabolism.
6. Steps 1–5 may not take longer than 60 s to prevent alterations in the metabolome.

   Pause Point: After this step the samples can be stored at –80°C and the sample preparation can be continued later.

   Since bacterial cells import and export a variety of metabolites, it is common to analyze the exometabolome as well. Important information can be gained in parallel to the endometabolome investigations, e.g., limitation for nutrients, accumulation of fermentation products or metabolites from altered pathways. Therefore, step 7 has to be performed in parallel to step 2.
7. Extracellular metabolite sampling. Filtrate 3 ml cell culture rapidly into a 2-ml tube through a sterile filter. Be aware of the fact that a part of the sample will always remain in the sterile filter (see Note 4). The filter can be discarded, while the supernatant has to be stored on ice and afterward frozen and stored at –20°C until measured. After thawing, the sample can be measured by ^{1}H-NMR (27) or by GC-MS.

3.1.2. Cell Disruption and Metabolite Extraction

1. For cell disruption thaw cells on ice, shake by hand and mix the cell solution ten times in alteration with a mechanical mixing device such as a Vortexer® to remove the cells from the filter.

   Critical Step: this step has to be carried out carefully, to ensure that all cells are washed from the filter.
2. Fill homogenizer tubes with 0.5 ml glass beads and 1 ml cell suspension (one sample will be aliquoted in 4–5 homogenizer tubes) to avoid a high ratio between cell biomass and glass beads (see Note 1).
3. The cells will be disrupted by two cycles in a Precellys® homogenizer for 30 s at 6,800 rpm.

   Caution: further cycles will possibly damage slightly instable metabolites.
4. Combine the supernatants derived from one sample in a 15-ml Falcon tube after cell disruption. Wash the glass beads once with 1 ml water p.a. quality. Combine the obtained washing

solution with the previous supernatant. To separate the metabolite containing solution from the remaining glass beads and the cell debris, centrifuge for 5 min at 4°C and 10,000 × *g*.

5. If available, use a slow stopping mode of the centrifuge to avoid a whirl up of the glass beads. Transfer the supernatant into a new 50-ml Falcon tube. Add water p.a. quality to an end concentration of organic solution (e.g., ethanol) of 10% (about 40 ml) to avoid a boiling process in the freeze-dryer. You can also use a vacuum rotation evaporator (e.g., Speed Vac®) to get rid of the extraction solvent. This is also a common method to concentrate metabolome samples which makes dilution of organic solvents unnecessary (see Note 13).

   Pause Point: The samples can now be stored at −80°C until lyophilization or vacuum evaporation. Caution: A loss of volatile compounds such as acetic acid, small alcohols, etc. during both evaporation processes cannot be avoided.
6. Lyophilization. Dry the sample by lyophilization at − 57°C and 0.053 mbar for 3–5 days to concentrate the sample and to remove solvents.
7. Dried samples should be again dissolved in a small amount of water p.a. quality and transferred into a more handy vial or reaction tube. Bring samples again to dryness and store at −80°C to −20°C.

3.2. GC-MS Metabolome Analysis

GC-MS metabolome analysis is a robust technique to achieve good quality data with a high degree of information. The methods described in Subheadings 3.2 and 2.2 were based on *S. aureus* metabolome investigations. The recently published and commercially available Agilent® Fiehn GC-MS metabolomics RTL library fulfills the requirements for starting with GC-MS metabolome analysis and readers which purchased this product should additionally follow detailed instructions made in the according manual.

3.2.1. Prepare GC-MS and Sample for Measurement

1. Check GC-MS function as described in Subheading 2.2.
2. Check injector system, fill wash solvents, empty waste, and check syringe for function.
3. Run at least one blank with your GC-MS method by injecting organic solvent. Check for impurities from the liner or other parts of the GC. Check baseline and replace injection liner if necessary. In that case, run at least three blank samples (derivatization reagent) to deactivate the new liner and to clean the system.
4. Run one sample with FAMEs for correct RI calibration per day (see Note 14).
5. Prepare quality control samples for your batch of samples.

6. Add 5 μl retention time locking solution (myristic acid d_{27}) to samples and QC samples (see Note 15), evaporate solvent again to complete dryness of sample (see Note 16).
7. Start derivatization with MeOx solution, add 60 μl to dry sample, close reaction tube, vortex, and heat for 90 min at 37°C in the heating block (see Note 17).
8. Add 120 μl MSTFA solution to the methoxyaminated sample, vortex, and heat for 30 min in heating block at 37°C (see Note 17).
9. Centrifuge sample to remove possible remaining salts or proteins, pipette supernatant into GC-MS vial.
10. Put the vial into the autosampler and start measurement of the batch.

3.2.2. Data Management and Data File Processing

1. Check all acquired data (a typical set is shown in Fig. 3) by overlay function for general quality, baseline, intensities, peak shapes and retention time drift.
2. Export files into needed format, e.g., netCDF or AIA.
3. Import data files into your software for peak finding and integration. Many free software packages are available e.g., XCMS² (28), metalign (29), metaquant (30), and tagfinder (31).
4. Perform statistics with QC samples and biological samples from one batch via normal principal component analysis (PCA) and/or partial least squares-discrimination analysis (PLS-DA) to prove the quality of the data set (32–34).
5. Peak list processing. To identify peaks use the commercial available EI spectra databases such as the NIST/EPA/NIH Mass Spectral Library (NIST 08), Wiley Registry™, 9th and 8th Editions Mass Spectral Library, or Agilent®Fiehn GC-MS metabolomics library or free database such as Golm metabolome database (http://csbdb.mpimp-golm.mpg.de/csbdb/gmd/gmd.html).

   Caution: For confidence in results run the pure compound under same conditions to prove the database hits. One option, if a peak still remains unknown but is of interest, is a decision tree (DT)-based prediction of the most frequent substructures based on mass spectral features and RI information which was developed out of the golm metabolome database. This approach could limit the number of possible metabolites and helps to identify the unknown compound (35).
6. Perform statistical analysis of results and visualization in metabolic context (see Subheading 3.4).

3.3. LC-MS Metabolome Analysis

In Subheadings 3.3.1 and 2.3, a guidance is given for successful maintenance of the HPLC-MS system as well as sample preparation. Manufacturer dependent details of the HPLC and MS parameters

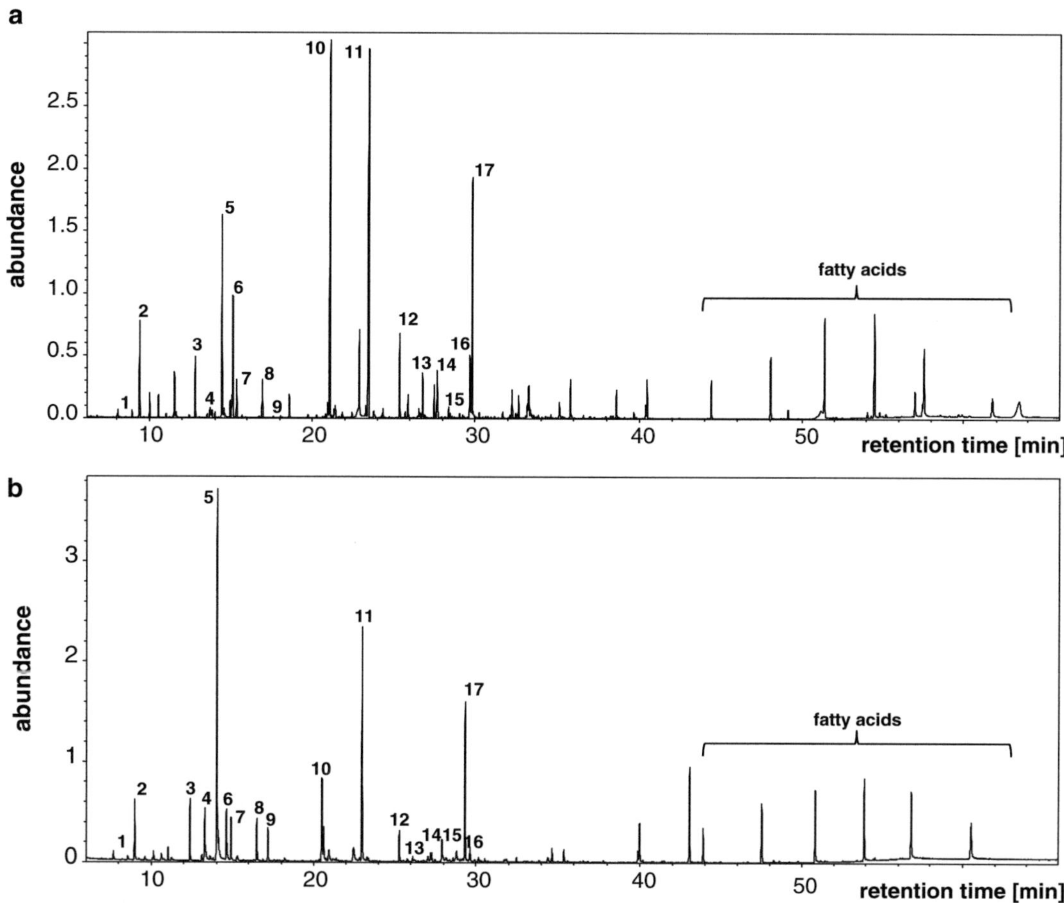

Fig. 3. GC-MS total ion-chromatograms (TIC) of bacterial metabolome samples, from (**a**) *Staphylococcus aureus* strain COL cultivated in chemically defined medium and sampled in stationary growth phase with the current protocol and (**b**) *Bacillus subtilis* 168 cultivated in complex Lauria-Bertani broth medium and sampled in stationary growth phase via a slightly altered protocol. Specific metabolites were identified by comparison of pure standard compounds and MS-library match and marked by numbers: (1) pyruvate, (2) alanine, (3) valine, (4) urea, (5) phosphate, (6) proline, (7) glycine, (8) serine, (9) threonine, (10) aspartate, (11) glutamate, (12) internal standard ribitol, (13) glutamine, (14) 2-phosphoglycerate, (15) ornithine, (16) histidine, and (17) lysine.

were excluded because of the wide variety of available products. The setup was chosen based on different criteria. An electrospray ionization source was used for soft ionization with relatively minimal fragmentation in combination with a high resolution mass detector such as a TOF-MS. This enables a dataset with less complexity through fragmentation but with high mass accuracy to identify metabolites based on the exact mass. With ion-pairing reagents, the use of the positive MS mode is not recommended because of complications from ion-suppression by the constant flow of positive charged tributylamine from the solvents. Nevertheless, the applied MS negative mode also covers nitrogen rich metabolites (preferable analyzed in positive mode) such as nucleotides and amino acid rich cell wall precursors based on negative ions formed by deprotonation on the phosphate groups.

3.3.1. Prepare LC-MS and Sample for Measurement

1. Before starting sample preparation the liquid chromatography mass spectrometry (LC-MS) system has to be checked and made ready for the analysis. Check LC-MS function as described under Subheading 2.3 item 4 and make sure enough disk space is available for batch data (typically raw data files can range from 100 MB to 2.5 GB per sample).
2. Resolve lyophilized samples in 100 μl cold p.a. quality water.
3. Centrifuge samples (10,000×*g*, 3 min, 0–4°C) to remove proteins or particulates with higher molecular masses.
4. Transfer supernatant of samples into HPLC glass vials and store them on ice or in the freezer until measured.

   Caution: fill level of the vial must be in accordance to the penetration depth of the injection needle. Therefore limited volume inserts (micro inlets) are necessary.
5. Put vials in the autosampler shortly before starting the run if no temperature-controlled (cooling) autosampler exists (see Note 18).
6. Use an injection volume of 25–50 μl (depending on the analytical system and sample concentration). The injection needle should be flushed with distilled water or other appropriate solvents before and after sample injection.
7. Start measurement of the batch.

3.3.2. Data Management and Data File Processing

1. Before starting data analysis the LC-MS dataset must be preprocessed, e.g., by internal calibration. Usually, the application software can do this step. For further peak detection, alignment and normalization as well as peak list generation and quantification (see Note 19) use instrument software or several available freeware packages, e.g., XCMS (28, 36), MZMine (37). Most packages include also statistical analysis functionality (38).
2. Peak list processing and identification of peaks. To our knowledge, there are no spectral libraries for LC-MS identification. Therefore pure standards have to be analyzed with the same method to get their retention time and accurate masses or fragmentation pattern. Nevertheless spiking a biological sample with a chemical standard may often be the only way for true identification. Some online databases (e.g., HMDB, metlin, Pubchem (39–41)) are freely available and provide mass searches, which supports metabolite identification.

   Remaining unknowns could be identified by the application of MS/MS approaches to determine fragmentation pattern or by using high resolution MS to elucidate the molecular composition. If enough of the sample is available, selective metabolite enrichment could be performed to gain pure material for structural information from NMR spectroscopy.

3. Perform statistical analysis of results and visualization in metabolic context (see Subheading 3.4).

   As shown in Fig. 4, the IP-LC-MS method is feasible to resolve the peaks for nucleoside mono-, di- and triphosphates. With increasing retention time more phosphate groups are present in the molecule. Details about the range of compounds analyzed by the IP-LC-MS method could be found in a recent *S. aureus* metabolome study (12).

4. Evaluation of sampling effectiveness and sample workup by determination of the Adenyl-Energy-Charge (AEC). As one control for accurate metabolome sampling, including fast sampling, quenching and metabolite extraction, the energy charge has to be determined for every biological sample (10, 12). Therefore the ion masses of AMP, ADP, and ATP have to be

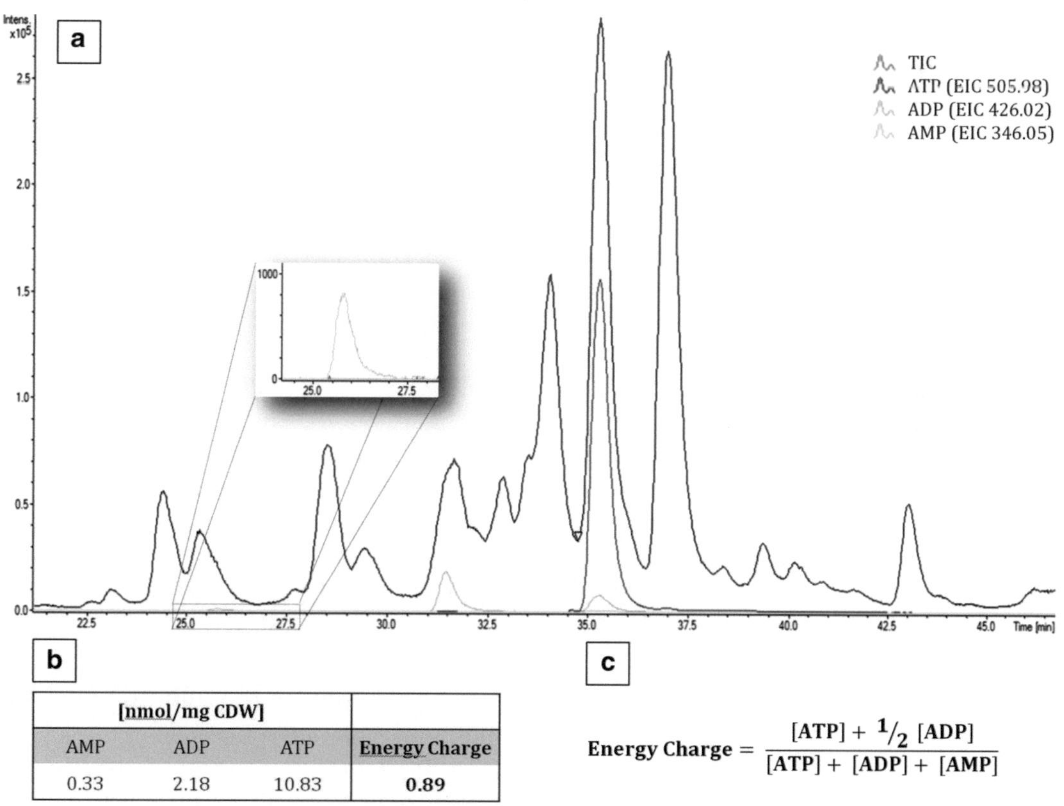

[nmol/mg CDW]			
AMP	ADP	ATP	Energy Charge
0.33	2.18	10.83	**0.89**

$$\text{Energy Charge} = \frac{[\text{ATP}] + \frac{1}{2}[\text{ADP}]}{[\text{ATP}] + [\text{ADP}] + [\text{AMP}]}$$

Fig. 4. (**a**) A part of a LC-MS total ion chromatogram (TIC) from a *Staphylococcus aureus* strain COL cell extract sampled in logarithmic growth phase in a chemically defined medium. Mass traces of adenosine nucleotides are represented as extracted ion chromatograms (EIC) with peak maxima at the following retention times: AMP at 25.8 min, ADP at 31.5 min and ATP at 35.3 min. (**b**) After quantifying the AMP, ADP, and ATP concentration, the adenylate energy charge (AEC) can be determined by the formula shown in (**c**) and values for AEC should be stable at around 0.80–0.95 for unstressed cells. CDW – cell dry weight.

extracted, integrated, and quantified via calibration curves for each metabolite. Unstressed and nutrition unlimited cells should have an AEC of 0.8–0.95 (42) (calculation see equation in Fig. 4c). If this value is below 0.8 it could be based on technical problems, e.g., slow sampling and/or unsuccessful quenching or based on biological reasons, e.g., cells were starved for their energy sources or stressed by other factors (see Note 20).

3.4. General Process of Data Extraction and Handling of Results

1. Choose your analysis for a global view or either targeted approach, e.g., metabolome profile comparison or focus on, for example, amino acids. The amount of work to align and integrate metabolite peaks is very different.
2. Use the resulting peak lists from GC-MS and LC-MS for statistical evaluation by importing the data into, for example, SIMCA or MATLAB. Different mathematical models can calculate the differences between samples and could also highlight the discriminating metabolites. Since this is a protocol description, in-depth reviews should be drawn in attention (32, 43).
3. A further challenge with all acquired metabolome data is to bring them into meaningful biological context. Different free software packages could help to visualize metabolite concentrations based on metabolic pathways maps (44, 45) and could give helpful indications of the metabolic processes.

4. Notes

1. Use gloves, safety equipment and work under a clean bench or fume hood if needed.
2. The choice of the extraction solution depends on the metabolites of interest. If a wide range of metabolites has to be analyzed, 60% (*w*/*v*) ethanol has to be used as an extraction solution. For special compound groups such as fatty acids, use a modified Bligh Dyer extraction solution [$EtOH/H_2O/CHCl_3$ (4/2/4)] (46). Caution: for each sample this extraction solution has to be prepared separate, based on phase separation.
3. Washing solution should be isotonic to the cultivation medium to avoid cell lyses while washing.
4. The filter has to be checked for chemical impurities such as chemical substances such as glycerol in the filter, which affect the analytics.
5. This reagent is toxic, please handle under fume hood.
6. Agilent® Fiehn GC-MS metabolomics Library is a useful tool. This database is based on a RTL method and is therefore suitable for other GC-MS methods also using RTL and RI calibration.

7. Tune interval is depending on MS type and should be done as recommended from the manufacturer.
8. The ion-pairing reagent sticks on MS parts and is hard to clean. The positive mode MS is strongly affected by the ions of tributylamine.
9. Several intermediates of, for example, the cell wall metabolism have higher molecular masses than 1,000 Da and are detectable with the LC method, also $[2M\text{-}H]^-$ and $[3M\text{-}H]^-$ of metabolites could be observed.
10. The required OD units are highly dependent on the analytical detection limit (analytical platform dependent).
11. Determine the Adenylate Energy Charge (see Subheading 3.3).
12. Tubes and the accordingly handling procedure for the cell disruption must be adapted to the existing homogenizer.
13. Check whether samples gave same results if dried with freeze-dryer or with a vacuum-extractor. We observed an increased conversion of glutamate to oxo-proline in *S. aureus* and *B. licheniformis* extracts dried with the vacuum-extractor.
14. Inclusion of FAME markers can guarantee a correct RI calibration. Examination of the retention times of the FAMEs is essential to judge if a RI calibration needs to be performed.
15. The RTL locking compound should always be included. Examination of the retention time of locking compound is essential to judge the relock of an RTL method.
16. A complete dry sample is required. Consider that derivatization reagent reacts with water.
17. Different amounts of derivatization reagents and temperature/ time combinations were used for metabolite derivatization with MeOx and MSTFA in the past. There is no common optimal setup for all metabolites. In addition to that, microwave irritation was used recently to shorten heating time in a significant manner (47, 48).
18. Be aware of sample temperature (max. 4°C) to avoid degradation of metabolites. An autosampler with temperature control is advisable.
19. Caution: despite the separation method via liquid chromatography a co-elution of compounds cannot be avoided at all. This effect may leads to ion-suppression, especially if salts or some high abundant metabolites were co-eluted. This effect can be ruled out by the continuously injection of a metabolite solution into the MS during sample measurement. If the intensity of the observed mass signals decreases during the chromatographic run, ion-suppression takes place (49).

20. Check in addition to AEC determination the GC-MS samples for content of nucleosides such as adenosine or guanosine. Elevated levels could probably be a hint for ongoing enzymatic degradation of nucleotides in your sample. As a proof-of-principle check of your protocol setup one sample that rests on the bench for 1 min or more before quenching should be utilized.

References

1. Fiehn O, (2002) Metabolomics – the link between genotypes and phenotypes. Plant Mol Biol 48, 155–171.
2. Oliver SG, Winson MK, Kell DB, and Baganz F (1998) Systematic functional analysis of the yeast genome. Trends in biotechnology 16, 373–378.
3. Bennett BD, Kimball EH, Gao M, Osterhout R, Van Dien SJ, and Rabinowitz JD (2009) Absolute metabolite concentrations and implied enzyme active site occupancy in Escherichia coli. Nat Chem Biol 5, 593–599.
4. Brauer MJ, Yuan J, Bennett BD, Lu WY, Kimball E, Botstein D, and Rabinowitz JD (2006) Conservation of the metabolomic response to starvation across two divergent microbes. P Natl Acad Sci USA 103, 19302–19307.
5. Zamboni N, and Sauer U (2009) Novel biological insights through metabolomics and 13C-flux analysis. Curr Opin Microbiol 12, 553–558.
6. Nakahigashi K, Toya Y, Ishii N, Soga T, Hasegawa M, Watanabe H, Takai Y, Honma M, Mori H, and Tomita M (2009) Systematic phenome analysis of Escherichia coli multiple-knockout mutants reveals hidden reactions in central carbon metabolism. Mol Syst Biol 5, 306.
7. Mashego MR, Rumbold K, De Mey M, Vandamme E, Soetaert W, and Heijnen JJ (2007) Microbial metabolomics: past, present and future methodologies. Biotechnol Lett 29, 1–16.
8. Durot M, Bourguignon PY, and Schachter V (2009) Genome-scale models of bacterial metabolism: reconstruction and applications. FEMS Microbiol Rev 33, 164–190.
9. Bolten CJ, Kiefer P, Letisse F, Portais JC, and Wittmann C (2007) Sampling for metabolome analysis of microorganisms. Anal Chem 79, 3843–3849.
10. Meyer H, Liebeke M, and Lalk M (2010) A protocol for the investigation of the intracellular Staphylococcus aureus metabolome. Anal Biochem 401, 250–259.
11. Donat S, Streker K, Schirmeister T, Rakette S, Stehle T, Liebeke M, Lalk M, and Ohlsen K (2009) Transcriptome and functional analysis of the eukaryotic-type serine/threonine kinase PknB in Staphylococcus aureus. J Bacteriol 191, 4056–4069.
12. Liebeke M, Meyer H, Donat S, Ohlsen K, and Lalk M (2010) A metabolomic view of Staphylococcus aureus and its serine/threonine kinase and phosphatase deletion mutants: involvement in cell wall biosynthesis. Chem Biol 17, 820–830.
13. Winder CL, Dunn WB, Schuler S, Broadhurst D, Jarvis R, Stephens GM, and Goodacre R (2008) Global metabolic profiling of Escherichia coli cultures: an evaluation of methods for quenching and extraction of intracellular metabolites. Anal Chem 80, 2939–2948.
14. Lisec J, Schauer N, Kopka J, Willmitzer L, and Fernie AR (2006) Gas chromatography mass spectrometry-based metabolite profiling in plants. Nat Protoc 1, 387–396.
15. Want EJ, Coen M, Masson P, Keun HC, Pearce JT, Reily MD, Robertson DG, Rohde CM, Holmes E, Lindon JC, Plumb RS, and Nicholson JK (2010) Ultra performance liquid chromatography-mass spectrometry profiling of bile acid metabolites in biofluids: application to experimental toxicology studies. Anal Chem 82, 5282–5289.
16. Villas-Boas SG, Mas S, Akesson M, Smedsgaard J, and Nielsen J (2005) Mass spectrometry in metabolome analysis. Mass Spectrom Rev 24, 613–646.
17. Cubbon S, Antonio C, Wilson J, and Thomas-Oates J (2010) Metabolomic applications of HILIC-LC-MS. Mass Spectrom Rev 29, 671–684.
18. Allwood JW, and Goodacre R (2009) An introduction to liquid chromatography-mass spectrometry instrumentation applied in plant metabolomic analyses. Phytochem Anal 21, 33–47.
19. Bajad SU, Lu W, Kimball EH, Yuan J, Peterson C, and Rabinowitz JD (2006) Separation and quantitation of water soluble cellular metabolites by hydrophilic interaction chromatography-tandem mass spectrometry. J Chromatogr A 1125, 76–88.
20. Buescher JM, Moco S, Sauer U, and Zamboni N (2010) Ultrahigh performance liquid

chromatography-tandem mass spectrometry method for fast and robust quantification of anionic and aromatic metabolites. Anal Chem 82, 4403–4412.

21. Roessner U, Luedemann A, Brust D, Fiehn O, Linke T, Willmitzer L, and Fernie A (2001) Metabolic profiling allows comprehensive phenotyping of genetically or environmentally modified plant systems. Plant Cell 13, 11–29.
22. Roessner-Tunali U, Urbanczyk-Wochniak E, Czechowski T, Kolbe A, Willmitzer L, and Fernie AR (2003) De novo amino acid biosynthesis in potato tubers is regulated by sucrose levels. Plant Physiol 133, 683–692.
23. Sangster T, Major H, Plumb R, Wilson AJ, and Wilson ID (2006) A pragmatic and readily implemented quality control strategy for HPLC-MS and GC-MS-based metabonomic analysis. Analyst 131, 1075–1078.
24. Wu L, Mashego MR, van Dam JC, Proell AM, Vinke JL., Ras C, van Winden WA, van Gulik W M, and Heijnen JJ (2005) Quantitative analysis of the microbial metabolome by isotope dilution mass spectrometry using uniformly 13C labeled cell extracts as internal standards. Anal Biochem 336, 164–171.
25. Bennett BD, Yuan J, Kimball EH, and Rabinowitz JD (2008) Absolute quantitation of intracellular metabolite concentrations by an isotope ratio-based approach. Nat Protoc 3, 1299–1311.
26. Pluskal T, Nakamura T, Villar-Briones A, and Yanagida M Metabolic profiling of the fission yeast S. pombe: quantification of compounds under different temperatures and genetic perturbation. Mol Biosyst 6, 182–198.
27. Liebeke M, Brozel VS, Hecker M, and Lalk M (2009) Chemical characterization of soil extract as growth media for the ecophysiological study of bacteria. Appl Microbiol Biotechnol 83, 161–173.
28. Benton HP, Wong DM, Trauger SA, and Siuzdak G (2008) XCMS2: processing tandem mass spectrometry data for metabolite identification and structural characterization. Anal Chem 80, 6382–6389.
29. Lommen A (2009) MetAlign: interface-driven, versatile metabolomics tool for hyphenated full-scan mass spectrometry data preprocessing. Anal Chem 81, 3079–3086.
30. Bunk B, Kucklick M, Jonas R, Munch R, Schobert M, Jahn D, and Hiller K (2006) MetaQuant: a tool for the automatic quantification of GC/MS-based metabolome data. Bioinformatics 22, 2962–2965.
31. Luedemann A, Strassburg K, Erban A, and Kopka J (2008) TagFinder for the quantitative analysis of gas chromatography–mass spectrometry (GC–MS)-based metabolite profiling experiments. Bioinformatics 24, 732–737.
32. Trygg J, Holmes E, and Lundstedt T (2007) Chemometrics in metabonomics. J Proteome Res 6, 469–479.
33. Lai L, Michopoulos F, Gika H, Theodoridis G, Wilkinson RW, Odedra R, Wingate J, Bonner R, Tate S, and Wilson ID (2010) Methodological considerations in the development of HPLC-MS methods for the analysis of rodent plasma for metabonomic studies. Mol Biosyst 6, 108–120.
34. Burton L, Ivosev G, Tate S, Impey G, Wingate J, and Bonner R (2008) Instrumental and experimental effects in LC-MS-based metabolomics. J Chromatogr B 871, 227–235.
35. Hummel J, Strehmel N, Selbig J, Walther D, and Kopka J (2010) Decision tree supported substructure prediction of metabolites from GC-MS profiles. Metabolomics 6, 322–333.
36. Smith CA, Want EJ, O'Maille G, Abagyan R, and Siuzdak G (2006) XCMS: processing mass spectrometry data for metabolite profiling using nonlinear peak alignment, matching, and identification. Anal Chem 78, 779–787.
37. Katajamaa M, Miettinen J, and Oresic M (2006) MZmine: toolbox for processing and visualization of mass spectrometry based molecular profile data. Bioinformatics 22, 634–636.
38. Sumner LW, Urbanczyk-Wochniak E, and Broeckling CD (2007) Metabolomics data analysis, visualization, and integration. Methods Mol Biol 406, 409–436.
39. Smith CA, O'Maille G, Want EJ, Qin C, Trauger SA, Brandon TR, Custodio DE, Abagyan R, and Siuzdak G (2005) METLIN: a metabolite mass spectral database. Ther Drug Monit 27, 747–751.
40. Ulrich EL, Akutsu H, Doreleijers JF, Harano Y, Ioannidis YE, Lin J, Livny M, Mading S, Maziuk D, Miller Z, Nakatani E, Schulte CF, Tolmie DE, Kent Wenger R, Yao H, and Markley JL (2008) BioMagResBank. Nucleic Acids Res 36, D402–408.
41. Wishart, DS, Knox C, Guo AC, Eisner R, Young N, Gautam B, Hau DD, Psychogios N, Dong E, Bouatra S, Mandal R, et al. (2009) HMDB: a knowledgebase for the human metabolome. Nucleic Acids Res 37, D603–610.
42. Atkinson DE (1968) The energy charge of the adenylate pool as a regulatory parameter. Interaction with feedback modifiers. Biochemistry-Us 7, 4030–4034.
43. Steuer R, Morgenthal K, Weckwerth W, and Selbig J (2007) A gentle guide to the analysis of metabolomic data. Methods Mol Biol 358, 105–126.

44. Gehlenborg N, O'Donoghue SI, Baliga NS, Goesmann A, Hibbs MA, Kitano H, Kohlbacher O, Neuweger H, Schneider R, Tenenbaum D, and Gavin AC (2010) Visualization of omics data for systems biology. Nat Methods 7, S56–68.
45. Scholz M, and Selbig J (2007) Visualization and analysis of molecular data. Methods Mol Biol 358, 87–104.
46. Bligh EG, and Dyer WJ (1959) A rapid method of total lipid extraction and purification. Can J Biochem Physiol 37, 911–917.
47. Kouremenos KA, Harynuk JJ, Winniford WL, Morrison PD, and Marriott PJ (2010) One-pot microwave derivatization of target compounds relevant to metabolomics with comprehensive two-dimensional gas chromatography. J Chromatogr B 878, 1761–1770.
48. Liebeke M, Wunder A, and Lalk M (2010) A rapid microwave-assisted derivatization of bacterial metabolome samples for GC/MS analysis. Anal Biochem 401, 312–314.
49. Coulier L, Bas R, Jespersen S, Verheij E, van der Werf MJ, and Hankemeier T (2006) Simultaneous quantitative analysis of metabolites using ion-pair liquid chromatography-electrospray ionization mass spectrometry. Anal Chem 78, 6573–6582.

Chapter 29

Metabolic Fingerprinting Using Comprehensive Two-Dimensional Gas Chromatography – Time-of-Flight Mass Spectrometry

Martin F. Almstetter, Peter J. Oefner, and Katja Dettmer

Abstract

Comprehensive two-dimensional gas chromatography–time-of-flight mass spectrometry (GC × GC–TOF-MS) is applied to the comparative metabolic fingerprinting of physiological fluids. Stable isotope-labeled internal standards plus norvaline serve as extraction standards and are added to the blanks, controls and patient samples prior to protein precipitation with methanol. The extracts are evaporated to complete dryness and derivatized in two steps using methoximation with methoxylamine hydrochloride (MeOx) and silylation with *N*-methyl-*N*-trimethylsily- trifluoroacetamide (MSTFA). Between derivatization steps a second internal standard containing odd-numbered, saturated straight chain fatty acids is added for quality control and to normalize retention time shifts. After GC × GC–TOF-MS analysis raw data are processed, aligned, and combined in one data matrix for subsequent statistical evaluation. Both a custom-made and the NIST 05 library are used to preliminarily identify significant metabolites. For verification purposes, commercial standards are run individually. Absolute quantification of selected metabolites is achieved by using a multipoint calibration curve and isotope-labeled internal standards.

Key words: Metabolic fingerprinting, Comprehensive two-dimensional gas chromatography electron ionization time-of-flight mass spectrometry, Physiological fluids, Alignment algorithms, Mass spectral libraries

1. Introduction

Metabolomics is the systematic study of small-molecule metabolite profiles in a biological system and their changes as a result of environmental, nutritional, genetic, and pathophysiological factors (1, 2). The ultimate objective is the quantitative analysis of the entire metabolome in a single run. The implementation of this task, however, is mostly impeded by the large number of chemically diverse metabolites

Michael Kaufmann and Claudia Klinger (eds.), *Functional Genomics: Methods and Protocols*, Methods in Molecular Biology, vol. 815, DOI 10.1007/978-1-61779-424-7_29, © Springer Science+Business Media, LLC 2012

present over a wide concentration range. Metabolic fingerprinting is a promising approach that tries to accomplish that feat. Samples are screened globally and classified upon their metabolite patterns directly yielding a snapshot of the physiological state.

Gas chromatography coupled to mass spectrometry (GC–MS) has emerged as a widely used tool for metabolomic investigations (3). Although one-dimensional capillary GC offers excellent chromatographic resolution, it cannot separate the multitude of metabolites present in extracts of physiological fluids and tissues. Comprehensive two-dimensional gas chromatography (GC×GC) combines two columns with orthogonal separation characteristics via a modulator leading to a multiplicative increase in peak capacity and a structured separation space (4). A scheme of a GC×GC setup is shown in Fig. 1. To maintain the separation accomplished in the first column, analytes eluting from the first column are retained in small adjacent fractions in the modulator, which are then released into the second column. In order to maintain first dimension separation each peak should be cut in at least three fractions. While one fraction of an analyte is separated in the second dimension column the next fraction is already sampled in the modulator. With peak widths of 6–25 s in the first column and 3–4 required modulations the separation in the second column can only last 2–8 s. Therefore, short narrow second dimension columns are used to achieve a fast second dimension separation. The generation of a schematic GC×GC chromatogram is shown in Fig. 2. A variety of modulators are available. The majority uses thermal modulation and are either cryogenic or heater-based or use a combination of both. Here, we use a dual-stage four jet modulator (Fig. 1). In order to accumulate and focus fractions of the first column effluent, the temperature in the modulator is decreased

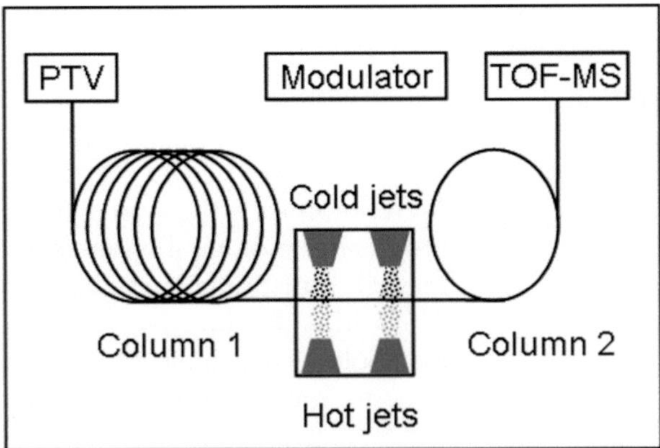

Fig. 1. Schematic setup of a comprehensive two-dimensional gas chromatograph coupled to a time-of-flight mass spectrometer (GC × GC–EI–TOF-MS). A thermal modulator is used to alternately cool and heat the incoming analytes and to release them periodically in packages onto the second column. *PTV*, programmed temperature vaporizer.

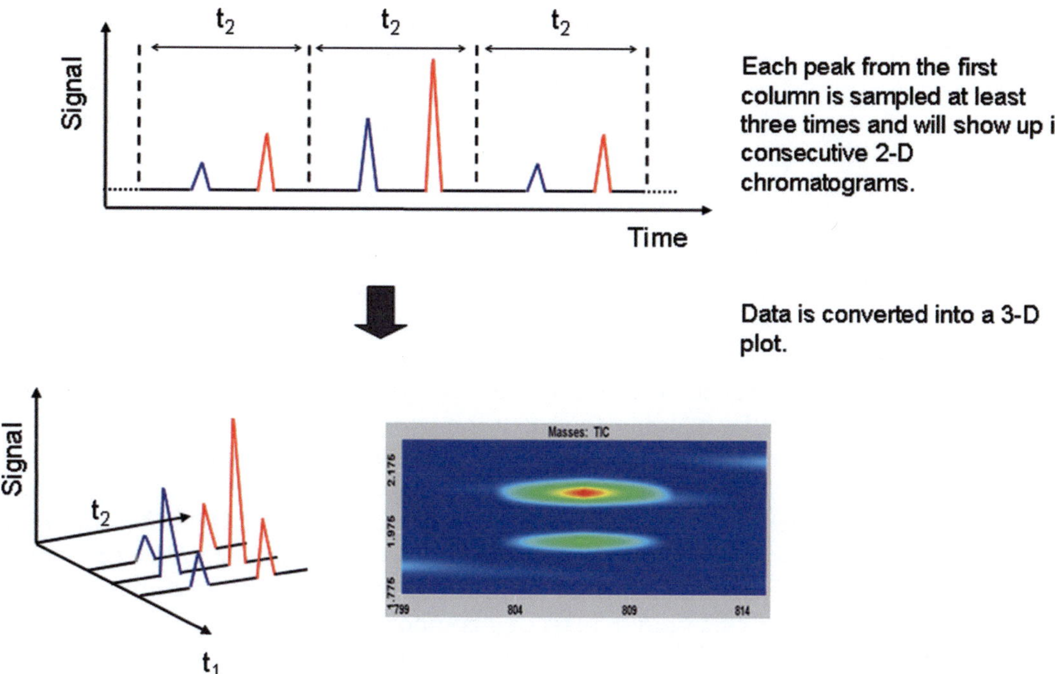

Fig. 2. Display of GC × GC chromatograms.

using cold nitrogen and the trapped fractions are released using hot air. The jets are used alternating. While the first cold jet is on, the effluent is focused in the first stage. To release the fraction the cold jet is turned off, the hot jet is turned on and, simultaneously, the second-stage cold jet is turned on. The fraction trapped in the first stage is released and together with "newly" arriving molecules transported to the second stage and focused by the second cold jet. Then, the second cold jet is turned off and the second hot jet is turned on to transfer the fraction onto the second-dimension column. Simultaneously, the first hot jet is turned off and the cold jet is turned on. With the dual-stage system one stage is always cold, thereby preventing the breakthrough of analytes. Thermal modulation also carries the benefit of creating narrow second dimension peaks and, thereby, increasing peak heights and, in turn, enhancing detection sensitivity (5).

GC × GC coupled to an electron ionization (EI)–time-of-flight mass spectrometer (TOF-MS) for identification and quantification is predestined for the characterization of metabolic fingerprints taking the complete (non-targeted) information from all experiments into account for subsequent statistical analysis. No prior knowledge of the metabolites is needed. If signals differentiate samples, an identification of the respective peaks is attempted. This approach enables the detection of possibly novel biomarkers.

In metabolomics, typically, numerous samples are analyzed and compared. It is important to recognize the same metabolites over many samples. To fully exploit the power of GC × GC–TOF-MS for metabolic fingerprinting, its data output requires reliable data processing, retention time (RT) correction, and alignment tools.

In the method presented here, the comparative metabolic fingerprinting analysis of physiological fluids, the metabolites of interest are first extracted using methanol followed by evaporation and a two-step derivatization to make the compounds more volatile. Raw data is being processed by the Leco ChromaTOF software. The spectra of trimethylsilyl (TMS) derivatives show a characteristic fragmentation behavior upon electron ionization and a typically abundant ion with *m/z* 73 that corresponds to the trimethylsilyl cation $[(CH_3)_3Si]^+$, while derivatives with at least two TMS groups always yield a strong signal at *m/z* 147 corresponding to the pentamethyldisiloxane cation $[C_5H_{15}Si_2O]^+$. The universally available area integrals for *m/z* 73 or 147 can be utilized as a quantitative measure for all peaks. If accessible, the area integral of the specific unique mass of a compound can also be used for quantification. The fragment mass spectra are then matched against a custom-made and the NIST library. The identifications are confirmed by comparison of chromatographic retention times and spectral match to commercially available standards. To ensure that metabolites of one metabolic fingerprint are consistent with their respective counterparts in the other fingerprints, raw data are processed and combined in one data matrix by alignment algorithms like the commercially available ChromaTOF Statistical Compare (SC) function or the in-house-developed INCA (6). The data matrix can now be subjected to multivariate statistical analysis to detect significant metabolites that are responsible for a clustering of groups.

2. Materials

2.1. Samples

Serum, plasma (collected using heparin or EDTA anticoagulants), cell culture media, milk, urine as well as cell and tissue extracts are all compatible with this method. In this work we focus solely on serum samples. If analysis must be delayed, samples should be stored at −20°C. Retained specimens are stored in the −80°C freezer for up to several years and as space permits.

2.2. Reagents and Internal Standards

1. Solvents: methanol (HPLC grade), isooctane, pyridine, and water (Milli-Q) (see Note 1).
2. Five millimolar EDTA solution in water (see Note 2).
3. Methoximation solution (20 mg/ml): add 200 mg methoxylamine hydrochloride (MeOx) (Sigma-Aldrich, Taufkirchen,

Germany) to 10 ml pyridine. Sonicate to get MeOx into solution. Store at room temperature in an amber vial away from moisture (see Note 3).

4. Trimethylsilylation: use *N*-methyl-*N*-trimethylsily- trifluoroacetamide (MSTFA, Macherey-Nagel, Dueren, Germany) (see Note 3).
5. Extraction standards: [$^2H_7$]*trans*-cinnamate, [2,2,4,4-$^2H_4$]citrate, [U-^{13}C]3-hydroxybutyrate, [U^{13}C]glucose, and norvaline are purchased from Sigma-Aldrich, [U-^{13}C]fumarate, [U-^{13}C]lactate, [U-^{13}C]pyruvate, and [U-^2H]succinate from Eurisotop (Saint-Aubin Cedex, France), and [2,3,3-$^2H_3$]malate from CDN Isotopes Inc. (Quebec, Canada). Prepare single stock solutions in methanol at a concentration of 100 mM. Prepare the internal standard mix by using 100 μl of each single stock and top of with methanol to a final volume of 10 ml resulting in a final concentration of 1 mM for each compound. Store solution at −80 to −20°C. If needed, other standards can be added for targeted or possibly untargeted analysis.
6. Internal derivatization standards: odd-carbon numbered fatty acids (C9–C19, Sigma-Aldrich) are used as internal derivatization standards. Prepare a stock solution in isooctane as described above with a final concentration of 1 mM for each odd-carbon numbered fatty acid (see Note 4). Store solution at −80 to −20°C.

2.3. Supplies

1. Amber glass vials with PTFE-lined caps for internal standard solutions.
2. Extraction of metabolites is carried out in 1.5-ml Eppendorf-cups.
3. Autosampler vials (12×32 mm; magnetic crimp caps) with 0.2-ml limited volume inserts.
4. Chemically inert SILTEC liner (Gerstel, Muehlheim, Germany).

2.4. Equipment

1. A Leco (St. Joseph, MI) Pegasus 4D GC×GC–TOF-MS instrument with an Agilent Technologies Model 6890 GC, a dual-stage, quad-jet thermal modulator, and a secondary oven coupled to a time-of-flight mass spectrometer providing unit mass resolution.
2. A PTV injector and MPS-2 Prepstation sample robot (Gerstel) for automated sample derivatization and handling. The robot is equipped with two agitators for sample incubation and two syringes of different volumes.
3. A 10-μl syringe is used for internal standard addition and sample injection, while reagents are added by means of a 250-μl syringe. Between adding steps, the syringes are washed five times with isooctane. Samples are kept in a cooled tray at 5°C.
4. An Rxi-5ms column (30 m×0.25 mm ID×0.25 μm film thickness) from Restek (GmbH, Bad Homburg, Germany) guarded

by a 3 m × 0.25 mm ID deactivated pre-column (Agilent, Palo Alto, CA) is used as the first-dimension column and a Rtx-1701 (2 m × 0.1 mm ID × 0.1 μm film thickness, Restek) as the second-dimension column.

5. GC–EI-MS mass spectral libraries, either custom made or commercially available (e.g. NIST05).
6. Heated evaporator module (CombiDancer, Hettich AG, Bäch, Switzerland) for drying down extraction solvents.

3. Methods

The most common steps in GC × GC–TOF-MS based metabolomics of physiological fluids comprise sample collection and storage, extraction of the metabolite fraction, derivatization, separation and detection of metabolites by GC × GC–TOF-MS, data processing, retention time correction and alignment of the raw data, discriminate analysis based on universal mass for all metabolites or on a compound-specific unique mass as a quantitative measure to determine the metabolites that differentiate samples, and identification of these metabolites (Fig. 3).

3.1. Stepwise Procedure

1. Label Eppendorf-cups and autosampler vials for each blank, control, and sample to be analyzed (see Note 5).
2. Add 20 μl of serum to the Eppendorf-cups.
3. Add 10 μl of extraction standard solution.
4. Add 20 μl of EDTA solution (see Note 2).

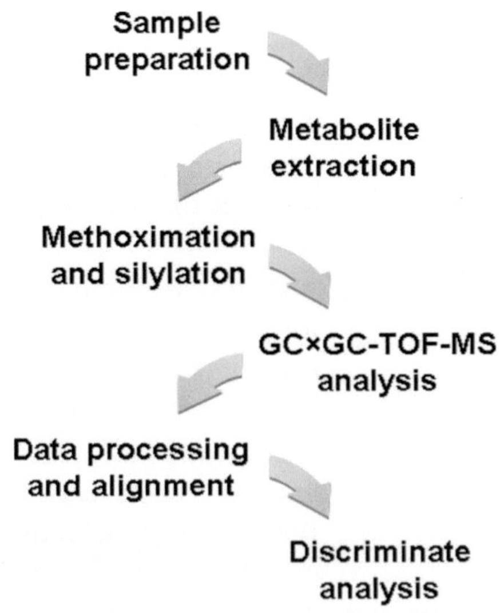

Fig. 3. Workflow of the complete process from sample preparation to discriminate analysis.

5. Add 80 μl of pure methanol.
6. Cap the cups and vortex for 5 min.
7. Centrifuge at 4°C and 3,375 ×g for 5 min.
8. Transfer the supernatant into a 2-ml glass vial with a 0.2-ml glass insert (see Note 6).
9. Re-extract the pellets twice with 50 μl of 80% methanol/20% water.
10. Combine the extracts and evaporate to complete dryness (see Note 7).
11. Close the vials with magnetic crimp caps for automated handling by the autosampler and place them in the instrument's cooled (5°C) tray.
12. Automated sample derivatization: add 50 μl of MeOx in pyridine and incubate at 60°C for 60 min, add 10 μl of internal derivatization standard solution containing odd-carbon numbered fatty acids, followed by 50 μl of MSTFA for 60 min at 60°C.
13. Inject 1.5 μl of the samples for GC×GC–TOF-MS analysis (see Note 8).

3.2. Instrument Operating Conditions

see Table 1 for instrument's operating conditions.

3.3. Data Analysis

1. For each patient and control sample, display the total ion chromatogram (TIC) in the ChromaTOF software. A representative GC×GC–TOF-MS chromatogram is shown in Fig. 4.
2. Perform baseline correction, deconvolution, and peak picking.
3. Select peaks above a given S/N value (e.g., 500:1).
4. Combine peaks in the second dimension using a spectral matching factor (e.g., 500) and an override of the allowed second-dimension retention time shift (e.g., 0.15 s early and 0.05 s late).
5. Integrate second-dimension subpeaks when exceeding a separate S/N (e.g., 50:1).
6. For tentative identification, match electron impact spectra against the NIST05 library, a custom-made or similar library (see Notes 11 and 12).
7. Apply Statistical Compare feature of ChromaTOF for alignment (steps 8–11); alternatively, peak lists for each chromatogram can be exported in csv file format and subjected to external alignment tools, e.g., INCA, tagfinder etc. (not shown here).
8. Add samples to the sample table and assign to their respective groups (control or patient).
9. Use scaling if shifts are present across the chromatograms.
10. Appoint values for retention time match criteria (max number of modulation periods apart, e.g., 1; max RT difference, e.g., 4 s)

Table 1
GC (Agilent 6890 model GC) and MS operating conditions

Oven	
Initial temp.	50°C
Initial time	0.2 min
Ramp	8°C/min to 265°C, hold for 10 min
Secondary oven offset	5°C relative to 1D column
Run time	37.08 min
Inlet	
Mode	Splitless, purge time 30 s, purge flow 20 ml/min
PTV injector	50°C for 0.5 min, 12°C/s to 250°C over 1 min
Pressure	31.1 psi
Liner	SILTEC liner from Gerstel
Injector	
Injection volume	1.5 μl
Column	
1D capillary column	Restek Rxi-5ms, 30 m × 0.25 mm × 0.25 μm with 3 m × 0.25 mm pre-column
2D capillary column	Restek Rtx-1701, 2 m × 0.1 mm × 0.1 μm
Carrier gas	Helium
Flow rate	1 ml/min
Mode	Constant flow
Outlet pressure	Vacuum
Modulator	
Thermal	Dual-stage, quad-jet
Temperature offset	15°C relative to GC oven
Modulation	4 s
Hot pulse time	0.6 s
Cool time	1.4 s
Transfer line	
Temperature	260°C
TOF-MS	
Tune	Auto tune (see Note 9)
Solvent delay	8 min (see Note 10)
Acquisition	*m/z* 40–600 at 200 spectra/s
Ion source	200°C
Electron energy	70 eV

and spectral match criteria (mass threshold, e.g. 50; minimum similarity match, e.g. 700). Set a separate *S/N* for peaks not found by the initial peak finding (e.g. 20:1) and define thresholds for analytes to keep for statistical evaluation (minimum number

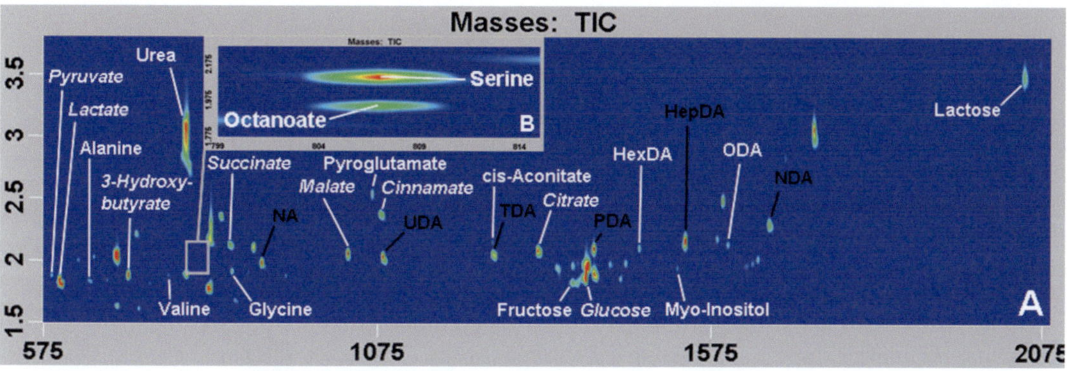

Fig. 4. (**A**) Total ion current (TIC) chromatogram of a human serum sample using GC × GC–TOF-MS. Major metabolites are labeled and the added extraction standards are *italicized* and derivatization standards are *printed in black* (**B**). The *insert* demonstrates the enhanced separation capacity of 2D-GC for the separation of octanoate and serine that could not be resolved in the first dimension. *NA* nonanoic acid, *UDA* undecanoic acid, *TDA* tridecanoic acid, *PDA* pentadecanoic acid, *HexDA* hexadecanoic acid; *HepDA* heptadecanoic acid, *ODA* octadecanoic acid, *NDA* nonadecanoic acid.

of samples or minimum percent of samples in a class that contain the analyte). Optionally, there is the possibility to fill the table with zero values if an analyte is not found. After assessing the appropriate values, start the alignment.

11. Export aligned peak table in csv file format.
12. Transform table to obtain a single data matrix. The generated matrix contains one column per measurement and one row per feature, where a feature corresponds to a set of aligned peaks. Each feature is characterized by an average of retention times, a unique mass and the areas detected for m/z unique in each sample (see Note 13). A flowchart of the entire process from raw 2D chromatograms to an appropriate data matrix is shown in Fig. 5.
13. Use the area integral of the unique mass of one of the extraction standard, e.g., for cinnamic acid m/z 212, which has been added to all samples at known concentration to normalize all peak areas prior to log-transformation.
14. The following steps can be accomplished by a number of software tools, e.g., Microsoft Excel, MATLAB, or the programming language R, just to name a few.
15. Perform Principal Component Analysis (PCA), Cluster Analysis, etc. as first steps for data analysis and to visualize the multivariate data. Furthermore, classification tools, such as support vector machine, can be applied (see Note 14).
16. Use a t-test to determine significant differences between the same metabolites in different samples. Make sure to correct for multiple testing (see Note 15).
17. Metabolites that differentiate samples need to be verified by running commercial standards.

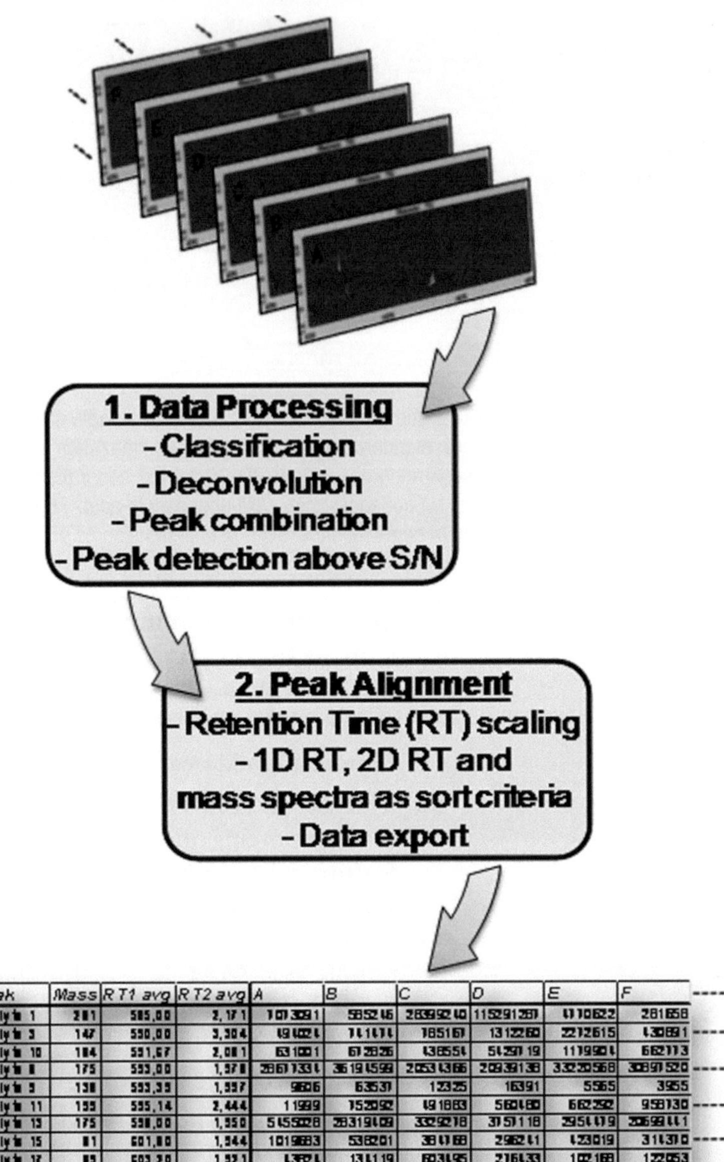

Fig. 5. Flowchart on how to compile a data matrix from the raw metabolic fingerprints for subsequent multivariate statistical analysis.

4. Notes

1. Avoid introducing excessive mass spectrometric background or chromatographic contaminants by using solvents that are HPLC grade or better.
2. Citrate and possibly other metabolites (e.g., *cis*-aconitate) form complexes with Ca^{2+} and Mg^{2+} ions in physiological fluids,

which are precipitated during methanol extraction. The chelating agent EDTA is used to sequester these metal ions and thereby release the metabolites from the divalent cation complexes.

3. Methoximation prevents cyclization and stabilizes carbonyl groups in the β position of sugars. MeOx is usually not suitable for use after prolonged storage. When in doubt, fresh MeOx should be made. Oxime derivatives are relatively stable and can be prepared many days in advance of trimethylsilylation. Be aware that oxime derivatives are formed as *cis/trans* isomers resulting in two chromatographic peaks.

   MSTFA is extremely moisture sensitive. TMS derivatives are volatile and degrade, even after only a few days. Trimethylsilylation is a technique where acidic hydrogens of functional groups are replaced with a $–Si(CH_3)_3$ group (see also Fig. 6). Make sure protic solvents are removed from the sample. They will react with MSTFA creating huge background signals and impede analyte derivatization. MSTFA should be stored at 4°C and removed from the bottle through the resealable septum with a syringe.

4. The long-chain fatty acids are not very soluble. The solutions should be vortexed frequently and sonicated mildly prior to freezing. Once frozen at –20°C the solutions are stable for several years.

5. Several blanks should be included in the measurement to account for contamination or false positives, which can derive from solvents, extraction solutions, impure internal standards,

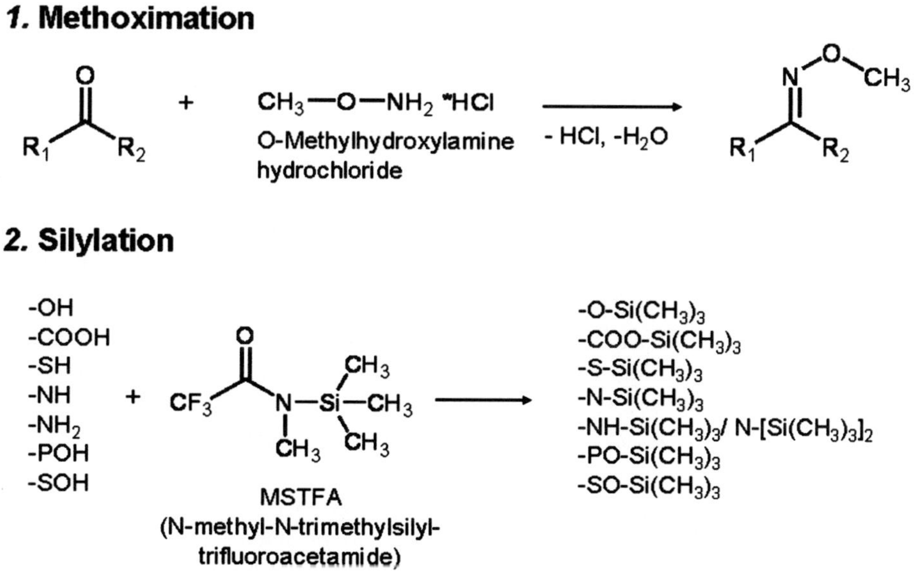

1. Methoxymation

$R_1C(=O)R_2$ + $CH_3—O—NH_2$ *HCl (O-Methylhydroxylamine hydrochloride) → (- HCl, -H_2O) $R_1C(=N–O–CH_3)R_2$

2. Silylation

-OH, -COOH, -SH, -NH, -NH_2, -POH, -SOH + MSTFA (N-methyl-N-trimethylsilyl-trifluoroacetamide) → -O-$Si(CH_3)_3$, -COO-$Si(CH_3)_3$, -S-$Si(CH_3)_3$, -N-$Si(CH_3)_3$, -NH-$Si(CH_3)_3$/ N-$[Si(CH_3)_3]_2$, -PO-$Si(CH_3)_3$, -SO-$Si(CH_3)_3$

Fig. 6. Derivatization is carried out in a two-step procedure; first the analytes undergo methoximation followed by trimethylsilylation.

dirty GC injection liners, or unconditioned columns. The blanks need to contain everything except the physiological fluid and undergo the complete process.

6. Be careful not to transfer any pellets at the bottom of the tube.
7. Do not over-dry the extract. This will result in poor recovery and a failed run.
8. Start by injecting blank samples followed by the control and patient samples in random order to avoid a systematic error. Distribute biological reference samples across the whole sample set to monitor for analytical variances across the measurement.
9. The mass spectrometer should be tuned for optimal sensitivity using perfluorotributylamine (PFTBA) as a reference.
10. Data are not collected during the first few minutes to exclude solvent, methoximation, and silylation artifacts.
11. Identification is achieved by comparing retention times and the mass spectrum of a peak with those of a library. There are many libraries of mass spectra available, for instance Wiley or NIST, both of which can be obtained commercially. The instrument software can be programmed to report a match similarity (0–100%) between a metabolite and the library entry. Note that the match quality may be lower when mass spectra acquired with a TOF mass analyzer are compared to spectra obtained with a quadrupole analyzer. To make sure that all peaks are identified correctly, the operator needs to review the comparison of mass spectra. Furthermore, it is useful to maintain a hard copy database of metabolites' retention times for a definite verification.
12. Be aware that degradation or rearrangement reactions can occur during derivatization and GC analysis. For example, ADP and ATP decompose during methoximation/silylation forming AMP, agmatine decomposes to putrescine, arginine and citrulline decomposition results in ornithine, while asparagine is converted to aspartate, and glutamate can partly lose H_2O and rearrange forming pyroglutamate, which is also produced from glutamine through loss of NH_3. It is well known that partial silylation can occur and result in more than one signal for an amino acid.
13. Evaluate the peak area and peak shape of extraction and derivatization standards to reveal potential problems during sample preparation and analysis. Extraction standards correct for analyte losses during the complete analytical process and can be used to reliably normalize the data for statistical analysis and for further quantification. Derivatization standards (odd-carbon numbered fatty acids) can be used for troubleshooting. They may help to reveal the sources of analytical variance, because they are added immediately before sample analysis. Low extraction standard recoveries in combination with sufficient recovery of

the derivatization standard are an indication of problems during sample preparation. The odd-carbon numbered fatty acids can also be used to correct for retention time shifts.

14. Data analysis can be performed using a variety of multivariate statistical tools, including unsupervised (e.g. PCA) and supervised techniques (e.g. support vector machine), which should be chosen according to the experimental setup and the hypothesis under testing.

15. In multiple testing the family wise error rate (FWER) is the probability that one or more false positives (type 1 error) occur. It can be controlled using for example Bonferroni or Westfall and Young correction. However, the controlling of the FWER is often too conservative. Instead the false discovery rate (FDR) can be determined. The FDR of a list of features is the expected relative frequency of false positives in it.

References

1. Dettmer, K., Aronov, P.A. and Hammock, B.D. (2007) Mass spectrometry-based metabolomics. Mass Spectrom Rev 26, 51–78.
2. Dettmer, K. and Hammock, B.D. (2004) Metabolomics – a new exciting field within the "omics" sciences. Environ Health Perspect 112, A396-397.
3. Roessner, U., Wagner, C., Kopka, J., Trethewey, R.N. and Willmitzer, L. (2000) Technical advance: simultaneous analysis of metabolites in potato tuber by gas chromatography-mass spectrometry. Plant J 23, 131–142.
4. Bertsch, W. (1999) Two-dimensional gas chromatography. Concepts, instrumentation, and applications – Part 1: Fundamentals, conventional two-dimensional gas chromatography, selected applications. J High Resol. Chromatogr. 22, 647–665.
5. Gorecki, T., Harynuk, J. and Panic, O. (2004) The evolution of comprehensive two-dimensional gas chromatography (GC×GC). J Sep Sci 27, 359–379.
6. Almstetter, M.F., Appel, I.J., Gruber, M.A., Lottaz, C., Timischl, B., Spang, R., Dettmer, K. and Oefner, P.J. (2009) Integrative normalization and comparative analysis for metabolic fingerprinting by comprehensive two-dimensional gas chromatography–time-of-flight mass spectrometry. Anal Chem 81, 5731–5739.

Index

A

B

C

Michael Kaufmann and Claudia Klinger (eds.), *Functional Genomics: Methods and Protocols*,
Methods in Molecular Biology, vol. 815, DOI 10.1007/978-1-61779-424-7, © Springer Science+Business Media, LLC 2012

D

E

F

G

R

S

T

U

V

W

Y

MIX
Papier aus verantwortungsvollen Quellen
Paper from responsible sources
FSC® C105338

If you have any concerns about our products,
you can contact us on
ProductSafety@springernature.com

In case Publisher is established outside the EU,
the EU authorized representative is:
Springer Nature Customer Service Center GmbH
Europaplatz 3, 69115 Heidelberg, Germany

Printed by Libri Plureos GmbH
in Hamburg, Germany